T0181886

Synthese Library

Studies in Epistemology, Logic, Methodology, and Philosophy of Science

Volume 462

Editor-in-Chief
Otávio Bueno, Department of Philosophy, University of Miami, Coral Gables, USA

Editorial Board Members
Berit Brogaard, University of Miami, Coral Gables, USA
Anjan Chakravartty, Department of Philosophy, University of Miami, Coral Gables, FL, USA
Steven French, University of Leeds, Leeds, UK
Catarina Dutilh Novaes, VU Amsterdam, Amsterdam, The Netherlands
Darrell P. Rowbottom, Department of Philosophy, Lingnan University, Tuen Mun, Hong Kong
Emma Ruttkamp, Department of Philosophy, University of South Africa, Pretoria, South Africa
Kristie Miller, Department of Philosophy, Centre for Time, University of Sydney, Sydney, Australia

The aim of *Synthese Library* is to provide a forum for the best current work in the methodology and philosophy of science and in epistemology, all broadly understood. A wide variety of different approaches have traditionally been represented in the Library, and every effort is made to maintain this variety, not for its own sake, but because we believe that there are many fruitful and illuminating approaches to the philosophy of science and related disciplines.

Special attention is paid to methodological studies which illustrate the interplay of empirical and philosophical viewpoints and to contributions to the formal (logical, set-theoretical, mathematical, information-theoretical, decision-theoretical, etc.) methodology of empirical sciences. Likewise, the applications of logical methods to epistemology as well as philosophically and methodologically relevant studies in logic are strongly encouraged. The emphasis on logic will be tempered by interest in the psychological, historical, and sociological aspects of science. In addition to monographs *Synthese Library* publishes thematically unified anthologies and edited volumes with a well-defined topical focus inside the aim and scope of the book series. The contributions in the volumes are expected to be focused and structurally organized in accordance with the central theme(s), and should be tied together by an extensive editorial introduction or set of introductions if the volume is divided into parts. An extensive bibliography and index are mandatory.

Wenceslao J. Gonzalez
Editor

Current Trends in Philosophy of Science

A Prospective for the Near Future

 Springer

Editor
Wenceslao J. Gonzalez
Center for Research in Philosophy
of Science and Technology (CIFCYT)
of the University of A Coruña
Faculty of Humanities and Information
Science
Ferrol (A Coruña), Spain

ISSN 0166-6991   ISSN 2542-8292   (electronic)
Synthese Library
ISBN 978-3-031-01317-1   ISBN 978-3-031-01315-7   (eBook)
https://doi.org/10.1007/978-3-031-01315-7

© The Editor(s) (if applicable) and The Author(s), under exclusive license to Springer Nature Switzerland AG 2022
This work is subject to copyright. All rights are solely and exclusively licensed by the Publisher, whether the whole or part of the material is concerned, specifically the rights of translation, reprinting, reuse of illustrations, recitation, broadcasting, reproduction on microfilms or in any other physical way, and transmission or information storage and retrieval, electronic adaptation, computer software, or by similar or dissimilar methodology now known or hereafter developed.
The use of general descriptive names, registered names, trademarks, service marks, etc. in this publication does not imply, even in the absence of a specific statement, that such names are exempt from the relevant protective laws and regulations and therefore free for general use.
The publisher, the authors and the editors are safe to assume that the advice and information in this book are believed to be true and accurate at the date of publication. Neither the publisher nor the authors or the editors give a warranty, expressed or implied, with respect to the material contained herein or for any errors or omissions that may have been made. The publisher remains neutral with regard to jurisdictional claims in published maps and institutional affiliations.

This Springer imprint is published by the registered company Springer Nature Switzerland AG
The registered company address is: Gewerbestrasse 11, 6330 Cham, Switzerland

Contents

1   From the Current Trends in Philosophy of Science
    to the Prospects for the Near Future   1
    Wenceslao J. Gonzalez

Part I   Philosophy of Medicine and Climate Change

2   Philosophy of Science Meets Medicine (Again):
    A Clearer-Sighted View of the Virtues of Blinding
    and of Tests for Blinding in Clinical Trials   15
    John Worrall

3   Environmental Decision-Making Under Uncertainty   45
    Joe Roussos, Richard Bradley, and Roman Frigg

Part II   Philosophy of Artificial Intelligence and the Internet

4   Artificial Intelligence and Philosophy of Science
    from the 1990s to 2020   65
    Donald Gillies and Marco Gillies

5   Whatever Happened to the Logic of Discovery?
    From Transparent Logic to Alien Reasoning   81
    Thomas Nickles

6   Scientific Side of the Future of the Internet as a Complex System.
    The Role of Prediction and Prescription of Applied Sciences   103
    Wenceslao J. Gonzalez

Part III   New Analyses of Probability and the Use of Mathematics in Practice

7   **From Logical to Probabilistic Empiricism: Arguments for Pluralism** 147
    Maria Carla Galavotti

8   **Instrumental Realism – A New Start for the Philosophy of Mathematics and the Philosophy of Science** 165
    Ladislav Kvasz

Part IV   Scientific Progress Revisited

9   **Scientific Progress and the Search for Truth** 191
    Philip Kitcher

10 **The Logic of Qualitative Progress in Nomic, Design, and Explicative Research** 207
    Theo A. F. Kuipers

Part V   Scientific Realism and the Instrumentalist Alternative

11 **Explicating Inference to the Best Explanation** 235
    Ilkka Niiniluoto

12 **Re-inflating the Realism-Instrumentalism Controversy** 261
    Stathis Psillos

Name Index 285

Subject Index 293

Contributors

Richard Bradley London School of Economics, London, UK

Roman Frigg London School of Economics, London, UK

Maria Carla Galavotti University of Bologna, Bologna, Italy

Donald Gillies University College London, London, UK

Marco Gillies Goldsmiths, University of London, London, UK

Wenceslao J. Gonzalez Center for Research in Philosophy of Science and Technology (CIFCYT) of the University of A Coruña, Faculty of Humanities and Information Science, Ferrol (A Coruña), Spain

Philip Kitcher Columbia University, New York, NY, USA

Theo A. F. Kuipers University of Groningen, Groningen, The Netherlands

Ladislav Kvasz Charles University of Prague, Prague, and Czech Academy of Sciences, Prague, Czechia

Thomas Nickles University of Nevada at Reno, Reno, NV, USA

Ilkka Niiniluoto University of Helsinki, Helsinki, Finland

Stathis Psillos National and Kapodistrian University of Athens, Athens, Greece

Joe Roussos Institute for Future Studies, Stockholm, Sweden

John Worrall London School of Economics, London, UK

Chapter 1
From the Current Trends in Philosophy of Science to the Prospects for the Near Future

Wenceslao J. Gonzalez

Abstract This text offers three aspects regarding the current trends in philosophy of science and prospects for the near future. First, the philosophical framework in which these most influential trends are situated. In these, there is novelty in terms of the topics covered and the approaches provided, along with continuity in other trends. Second, this initial chapter affords the main lines of research followed in the other chapters offered here, which are articulated in the five thematic parts in the book: (I) Philosophy of Medicine and Climate Change; (II) Philosophy of Artificial Intelligence and the Internet; (III) New Analyses of Probability and the Use of Mathematics in Practice; (IV) Scientific Progress Revisited; and (V) Scientific Realism and the Instrumentalist Alternative. Third, the origin of this book and its link with the Center for Research in Philosophy of Science and Technology (CIFCYT) of the University of A Coruña.

Keywords Current · Trends · Philosophy of Science · Prospects · Near · Future · Framework · Lines of research · Book's origin

1.1 Philosophical Framework

There is currently a flow of trends in philosophy and methodology of science, including new research topics and novel contributions to problems that have long been discussed. To a large extent, the frame drawn in "Novelty and Continuity in Philosophy and Methodology of Science" (Gonzalez, 2006) continues to be in place and is developing in new directions. Thus, (1) the naturalist approaches (cognitive,

I am grateful to John Worrall and Thomas Nickles for their comments on this paper.

W. J. Gonzalez (✉)
Center for Research in Philosophy of Science and Technology (CIFCYT) of the University of A Coruña, Faculty of Humanities and Information Science, Ferrol (A Coruña), Spain
e-mail: wenceslao.gonzalez@udc.es

© The Author(s), under exclusive license to Springer Nature Switzerland AG 2022
W. J. Gonzalez (ed.), *Current Trends in Philosophy of Science*, Synthese Library 462, https://doi.org/10.1007/978-3-031-01315-7_1

biological, etc.), (2) the social concern about science, which commonly includes the relation to technology, (3) the interest in scientific realism, either in the general perspective of science or in some particular sciences,[1] and (4) the philosophico-methodological views based on theories of probability (mainly Bayesianism) have their role, which seems less intense now in the case of Bayesianism. In addition, there is the emergence of new issues, such as those related to updates of Artificial Intelligence — like the new advances in machine learning — and the Internet understood in the broad sense, as a network of networks.

Where there has been a clear expansion is in the case of scientific realism. The philosophical framework was expanded with the contributions made in the volume *New Approaches to Scientific Realism* (Gonzalez, 2020b), where the recent directions can be found in "Novelty in Scientific Realism: New Approaches to an Ongoing Debate" (Gonzalez, 2020c). In addition, there is a greater prominence of methodological pluralism as opposed to monist, reductionist, universalist or imperialist methodological prospects. This relevance of methodological pluralism has been noticeable at least since 2006 (Kellert et al., 2006) and from various philosophical angles.[2] There is also the renewed commitment to methodological pragmatism, either as complementary to other methodological approaches (naturalism, realism, etc.) or as opposed to other methodological alternatives (cf. Gonzalez, 2020d).[3]

Situated within this contemporary philosophical framework, the main aim of this book is to offer especially relevant topics of the philosophy of science today, presented by philosophers of international standing, conceptions that may also have a projection in the near future. Thus, the book seeks to provide new perspectives, to broaden the field of philosophy of science, or to renew themes that have had a great impact on the profession. In addition to the current situation, the perspective of the near future is also interesting, so the volume wants to draw attention to topics that will receive more attention in the coming years at least in epistemology, methodology of science and ontology of science.[4]

---

[1] The interest in scientific realism continues to grow and so does how to measure its presence among scientists and philosophers of science. Thus, the conceptions of researchers working in seven scientific disciplines and philosophers of science attentive to the contributions of the history of science have been investigated regarding a number of aspects of scientific realism. Among other things, they have found: "(i) that natural scientists tended to express more strongly realist views than social scientists, (ii) that history and philosophy of science scholars tended to express more antirealist views than natural scientists, (iii) that van Fraassen's characterization of scientific realism failed to cluster with more standard characterizations, and (iv) that those who endorsed the pessimistic induction were no more or less likely to endorse antirealism" (Beebe & Dellsén, 2020, 336).

[2] See, for example, Giere (2006).

[3] On methodological pragmatism, see the recent contribution of a pioneer: Rescher (2020).

[4] In addition to these thematic areas, the philosophy of science includes the semantics of science, the logic of science, the axiology of research and the ethics of science. All of them are relevant when dealing with topics such as scientific prediction, cf. Gonzalez (2015).

Following these lines, which include contributions to some "neglected" disciplines in the current philosophy of science,[5] this book is structured in five thematic blocks: (I) Philosophy of Medicine and Climate Change; (II) Philosophy of Artificial Intelligence and the Internet; (III) New Analyses of Probability and the Use of Mathematics in Practice; (IV) Scientific Progress Revisited; and (V) Scientific Realism and the Instrumentalist Alternative. Thus, this volume addresses such relevant issues as the methodological validity of medical evidence or decision-making in situations of uncertainty; recent advances in Artificial Intelligence and the future of the Internet; current forms of empirically based methodological pluralism and new ways of understanding mathematics within scientific practice; and the revision of the approaches to scientific progress based on the experiences accumulated in recent decades.

1.2   Main Lines of Philosophical Research

Within Part I, which addresses the philosophy of medicine and climate change, two issues of great current relevance, John Worrall analyzes double-blind clinical trials in his paper, entitled "Philosophy of Science Meets Medicine (Again): A Clearer-Sighted View of the Virtues of Blinding and of Tests for Blinding in Clinical Trials." These are the cases when neither the participants themselves nor the administering clinicians in the trial know to which arm — either experimental or control — any particular participant has been assigned. The methodological aim of double-blinding is to control possible biases in the clinical trial, which might otherwise affect the trial's result. Thus, this type of methodological process in medicine seems to be virtuous. In this regard, if blinding is methodologically virtuous, then so should be testing whether blinding was retained throughout the trial's course.

However, Worrall points out that, de facto, end-of-trial tests for blinding are relatively seldom performed. Furthermore, when they are performed, it often happens that the blinding has been lost. In addition, leaders of Evidence-Based Medicine have become cool or even outright negative about tests for blinding, which in some cases leads to questioning blinding itself. Worrall's paper investigates this unexpected situation. His analysis follows the point of view of the general account of evidence supplied by philosophy of science. It argues that the correct response to the problems associated with blinding is not a negative view of end-of-trial tests for blinding. His proposal is a nuanced but still positive view on this methodological process in medicine.

Issues related to climate change are discussed in the paper "Environmental Decision-Making Under Uncertainty," which Roman Frigg has prepared with Joe Roussos and Richard Bradley. How can we apply science properly in situations of

---

[5]On this issue, see Gonzalez (2013). These "neglected" disciplines include the sciences of the artificial in the sense used by Herbert Simon (1996).

extreme weather conditions, such as hurricanes or typhoons? They occur relatively frequently, there are areas more prone to receive their impact and they cause immense damage to people and material goods. How can we provide knowledge contextualized to specific cases that will facilitate subsequent decision-making by public or private agents, who have to deal with potential or actual damage from these atmospheric phenomena? This paper deals with the severe uncertainty of those who have to make decisions in the face of these phenomena. It does so by considering a "confident-approach" in the case study of insurance pricing using hurricane models.

This chapter on decision making for environmental problems can be considered from the perspective of the existence of diverse philosophico-methodological aspects. First, there is *basic science*, which broadens our knowledge of atmospheric phenomena — including climate change phenomena — through explanation and prediction. Second, there is *applied science*, which seeks to solve concrete problems[6] — as is the case of the hurricanes that hit the coasts of the United States — through predictions that anticipate the possible future and prescribe patterns on how to face these types of phenomena that can have quite different levels of intensity. Third, there is *application of science*, when the use of scientific knowledge faces specific contexts with particular circumstances or features.[7] This includes a higher level of variability with respect to the previous case.[8] If the concern is the impact of the hurricane, then one must consider what to do when it is still on its particular coast and when it reaches the nearby inhabited area, since it can cause rivers to overflow (as in the case of Katrina in New Orleans) or completely destroy homes, depending on where it passes through.

Fourth, there is a distinction between scientific advice and political decision. This leads to the *public administration of the knowledge provided by scientists*.[9] This kind of public management with decisions on many matters (such as environment, health, economic, social, etc.), insofar as they are policy-making with politically motivated goals, are in principle outside the general framework of the scientific research mentioned (basic, applied and of application) and belong to a different sphere of

---

[6]Regarding the features of applied science, see Niiniluoto (1993).

[7]On the characteristics of application of science, see Gonzalez (2015, v, 4, 18, 33, 40, 70–71, 151n, 321 and 325). This variability of contexts of use for the application of science and the accumulation of particular possible circumstances to consider in it involves a *pragmatic complexity*, which adds to the already known structural complexity and dynamic complexity of basic science and applied science.

[8]This variability, directly modulated by the specific elements of the particular context of use, is usually the case in medicine. It can be seen when comparing the activity in the hospitals with respect to what is learned in medical schools, especially when treatment must be given to patients with a differentiated medical history and may have other pathologies in addition to those that caused their admission to the hospital, as we have seen during the Covid-19 pandemic with the diversity of therapies used.

[9]This issue forms part of the topics of the "external" philosophy of science, within the framework of studies on science, technology and society. On this subject, see Gonzalez (2005).

human reality.[10] Although they can be based on the advice of scientists as experts, this practical activity — international, national or regional — amounts to politics or policy with consequences for the society. In this regard, rather than in the general scope of science — due to the conceptual framework used and the institutional bases to make decisions (public or private organizations) — they are in the specific territory of politics or policy management. Thus, the decisions are made outside of scientific institutions, such as governments, parliaments, multi-national organisms, etc.[11]

All these philosophico-methodological aspects mentioned above are also pertinent when we move from medicine and environmental studies to questions related to the artificial that have social importance and where technological support is crucial. In this regard, Part II begins with the join paper by Donald Gillies and Marco Gillies entitled "Artificial Intelligence and Philosophy of Science from the 1990s to 2020." Following a path of "internal" factors (cf. Gonzalez, 2017), the authors highlight the relevance of AI for philosophico-methodological issues such as induction, confirmation and probability. The chapter examines the implications for philosophy of science set out in the book *Artificial Intelligence and Scientific Method* (Gillies, 1996) and the advances made in AI since 1996, mainly in machine learning.

Since this chapter — where the science of the artificial sciences and technology intersect — has a clear historical component, there is a philosophico-methodological comparison between the machine learning programs in the 90's and the much more powerful machine learning programs available in 2020. This new version of machine learning is particularly relevant for the development of Virtual Reality (VR). In this regard, the explicit recognition of the advances made since 1996 in the field of AI allows a comparison with the philosophico-methodological conclusions reached in the previous book, which leads to an update based on the new achievements in AI.

Thomas Nickles also focusses on Artificial Intelligence in his paper "Whatever Happened to the Logic of Discovery? From Transparent Logic to Alien Reasoning." This chapter discusses recent developments in deep learning, artificial neural networks (ANNs), machine speed, and Big Data. He considers if these developments will revolutionize scientific discovery across many scientific fields. Nickles investigates the claim that deep learning fueled by Big Data is providing an actual methodological revolution across the sciences, one that overturns traditional methodologies of science. From the point of view of rationality, these new approaches

---

[10]They are social phenomena as political decisions or processes of management regarding the policy making. In this regard, these types of actions, insofar as they are political decisions of governments or policy patterns of private organizations (like big corporations or large companies), can be analyzed by specialized disciplines, such as political and administrative science, international relations studies, etc.

[11]This fourth aspect can be illustrated with the management of the Covid-19 pandemic in recent months. Thus, on the basis of recommendations that international organizations, scientific societies and committees of scientists in each country have made to the authorities, public actions have been managed with enormous variations from one case to another due to political reasons. See, in this regard, The Economist (2020). On the types of design for institutions, the democratic innovations they introduce and international relations see Gonzalez (2021), where the variations are analyzed.

bring unknown forms of rationality, so he addresses the challenges of alien reasoning and the black-box problem for these novel computational methods.

Nickles considers the black-box problem or explainable AI (XAI) problem for deep learning. This is the difficulty of understanding how or why the device of AI arrived at its conclusions, which are becoming more incomprehensible for us. He addresses this opacity issue through several responses to the problem of achieving XAI. His present position involves several elements. Among them are, on the one hand, that if we are at a major turning point in scientific research, it is not the one initially advertised; and, on the other, that the alleged methodological revolution for new discoveries is also self-limiting in important ways.

Along with Artificial Intelligence, the Internet holds a preeminent position today and for the future. Wenceslao J. Gonzalez analyzes it in the chapter "Scientific Side of the Future of the Internet as a Complex System. The Role of Prediction and Prescription of Applied Sciences." This future of the network of networks has many faces and the focus here is on the scientific side, which is complementary to the technological facet and the social dimension. In this regard, the paper considers several steps relevant to the possible future of the Internet — in the broad sense — as a complex system:

(i) Its ontological framework is articulated in three major layers: the technological infrastructure (Internet *sensu stricto*); the Web; and cloud computing, practical applications (apps) and the "mobile Internet." They develop designs within artificial environments, whose configuration is analyzed in terms of dualities of a complex system (such as structure and dynamics, internal and external perspectives, epistemological and ontological factors). (ii) The analysis of the Internet as a whole requires the scientific side, the technological facet and the social dimension. In the scientific side, applied science and application of science are the central aspects. This involves the sciences of the Internet and other sciences related to the network of networks. (iii) To deal with the problems of epistemological and ontological complexity of the Internet in the broad sense, an internal and external perspective are needed. These affect the structural complexity, the dynamic complexity and the social dimension of the network of networks. (iv) The role of prediction and prescription concerns above all applied science that deals with complexity. The task of prediction is studied in the face of the complexity of scientific activity and the social dimension. The task of prescription is analyzed when facing the complexity of the Internet from the scientific side. Thereafter, this leads to the question of evaluation and meta-evaluation of future studies. (v) The coda finally reflects on the question of how to deal scientifically with the future of the Internet in a broad sense.

Part III looks towards new analyses of probability and the role of mathematics in practice. In her paper, Maria Carla Galavotti characterizes the move "From Logical to Probabilistic Empiricism: Arguments for Pluralism," which leads to a methodological pluralism based on empiricism and probabilism. Since the 1960s, the conception inherited from logical empiricism was gradually superseded by a new way of thinking characterized by a progressive opening to the pragmatic components of science. Thus, (a) the historical turn opened the field to sociological components, giving a new perspective to the formation of scientific knowledge, and (b) the semantical view of theories emphasized the role of models rather than scientific theories and highlighted aspects of experimentation. Patrick Suppes took this direction with his "probabilistic empiricism," where the crucial role is played by probability and statistical methods in science.

Galavotti points out that this approach brought about a pluralistic attitude regarding scientific methodology and extended the spectrum of disciplines of interest for philosophers of science. Nowadays, pluralism is a widespread position that different authors develop in different ways. In Suppes' view — probabilistic empiricism — there is a strict link between pluralism and the statistical methodology for forming hypotheses and experimentation. Galavotti maintains the desirability of adopting a bottom-up epistemological approach that assigns crucial importance to the context. This is in tune with Suppes' constructivist and pragmatical attitude. She considers that pluralism and the adoption of a bottom-up approach are needed following two examples: (1) cluster analysis and (2) the use of probability and statistics in courtrooms, which is discussed in the light of a controversial legal case.

Regarding the role of mathematics in practice, Ladislav Kvasz focuses on a type of scientific realism which emphasizes the importance of scientific instruments for the acquisition of scientific knowledge. Thus, in his paper "Instrumental Realism — A New Start for Mathematics and Scientific Practice," he maintains that our epistemic access to reality is often indirect, because it is mediated by the use of instruments. In this regard, the development of science is commonly accompanied by the introduction of new instruments. They open up access to phenomena — mainly in the natural world — to which we had formerly lacked epistemic access.

According to Kvasz, there is a second effect: the knowledge gained through the use of new instruments usually leads to the refinement of our scientific theories. But it also happens that sometimes the theories require their correction due to new experimental results, which do not support such theories and force scientist to engage in the correction of those scientific theories. The results that are obtained by means of particular instruments are then integrated or synthesized into a practice, which includes a unification of instrumental, linguistic and theoretical components. He distinguishes four kinds of synthesis: instrumental, relational, compositional, and deductive. Kvasz considers that all four kinds of synthesis require idealization, which in his instrumental realism approach is understood in terms of linguistic reduction.

Thereafter, Part IV revisits the topic of scientific progress. In his new contribution to this important issue, Philip Kitcher emphasizes the pragmatic component of his philosophico-methodological approach.[12] Thus, in his paper "Scientific Progress and the Search for Truth," he argues that scientific progress is better understood as *pragmatic* progress: "progress from" rather than "progress to" (with the goal of finding the complete true story about nature or the complete truth about the

---

[12] This particular emphasis on the pragmatic component is best appreciated when compared to the greater relevance of scientific realism in his systematic book on the subject: Kitcher (1993). On Kitcher's intellectual trajectory, see Gonzalez (2011).

fundamental structure of nature). For Kitcher, progress consists then in overcoming the difficulties and the limits of the current situation. He also considers that this view should be elaborated through explicit recognition of the role the sciences play in human life.

Besides the epistemological and methodological aspects, scientific progress requires ethical values, which inevitably frame and constrain the progress of the sciences: the problems that are worth tackling (*significant questions*) and judgments of significance are always subject to evaluation. Kitcher's approach emphasizes the practical deliverances of scientific inquiry. But basic research has also a place because the route to practical success often runs through studies with no pragmatic payoff. His analysis considers that the organization of the world is largely shaped by human goals and values. Thus, human cognition and action have a role regarding the structure of nature as it is.

Theo Kuipers discusses scientific progress based on two central ideas: (1) claims that scientific progress has been made can be justified, and (2) scientific progress in different areas has a common structure. Thus, in his paper "The Logic of Qualitative Progress in Nomic, Design, and Explicative Research," he proposes two general sufficient conditions for making justified claims of qualitative progress, conditions that are presented in their strict form and assuming a specific target. Thereafter, they are applied to theory-oriented research aiming at nomic truth approximation and to design research aiming at a certain product or process.

Kuipers then generalizes the two conditions and assumes that some specific constraints have to be satisfied, instead of assuming a specific target. Again, they are applied to theory-oriented science, but this time only as far as it primarily aims at empirical progress in the light of constraints provided by evidence. He considers that such empirical progress is functional for nomic truth approximation. Thereafter, the generalized conditions are applied to concept explication, undertaken in philosophy and science. In this case, the constraints are provided by evident examples and conditions of adequacy, leading to explicative progress. He also revisits progress in design research in the generalized perspective.

Part V completes the thematic framework of the book, addressing issues of scientific realism and the alternative of instrumentalism. Ilkka Niiniluoto focuses on "Explicating Inference to the Best Explanation," which he considers a pattern of everyday and scientific reasoning. IBE assumes that a hypothesis is accepted if it gives a better explanation of the known evidence than any alternative hypothesis. Since Harman's proposal (Harman, 1965), this subject has received various formulations. Moreover, it has attracted — and still attracts — the attention of many philosophers of science and in different directions.[13] Historically, Charles S. Peirce had a similar idea with "hypothetical reasoning" conceived as a special kind of ampliative reasoning, in addition to the processes of deduction and induction.

Niiniluoto points out that this sort of "hypothetical reasoning" proceeds backward from effects to causes and "abduction" leads from surprising facts to explanatory

---

[13] Among the recent analysis are Rescher (2019) and Bird (2020).

theories. In this regard, there are contemporary authors who treat Inference to the Best Explanation and abduction as synonyms. Meanwhile, some scholars argue that Peirce's abduction is restricted to the generation of explanatory hypotheses, but it does not cover their comparison and selection. Commonly, the discussion is concerned with the definition of "the best explanation" in science. Niiniluoto takes into account the possibility of analyzing the nature of explanatory reasoning by the Bayesian probabilistic approach, which connects with a line of research that has been particularly influential in recent decades.

Stathis Psillos analyzes central aspects of the trajectory followed by the philosophico-methodological divergence between scientific realism and the instrumentalist approach, to address the current situation between both in his paper: "Reinflating the Realism-Instrumentalism Controversy." In the 50's and 60's there was a denialist stance, according to which theoretical entities do not exist and the dispute between scientific realists and methodological instrumentalists should focus on whether or not theoretical terms have meaning. Then came the constructivist empiricism of van Fraassen, clearly critical of the positions of scientific realism, and a compatibilism of authors such as Ernst Nagel, who tend to underestimate the ontological differences between the two approaches and think of differences related to language.

After a period of decline in the compatibilist position, two new positions appeared. In the first, Howard Stein argued that there is no significant difference between a cogent and enlightened realism and a sophisticated instrumentalism (cf. Stein, 1989, 47–65, especially, 61); and, in the second, Kyle Stanford maintained a new compatibilism between an instrumentalism in historical terms and some versions of scientific realism.[14] Psillos, after reviewing the main arguments for compatibilism between scientific realism and instrumentalism, argues against Stein and Stanford that the difference between realism and instrumentalism (even when they are taken to be 'sophisticated') is deep and philosophically significant. Furthermore, he argues that there are still important and compelling reasons to prefer scientific realism to instrumentalism.

1.3   Origin of This Book

Several of the chapters have their origin in the papers of the *Conference on the Philosophy of Science: Seven perspectives* (*Jornadas sobre Filosofía de la Ciencia hoy: Siete perspectivas*), of the University of A Coruña (Campus of Ferrol), scheduled for March 12 and 13, 2020. This was the *XXV Conference on*

---

[14] "This more sophisticated realist has just granted everything that her historicist opponent was concerned to assert in the first place: after all, the historicist critic's central commitment was to the idea that we are in the midst of an ongoing historical process in which our theoretical conceptions of nature will continue to change just as profoundly and fundamentally as they have in the past" (Stanford, 2015, 875).

Contemporary Philosophy and Methodology of Science, which is held annually. The conference was preceded by the award of the Doctorate *Honoris Causa* to John Worrall (London School of Economics), an act celebrated on March 11th, 2020.[15] This ceremony was chaired by the Rector of the University of A Coruña, who was accompanied by the Rector of the University of Vigo and a Vice-Rector of the University of Santiago de Compostela, together with other authorities.

After a reasonable time to prepare the final versions of the texts, a publication was forthcoming. Thus, a book or a monographic issue of an indexed philosophical journal is published for each of these conferences. This group of publications is called the *Gallaecia Series: Studies in Contemporary Philosophy and Methodology of Science*.[16] In addition to the papers of the conference mentioned, there are several new texts to complete the picture of current trends in the philosophy of science and what is expected in the coming years. This makes the book thematically more complete than the XXV congress and the structure of the book is richer.

Let me express my recognition to the persons and institutions that cooperated in the original event, organized by the Center for Research in Philosophy of Science and Technology (CIFCYT) at the University of A Coruña. First, my appreciation to the guest speakers of the conference: John Worrall, Philip Kitcher, Donald Gillies, Maria Carla Galavotti, Theo Kuipers, Thomas Nickles, and Roman Frigg. I am very pleased with the efforts they made regarding their papers. Second, my acknowledgement to the organizations that gave their support: the University of A Coruña (especially to the Rector of the University and the Vice-rector for Economics and Strategic Planning), the Santander Bank, the Foundation Philosophy of Science and Technology, and the Society of Logic, Methodology, and Philosophy of Science in

---

[15] There is a publication with the *laudatio* of the promoter of this recognition and the masterly lecture of Professor Worrall. Cf. Gonzalez (2020a) and Worrall (2020). See also Worrall (2006).

[16] This collection includes the following titles: *Progreso científico e innovación tecnológica* (1997); *El Pensamiento de L. Laudan. Relaciones entre Historia de la Ciencia y Filosofía de la Ciencia* (1998); *Ciencia y valores éticos* (1999); *Problemas filosóficos y metodológicos de la Economía en la Sociedad tecnológica actual* (2000); *La Filosofía de Imre Lakatos: Evaluación de sus propuestas* (2001); *Diversidad de la explicación científica* (2002); *Análisis de Thomas Kuhn: Las revoluciones científicas* (2004); *Karl Popper: Revisión de su legado* (2004); *Science, Technology and Society: A Philosophical Perspective* (2005); *Evolutionism: Present Approaches* (2008); *Evolucionismo: Darwin y enfoques actuales* (2009); *New Methodological Perspectives on Observation and Experimentation in Science* (2010); *Scientific Realism and Democratic Society: The Philosophy of Philip Kitcher* (2011); *Conceptual Revolutions: From Cognitive Science to Medicine* (2011); *Freedom and Determinism: Social Sciences and Natural Sciences* (2012); *Creativity, Innovation, and Complexity in Science* (2013); *Bas van Fraassen's Approach to Representation and Models in Science* (2014); *New Perspectives on Technology, Values, and Ethics: Theoretical and Practical* (2015); *The Limits of Science: An Analysis from "Barriers" to "Confines"* (2016); *Artificial Intelligence and Contemporary Society: The Role of Information* (2017); *Philosophy of Psychology: Causality and Psychological Subject* (2018); *Methodological Prospects for Scientific Research: From Pragmatism to Pluralism* (2020); and *Language and Scientific Research* (2021). The table of contents are available in the webpage of the Center for Research in Philosophy of Science and Technology (CIFCYT): https://cifcyt.udc.es/coleccion-gallaecia/ (accessed on 19. 2. 2022).

Spain. Third, I am grateful to the support team of the conference and the technical secretary, Jessica Rey, which had to make an extra effort because of the circumstances due to the diffusion of Covid-19.

I am also grateful that Ilkka Niiniluoto, Stathis Psillos and Ladislav Kvasz have accepted the invitation to publish in this book, to expand the thematic framework of the volume. I would also like to thank Otávio Bueno for his positive response to this publication at *Synthese Library* and for his suggestions. In addition, let me point out that I am thankful to Amanda Guillan, Alba Garcia Bouza and Ana María Alonso for their contribution to the edition of this book. Finally, my gratitude to the Centre for Philosophy of Natural and Social Sciences (CPNSS) of the London School of Economics, where I have carried out a substantial part of the research that I publish here.

Ferrol, 19. 2. 2022

References

Beebe, J. R., & Dellsén, F. (2020). Scientific realism in the wild: An empirical study of seven sciences and history and philosophy of science. *Philosophy of Science, 87*(2), 336–364.

Bird, A. (2020). Scientific realism and three problems for inference to the best explanation. In W. J. Gonzalez (Ed.), *New approaches to scientific realism* (pp. 48–67). De Gruyter.

Giere, R. N. (2006). *Scientific perspectivism.* The University of Chicago Press.

Gillies, D. A. (1996). *Artificial Intelligence and scientific method.* Oxford University Press.

Gonzalez, W. J. (2005). The philosophical approach to science, technology and society. In W. J. Gonzalez (Ed.), *Science, technology and society: A philosophical perspective* (pp. 3–49). Netbiblo.

Gonzalez, W. J. (2006). Novelty and continuity in philosophy and methodology of science. In W. J. Gonzalez & J. Alcolea (Eds.), *Contemporary perspectives in philosophy and methodology of science* (pp. 1–28). Netbiblo.

Gonzalez, W. J. (2011). From mathematics to social concern about science: Kitcher's philosophical approach. In W. J. Gonzalez (Ed.), *Scientific realism and democratic society: The philosophy of Philip Kitcher* (Poznan studies in the philosophy of the sciences and the humanities, pp. 11–93). Rodopi.

Gonzalez, W. J. (2013). From the sciences that philosophy has 'neglected' to the new challenges. In D. Dieks, W. J. Gonzalez, T. Uebel, M. Weber, & G. Wheeler (Eds.), *New challenges to philosophy of science* (pp. 1–6). Springer.

Gonzalez, W. J. (2015). *Philosophico-methodological analysis of prediction and its role in economics.* Springer.

Gonzalez, W. J. (2017). Artificial Intelligence in a new context: 'Internal' and 'external' factors. *Minds and Machines, 27*(3), 393–396. https://doi.org/10.1007/s11023-017-9444-3. Accessed 6 Oct 2017.

Gonzalez, W. J. (2020a, March 11). Recognition of John Worrall as Doctor *Honoris Causa.* In *John Worrall*, Lectio Doctoralis, Universidade da Coruña, pp. 43–62.

Gonzalez, W. J. (Ed.). (2020b). *New approaches to scientific realism.* De Gruyter.

Gonzalez, W. J. (2020c). Novelty in scientific realism: New approaches to an ongoing debate. In W. J. Gonzalez (Ed.), *New approaches to scientific realism* (pp. 1–23). De Gruyter. https://doi.org/10.1515/9783110664737-001

Gonzalez, W. J. (2020d). Pragmatism and pluralism as methodological alternatives to monism, reductionism and universalism. In W. J. Gonzalez (Ed.), *Methodological prospects for scientific research: From pragmatism to pluralism* (Synthese library, pp. 1–18). Springer. https://doi.org/ 10.1007/978-3-030-52500-2_1

Gonzalez, W. J. (2021). Tipos de diseño, innovaciones democráticas y relaciones internacionales. In A. Estany & M. Gensollen (Eds.), *Diseño institucional e innovaciones democráticas* (pp. 37–52). Universidad Autónoma de Barcelona-Universidad Autónoma de Aguascalientes.

Harman, G. (1965). The inference to the best explanation. *The Philosophical Review, 74*(1), 88–95.

Kellert, S. H., Longino, H. E., & Waters, C. K. (Eds.). (2006). *Scientific pluralism* (XIX Minnesota studies in the philosophy of science). Minnesota University Press.

Kitcher, P. (1993). *The advancement of science: Science without legend, objectivity without illusions*. Oxford University Press.

Niiniluoto, I. (1993). The aim and structure of applied research. *Erkenntnis, 38*(1), 1–21.

Rescher, N. (2019). Does the inference to the best explanation work? In N. Rescher (Ed.), *Philosophical clarifications: Studies illustrating the methodology of philosophical elucidation* (pp. 145–154). Palgrave Macmillan.

Rescher, N. (2020). Methodological pragmatism. In W. J. Gonzalez (Ed.), *Methodological prospects for scientific research: From pragmatism to pluralism* (Synthese library, pp. 69–80). Springer.

Simon, H. A. (1996). *The sciences of the artificial* (3rd ed.). The MIT Press (1st ed. in 1969, and 2nd ed. in 1981).

Stanford, K. (2015). Catrastophism, uniformitarianism and a scientific realism debate that makes a difference. *Philosophy of Science, 82*(5), 867–878.

Stein, H. (1989). Yes, but … some skeptical remarks on realism and anti-realism. *Dialectica, 43*(1–2), 47–65.

The Economist. (2020, September 26). Why governments get it wrong? Section *Leaders*, p. 9.

Worrall, J. (2006). Why randomise? Evidence and ethics in clinical trials. In W. J. Gonzalez & J. Alcolea (Eds.), *Contemporary perspectives in philosophy and methodology of science* (pp. 65–82). Netbiblo.

Worrall, J. (2020, March 11). Evidence and ethics in 'Evidence-Based-Medicine:' Epistemology and equipoise. In *John Worrall*, Lectio Doctoralis, Universidade da Coruña, pp. 63–77.

Part I
Philosophy of Medicine and Climate Change

Chapter 2
Philosophy of Science Meets Medicine (Again): A Clearer-Sighted View of the Virtues of Blinding and of Tests for Blinding in Clinical Trials

John Worrall

Abstract A clinical trial is double-blind if neither the administering clinicians nor the participants know to which arm of the trial – experimental or control – any particular participant has been assigned. Double-blinding controls for various possible biases that might otherwise affect the trial's result; and hence seems to be an unambiguous methodological virtue. And if blinding is a virtue, then so also, it would seem, is testing that blinding was retained throughout the trial's course. As a matter of fact, however, end-of-trial tests for blinding are relatively seldom performed and, when they are performed, frequently find that blinding has been lost. Rather than decrying this situation, as might have been expected, leaders of Evidence-Based Medicine have become cool, or outright negative, about tests for blinding (and occasionally even about blinding itself). This paper investigates this *prima facie* mysterious situation from the point of view of the general account of evidence supplied by philosophy of science. It argues that, although interesting and unexpected complexities and difficulties are associated with blinding, the correct response to them is not a negative view of end-of-trial tests for blinding, but rather a nuanced but still positive view.

Keywords Clinical trials · Evidence · Blinding · End-of-trial tests for blinding

2.1 Introduction

My perspective on philosophy of science is one of the very few things that I share with John Locke. Locke believed that philosophers in general should restrict themselves to being, in his famous phrase, "underlabourers to the sciences" ("Epistle to the reader" in his (1689)). I completely agree – there is no special philosophical way

J. Worrall (✉)
Department of Philosophy, Logic and Scientific Method, London School of Economics,
London, UK
e-mail: j.worrall@lse.ac.uk

© The Author(s), under exclusive license to Springer Nature Switzerland AG 2022
W. J. Gonzalez (ed.), *Current Trends in Philosophy of Science*, Synthese Library
462, https://doi.org/10.1007/978-3-031-01315-7_2

of knowing: just logic (and mathematics) on the one hand, and the empirically-based sciences on the other. Hence, the only legitimate role for a philosopher is as an applied logician underlabouring in science: clarifying concepts and especially analysing arguments that are of relevance to science. So, when I am asked "what's new in philosophy of science?", I say "Nothing, or, at least, nothing at the fundamental level; it's just one damned underlabourer's job after another". The only thing in my view that has really changed during my 50 years or so in philosophy of science is that it now supplies underlabourers to a much wider range of sciences than it used to.

I am proud, for example, of having played a small role in extending the range of philosophy of science to include medicine.[1] Philosophy of medicine has become in recent years a small but energetic sub-field of philosophy of science. One focus of my work there has been Evidence-Based Medicine (EBM) and the hierarchies of evidence that it has spawned. Now, it surely ought to seem astounding to any right-minded person that there is an explicit Evidence-Based Medicine movement. *Of course*, medicine, like any rational discipline, should be based on the evidence. What else? And, indeed, the novelty and interest of the EBM approach lies, not in its general claim that medicine should be based on evidence, which is beyond sensible criticism, but rather in its specific claims about what exactly counts as evidence and, especially which kinds of evidence count as particularly strong or telling. These specific claims can be, and have been, criticised. EBM is seen as endorsing the view that evidence from clinical trials that have been randomized (division of participants into experimental and control arms made by some hidden random process) provides the "gold standard". But in fact, many in medicine, influenced by EBM, hold that, when it comes to evidence from a single trial,[2] the absolute pinnacle, the "gold plus" or "platinum" standard evidence is that supplied by trials that are not only randomized but also *double blind*: meaning of course that neither the clinicians involved in the trial nor the participants know to which arm, experimental or control, any individual participant has been assigned.[3] This paper does some underlabouring on arguments that have carried weight in medicine concerning the virtues of blinding and of testing to see that blinding has been retained during the course of a trial. Of course, "underlabourer" need not imply "underling". Philosophers of science are good at spotting detritus amongst the wonderful structures produced by science and encouraging its removal. In particular, they are good at identifying invalid or confused arguments that have nonetheless carried some persuasive weight in science. This is exactly what I try to do in this paper.

---

[1] Following the lead of Peter Urbach, see Urbach (1993) and, e.g., Worrall (2006, 2007a, b, 2010).

[2] EBM evidence hierarchies generally rank evidence from systematic reviews or meta-analyses (which attempt to amalgamate the results from many individual trials) as higher than evidence from any single trial.

[3] I shall count a trial as double blind only if everyone involved in the trial is blind to treatment arm for any particular participant, this includes both the participants and *all* the scientists: whether attending physicians, outcome assessors or data analysts. This cuts through the confusion noted, e.g., in Schulz and Grimes (2002, 697).

So, the most telling evidence for the effectiveness of some treatment, according to many in medicine (influenced by the original EBM position), is evidence provided by trials that are (a) randomized and (b) double-blind. My previous work in philosophy of medicine has concentrated largely on randomization and has argued that while randomizing has some epistemic virtues, it is far from the *sine qua non* of scientific validity that many influenced by EBM have taken it to be (see the references in footnote 1). However, no one could deny that RCTs may sometimes provide important evidence of effectiveness and such evidence definitely seems to carry extra weight if the trial was performed double-blind. The virtues of blinding will be examined in some detail soon, but roughly (and partially): if a clinical trial, for instance, is *not* blind, and if (as may be the case) the clinicians involved have a vested interest in achieving a "positive" result, then they may (perhaps subconsciously) bestow special care on the participants whom they know to be taking the experimental treatment, and those participants may improve because of that auxiliary care and the expectations raised by it, rather than because of any effect of the treatment under trial. More significantly, if the trial is not blind and the outcome involved is "subjective" ("participant showed some improvement in her symptoms") rather than "objective" ("participant died"), then an unblinded clinician (acting as outcome assessor) may – again perhaps subconsciously – be more inclined to declare a positive outcome for a participant that she knows to be receiving the experimental treatment.[4] Either of these possibilities, if actualised, would clearly bias the outcome: pushing that outcome toward a "positive" result whether or not the experimental treatment is actually better than the control treatment. If, on the other hand, it is the participants who are not blinded, then those knowingly taking the control treatment (let's suppose it's a placebo) may well, for example, break the trial-protocol and seek extra treatment for their condition outside the trial. This possibility, if actualised, would again bias the trial outcome but now in the opposite direction: because a number (or a disproportionate number) of participants on the placebo arm took concomitant therapies, a "negative" result might well be produced (meaning that the experimental treatment would be taken as "failing to outperform placebo") even though the experimental treatment is in fact more effective than placebo (but not more effective, or not much more effective, than placebo plus the concomitant treatment).

Double-blinding avoids these potential biases, as well as others that will be considered shortly, and hence appears to be a clear epistemic virtue. If the trial starts out unblinded (meaning that treatment allocation was not successfully concealed), then the scope for such biases to intrude is maximal; but, if a trial becomes unblinded

---

[4]This is far from being a merely theoretical possibility. Both Sackett (2011) and Schulz and Grimes (2002) cite the celebrated case of an RCT carried out on a triple concoction of cyclophosphamide, prednisone, and plasma exchange as a treatment for multiple sclerosis (Noseworthy et al. 1994). Sackett reports (2011, 675–676) that, in this trial, 'participants were periodically examined by two groups of neurologists, one group blind to their treatment groups and the other unblind. The blind neurologists found no effect ... on patient-outcomes, but the unblind neurologists concluded that [the] triple therapy was effective.'

at any stage before its end, then there is some scope for biases to intrude. It seems clear, then, that the epistemic ideal is that a trial should begin, and remain, double-blind.[5] And if so, the importance of *testing whether blindness was maintained throughout the trial* also seems clear: the failure of such a test may necessitate a re-evaluation of the weight of evidence produced by the trial, because it raises the possibility that biases have affected the outcome.

However, if blinding trials and testing for the retention of blinding at the end of the trial form the epistemic ideal, then it is an ideal more often honoured in the breach than the observance. Ney et al. (1986, 120) complained some time ago that "Although current medical science often relies on the double-blind trial to determine the value of a medication, there is very little evidence that the double-blind trial is blind for anybody, except those who read the report." This view has been endorsed by recent, more systematic research. For instance, a study by Fergusson et al. (2004) looked at 191 published reports of placebo-controlled trials and found that only 13 (8%) reported end-of-trial tests for blinding, and of those 13, blinding was discovered to have been lost in 9 of them (60%). Boutron et al. (2005) undertook a study that looked at 82 trial reports; while 54 of those trials tested for blinding of participants, in 22 of them blinding was reported to have been lost. As is reflected in the fact that these two studies came to quantitatively different (albeit similarly worrying) conclusions, there are issues about the representativeness of their samples of trial reports. A study by Hrobjartsson et al. (2007) not only was much larger but also made explicit and systematic attempts to take a representative sample of trials. This study ended up looking at 1599 trials and found that, of these, only 31 (2%) reported undertaking end-of-trial tests for blinding (of "key" trial personnel – whether clinicians, participants or outcome-assessors); of those 31 trial reports, 14 recorded that blinding was retained, 7 that blinding had not been retained, and in the remaining 19, the results of the end-of-trial test were either unclear or unreported.

So, this most extensive and rigorous study found that end-of-trial blinding was tested-for and blinding clearly discovered to have been retained in only 14 out of a

---

[5] As we shall see, the "remains blind" condition is important. Schulz and Grimes (2002, 696), for example, complain that: "many medical researchers confuse the term blinding with allocation concealment... [In fact] the term blinding refers to keeping trial participants, investigators .... or assessors ... unaware of an assigned intervention, so that they are not influenced by that knowledge." The view turns out to be controversial however: the medical statistician, Stephen Senn, had earlier complained (2004) about investigators who, "consider a trial to be double-blind when the patient, investigators, and outcome assessors are unaware of the patient's assigned treatment throughout the conduct of the trial." But Senn states, they are "quite wrong to do so." The reason they are wrong, according to Senn, is that "The whole point of a successful double-blind trial is that there should be un-blinding through efficacy. That is to say that there should be no incidental reasons, apart from efficacy, as to why the treatments are distinguishable but that the treatments should reveal themselves through efficacy. If the treatments are not distinguishable at all, then the treatments have not been proved different." I will consider Senn's claim in detail later. But note for now that, in whatever way and at whatever stage it may be broken, breaking blind inevitably opens up a trial to *possible* bias.

total of 1599 trials. The study's conclusion (Hrobjartsson et al., 2007, 659) – that "There is an urgent need for improving the methods of evaluating the success of blinding" – seems, therefore, amply justified. Since making sure that blinding is instituted and retained in a trial is a matter of avoiding biasing the trial's outcome, and since a central driving force behind EBM was the elimination of bias (or confounding) in trials, it might be expected that the leaders of EBM would be at the forefront of those endorsing Hrobjartsson et al's plea for "improving the methods of evaluating the success of blinding". The facts are surprisingly at odds with this expectation.

David Sackett, who was often called "the father of EBM", writing in direct response to the study by Hrobjartsson et al., explained why "we vigorously test for blindness before our trials, but not during them and never at their conclusion" (Sackett, 2004, 1136). Later, in his paper (Sackett, 2011), he went so far as to describe testing for blindness at the end of a trial as "playing a mug's game". Although they don't express it quite so colourfully, Sackett's negative view is shared by other leading proponents of EBM. Schulz and Grimes, for example, in an influential (2002) paper in *The Lancet* had earlier "question[ed] the usefulness of tests of blinding in some circumstances".

This negative view has had important practical consequences. Schulz was one of the authors of the CONSORT guidelines for clinical trials (*www.consort-statement. org*). These guidelines are endorsed by prominent EBM-ers and widely accepted within medicine, to the extent that satisfying them is effectively a necessary condition for a trial report to be published in a "top" medical journal. The guidelines – rather oddly – govern the reporting, rather than the conduct, of trials. But reporting may well have knock-on effects concerning conduct. From their first appearance in 2004, the CONSORT guidelines were lukewarm both about the value of testing for blindness and even about blinding itself. Recommendations concerning the reporting of both were conditional. The checklist entry for blinding specifies only "If [blinding is] done, [an ideal trial report should state] who was blinded after assignment to interventions (for example, participants, care providers, those assessing outcomes)": hardly a ringing endorsement of what seems a great epistemic virtue. And the entry concerning end-of-trial tests for blindness is again conditional and therefore lukewarm at best: "If [tests for blindness are] done, [an ideal trial report should state] how the success of blinding was evaluated". A 2010 revision of the guidelines (still in force) went further – the recommendation about blinding remains in conditional form, but mention of end-of-trial tests of blindness is dropped entirely. As its authors reported in one of several papers introducing the revised guidelines, in CONSORT 2010, "we have eliminated mention of how the success of blinding (masking) was assessed" (Schulz et al., 2010, 701). So there is currently no incentive for an investigator to report on whether or not blind was maintained in her trial and hence no incentive for her to institute a test to check if it was maintained. Indeed Sackett at least, as just noted, explicitly advises against making such a check.

This seems baffling. Double-blinding appears to be a clear evidential virtue and, if that's true, then it surely follows that there is also virtue in performing tests to check if a trial was in fact double-blind throughout its course. And yet some of the staunchest advocates of EBM feel that such tests are unnecessary and even that performing them may be a "mug's game"! Why?[6]

In the next section, I analyse the arguments that have traditionally been taken as establishing that double-blinding a trial, both initially and throughout its course, adds weight to the evidence it provides. Then I analyse the claims presented by Schulz and Grimes, Sackett and others to underwrite their negative view of end-of-trial-tests for blinding. Their views ultimately evaporate as based on confusion but this is more than a mere clarificatory exercise. First, it turns out that some valid and interesting points underlie their analyses and show that the issues both of blinding and of end-of-trial tests for blinding are more complex than might initially be thought. But secondly, once these valid points are clearly articulated, it becomes evident that the correct conclusion from them is a nuanced but still positive view of the value of end-of-trial tests This in turn potentially has practical consequences: it mandates a re-revision of the CONSORT guidelines (as explained in the final section of this paper – Sect. 2.4).

2.2 The Traditional View: Evidence from Double-Blinded Trials Is More Telling

Despite some claims in philosophy of science,[7] the *fundamental* principles involved in weighing evidence are, I believe, both universal and simple. There is, in particular, one principle that is common to all serious accounts of confirmation and that, on its own, sheds a great deal of light on issues concerning evidence from clinical trials. The principle states that *evidence e tells more strongly in favour of a theory T if e not only conforms with T* (in the simplest case, follows from it deductively, given accepted auxiliaries and initial conditions), but also, at the same time, *tells against* plausible theories that rival T. The basic idea is that evidence tells more strongly in favour of a theory to the extent that it not only conforms with the theory but also picks out that theory from its competitors – in the ideal case, the evidence picks out the theory as its only currently available explanation.

The principle lies behind Popper's idea that theories are particularly strongly confirmed ("corroborated" was the term he preferred) by passing "severe tests", but

---

[6]Jeremy Howick first made me aware of this problem and gave me important guidance to the literature on it. We discussed the issues at length during the course of his doctoral studies at LSE. His own treatment of, and attempt to resolve the problem are contained in chapter 6 of his (2011) – that chapter being itself a revised version of part of his PhD thesis. Unsurprisingly there are a number of overlaps between his treatment and the one developed in this paper. But also a number of differences of approach and content (some of which I will allude to later). This paper is how chapter 6 of Jeremy's book would have read had I been writing it.

[7]The most prominent advocate of the view that methodological principles are not universal, but instead have a complex and historically-evolving character, is Larry Laudan. For references to his work and my rebuttal of his claims see Worrall (1988).

it is especially clearly underwritten by the Bayesian account of confirmation. On this approach, the degree to which the discovery that potential evidence e actually holds supports the theory T is measured by the ratio p(e/T &B)/p(e/B), where B is background knowledge. In the simplest (positive) case where T, against the background B, deductively entails e, the numerator in this ratio (sometimes called "the Bayes factor") takes the value 1 and so the ratio reduces to 1/p(e/B). The denominator here measures the "prior probability" of the evidence e; so, remembering that probabilities vary between 0 and 1, the basic idea is that the Bayes factor is higher, that is, e tells more strongly in favour of T, the less likely e is otherwise; where "otherwise" means "on the assumption that some theory other than T were true". More precisely, if background knowledge B allows only a finite range of theories T_i of which T_n (=T) is the theory under consideration, then

$$p(e) = \sum_{i=1,...,n} p(T_i) \cdot p(e|T_i)$$

That is, p(e) is the sum of the degrees to which e would be expected on each of the various rival theories, weighted by the plausibilities (prior probabilities) of those theories. In particular, p(e) is the smaller, and so 1/p(e) (and with it the Bayes factor measuring strength of confirmation) is the larger, the fewer plausible theories, aside from T itself, that attribute significant probability to e. Intuitively: the more unlikely evidence e would be, in the light of plausible theories other than T, the more the fact that T correctly predicts e tells in favour of it. Howson and Urbach suggest that this reflects "the everyday experience that information that is particularly unexpected or surprising unless some hypothesis is assumed to be true, supports that hypothesis with particular force" (2004, 97).

This simple principle underwrites the whole process of controlling clinical trials. Suppose that a trial produces a "positive result". This certainly *might* be explained by the greater effectiveness of the experimental treatment, but of course lots of alternative explanations are possible. For example, those in the control group may have been older, less fit, had more concomitant conditions ... than those in experimental group. Every time such an alternative is "controlled for", that is, every time the experimental and control groups are made similar with respect to such a "possible confounder", one alternative explanation of the positive result is undermined and the strength carried by the trial result in favour of the effectiveness of the experimental treatment is increased – in accordance with this basic principle.

But even if there is strong evidence that the two groups in a trial were initially similar in respect of relevant possible confounders, the evidence for effectiveness from an observed "positive" result becomes weaker if differences between the two groups emerged *during the trial* – differences that might plausibly have had an impact on the outcome. In case randomization has been used to create the control and experimental groups, such differences are often referred to as "post-randomization differences" or "post-randomization confounders".

If a trial is performed without blinding, then it is open to a range of post-randomization confounders (or "biases"); whereas blinding a trial provides a safe-guard against such confounders. Suppose, for example, that the participants in a placebo-controlled trial know to which treatment they have been assigned; then those on the experimental treatment might well develop higher expectations of a good outcome than those on placebo; this difference would introduce an "expecta-tion bias", which might, independently of any "characteristic" feature of the exper-imental treatment, lead to a better outcome, at least in trials on outcomes such as pain relief that are known to be subject to the placebo effect.[8] Blinding participants as to which arm of the trial they are on effectively eliminates the possibility of expectation bias; and hence, in accordance with our basic principle, increases the evidential weight of any observed 'positive result' in favour of the theory that the experimental treatment is effective.[9]

Similarly, at any rate in the case of a trial where the outcomes are "subjective" (perhaps dependent in part on what response a participant reports to the clinician), blinding eliminates potential "response bias": if the participant knows that s/he has been given the experimental treatment and is inclined to please the clinicians, then s/he may be more inclined to report an improvement in her/his condition. If the participant knows that s/he is, instead, on the placebo arm of a trial, then s/he may be more inclined to drop out of the trial: blinding reduces the possibility of "attrition bias". Moreover, a participant who knows s/he is on the experimental arm may feel a greater incentive to stick to the protocol when not being directly observed than a participant who knows s/he is on the placebo arm (after all, the latter participants may well share the common, but sometimes mistaken, view that taking a placebo is the equivalent of doing nothing) – in other words, blinding helps avoid "non-compliance bias". Finally, as noted earlier, if a participant knows s/he is on the placebo arm of a trial, there is clearly a greater incentive to seek treatment for her/his condition outside the trial; if s/he is blind to treatment then any such differential incentive disappears – blinding undermines the possibility of "co-intervention bias".[10]

If, on the other hand, it is the clinicians who are not blind and instead know which treatment each participant is given, then, because they often expect, or hope for, a result that favours the experimental treatment, they may communicate their expec-tations of success to those participants whom they know to be on that treatment; and those participants may then experience a greater expectation/placebo effect than

---

[8] The placebo effect being in general stronger if the person taking it believes that the treatment has some specific or characteristic effect beyond placebo (see, for example, Benedetti 2014).

[9] Of course, other methodological factors – such as the size of the study population – also play a role. The claim is not that if a trial was performed double-blind then its result is automatically telling; but only that a trial that is double-blind carries more weight than an otherwise methodologically similar one that is not double-blind.

[10] Of course, it may just happen that, in a particular trial, more people on the placebo arm, despite not knowing this, took concomitant treatment; but, unlike when the trial becomes unblinded, there would be no systematic reason to believe that they might.

those who are taking the placebo. Hence blinding controls against another form of expectation bias ("communication of expectation bias"). Also, an unblinded clinician may (perhaps unconsciously) give better, more solicitous treatment to those whom s/he knows to be on the treatment arm, and those participants may improve because of that auxiliary treatment rather than because of the intervention under test. Conversely, a clinician who knows that a participant is taking the placebo, and who prioritises the welfare of the participants over the scientific validity of the trial, may prescribe subsidiary treatment. Blinding eliminates systematic "differential subsidiary treatment bias". Finally, where the outcome at issue is somewhat subjective, then, as already noted, a clinician who would prefer a "positive" result may be (perhaps sub-consciously) more inclined to judge that a participant's symptoms have improved if s/he knows that that participant has been given the experimental treatment. Blinding helps avoid the possibility of "observer (or reporter) bias".

And, in all these cases, by avoiding a potential bias, blinding rules out another possible explanation of the result. Hence, in accordance with the basic principle identified above, it increases the weight of evidence produced by the trial.

So, what's not to like about double-blinding? And, if blinding is such an obvious epistemic virtue, then:

2.3 How Could Testing for Blindness at the End of a Trial Possibly Be a "Mug's Game"?

Why, then, do some of the leading proponents of EBM sometimes at least seem to give the impression that they are rather lukewarm about blinding? And, more particularly, why are they at best cool, or even outright negative about testing for blindness at the end of a trial? How could it possibly be considered a "mug's game" to check if blinding has been maintained throughout a trial? The period during which biases can creep in to affect the trial outcome is clearly longer the earlier blinding is lost (and so is maximal if initial treatment allocation is not concealed), but if blinding is lost at any stage before the trial ends then the possibility arises of identifiable biases affecting the result. End-of-trial tests seem, then, very important: if such a test indicates that blinding has been lost, then questions need to be asked about whether any of the biases that lack of blinding makes possible has been actualized and hence has affected the evidential weight properly carried by the trial result.[11]

---

[11] This is true however the trial works out. A "positive result" might seem to tell in favour of the effectiveness of the experimental treatment under trial (or rather its greater effectiveness than the control treatment) but post-randomization confounders may have instead been responsible. However, some such possible confounders, as we saw, can tell in the opposite direction: for example, concomitant treatment bias (participants who know they are on placebo seeking other treatments for their condition) can help produce a "negative" result even if the experimental treatment is effective (again, more accurately, more effective than control).

The "mug's game" claim, it seems, cannot be correct. The treatments both of Schulz and Grimes and of Sackett are indeed, as we shall see, based on confusion in this respect. However, underlying those treatments and more or less explicit in them (particularly in the treatment by Schulz and Grimes) are three related important and under-recognised points. These show that the issues of blinding and of tests for blinding are not as straightforward as they might initially seem. Having first explained those underlying insights (Sect. 2.3.1), I then (Sect. 2.3.2) show how Schulz and Grimes and Sackett misrepresent them, fall into some confusion and end up endorsing their unjustifiably negative 'mug's game' view of end-of-trial tests for blinding. In Sect. 3.3, I articulate the more nuanced, but still positive, account of the significance of blinding and of end-of-trial tests for blinding that is really justified by the insights outlined in Sect. 2.3.1. In the final part (Sect. 2.4) I spell out the re-modifications of the CONSORT guidelines that this more nuanced account underwrites.

2.3.1   Blinding and Its Possible Loss Are More Complex Issues Than Might Appear: The Insights Underlying the Analyses of Schulz and Grimes and of Sackett

2.3.1.1   Loss of Binding May Not Result from Methodological Sloppiness

However virtuous double-blinding may be from an epistemic point of view, it is not always easy to achieve in practice; and, moreover, once achieved, is not always easily maintained. Loss of blinding may not be the result of any methodological defect in the conduct of the trial; instead, however strong the clinicians' commitment to the ideal of a double-blinded trial might be, their trial may become unblinded through circumstances beyond their control.

The natural home of double-blinding is the drug trial where it is *relatively* easy to make the control and experimental treatments observably indistinguishable both for the clinicians and participants. This makes allocation concealment, and hence initial double-blinding, reasonably simple; though even here, as we shall soon note, keeping the trial blind after the initial allocation may be an altogether different and more difficult matter. In non-pharmacological trials, even initial double-blinding may be difficult – indeed, in some cases, it may be impossible. To take one minor example, trials have been performed on the relative analgesic effects of real and sham acupuncture. The sham procedure involves an implement sometimes called the 'Streitberger needle' featuring a sheath into which the acupuncture needle can retract instead of penetrating the skin. The whole idea is to give the visual impression that the needle has penetrated the skin in the attempt to keep *participants* blind as to which of real or sham acupuncture they receive. But it is clearly difficult in such a trial to blind the administering acupuncturists (though it will be possible, and will clearly be a good idea, to try to ensure that those clinicians assessing outcomes are blind to interventions). To take another, even clearer, example: there is obviously no way that either participants or clinicians can be blinded in trials comparing a vigorous exercise programme with a course of antidepressant pills for the treatment of mild depression (though again outcome-assessors may, and should, be blinded).[12]

Even in its natural home of drug trials, although relatively easily achieved at the outset, double-blinding often cannot be retained as the trial progresses. Laying aside the possibility of outright malpractice *via* gaining access to the randomization code, there are two main ways in which blinding may be lost during the course of an initially blind trial: (a) through tell-tale side-effects and/or (b) through large positive clinical effects. To take an extreme case of the first: suppose that, in a placebo-controlled trial of, let's say, a treatment for migraine, several participants notice that their urine has become significantly discoloured; if so, then those participants and the clinicians involved will very likely, and very probably correctly, conclude that they are not taking the sugar pill, but are instead on the experimental arm of the trial. Suppose, alternatively, that the trial is an "active" one – that is, one in which the control is the currently accepted best treatment for the condition at issue. Because that treatment's side-effect profile will generally be well-known, if a participant experiences some different side-effect, then clinicians, and probably participants too, will likely, and probably correctly, conclude that they are being given the experimental treatment. On the other hand, those participants who had been taking the current treatment beforehand, and who have a similar experience as when knowingly taking the active control treatment, will likely, and probably correctly, conclude that they are in the control group. (And the clinicians who will be monitoring the participants' side-effects are likely to make the same inference.)[13]

So, double-blinding may be lost because of side-effects; it may also be lost because of a clear positive effect of the experimental treatment (though as we shall see, we should strictly speaking talk about "*apparent* positive effects"). To take an unrealistic "philosopher's" example – one that, nonetheless, makes the point: suppose that a placebo-controlled trial is being conducted on an experimental treatment for the common cold; all participants, who are being treated at the same time, have heavy colds of recent onset, but, within 5 minutes of administering the treatments, half of the participants have fully recovered while the other half continue sneezing and snuffling. In the light of background knowledge of the natural history of colds

---

[12] A supplementary insight is that such trials should not be dismissed as incapable of producing "real" evidence of (comparative) effectiveness just because they were not double-blind trials. Much, as we will reflect in Sect. 2.3.1.3, depends on the (apparent) size of the treatment effect. (I was introduced to the Streitberger needle example by Jeremy Howick.)

[13] It is not quite true that unblinding through side-effects is always outside of the investigators' control. If side-effects can be predicted (or have emerged in earlier-phase trials) then a so-called "active placebo" can, in principle, be developed to reinstate control over blinding. Suppose, for example, that an experimental treatment is known to have the side-effect of discolouring the urine, then an agent can be added to the placebo substance that simply has the effect of discolouring the urine in a similar way. Such "active placebos" are sometimes used, but, where the side-effect is adverse, rather than neutral (persistent headache, say) there are clear ethical obstacles to the addition of a substance to the "placebo" designed to produce that adverse effect. Moreover, placebos cannot, by definition, be designed to be "active" with respect to *unexpected* side-effects.

(and background knowledge about the common cold's positive, but slight suscepti-
bility to the placebo effect), it would be difficult not to form a view as to which
participants had been given the experimental treatment, and which were the controls;
and difficult to conceive that that view would be (at all substantially) incorrect.
Certainly those participants themselves who experience this "miracle cure" and who
presumably have all had colds before, will infer that they have been given the
experimental treatment. The "hard-line" classical statistician would hold that there
is no real "objective" evidence until the (properly-powered) trial has ended (let's
assume that the trial design specified that treatment should continue for a week), the
data have been analysed and a statistically significant result has been declared – until
then we should speak of an *apparent* large positive effect. (This is no doubt why
Schulz and Grimes, as well as Sackett, refer, as we shall see, to clinicians' "hunches"
about treatment allocation in such circumstances.) But, as not infrequently happens,
classical statistics seems to be out of line with educated scientific common sense.
Certainly, if, as in this far-fetched case, the effect is marked enough (sadly seldom
the case in recent clinical trials), then, given their knowledge of the natural history
and of the likely extent of any placebo effect, it will be difficult to prevent clinicians
and participants themselves making conjectures about which participants are on the
treatment that is going to turn out to be effective. Of course, those conjectures *may* be
wrong; but it seems, in a case like this, extraordinarily unlikely that they are.[14]

In sum, as Schulz and Grimes write (2002, 698–99):

> Disproportionate levels of side-effects might provide strong hints as to the intervention.
> Irrespective of the painstaking efforts to do double-blinded trials, some interventions have
> side-effects that are so recognisable that their occurrence will unavoidably reveal the
> intervention received to both the participants and the health-care providers. Even more
> fundamental than the hints from adverse effects are the hints from clinical outcomes.
> Researchers usually welcome large clinical effects ...If they arise, health-care providers
> and participants would likely deduce ... that a participant with a positive outcome received
> the active (new) intervention rather than the control (standard).

2.3.1.2   Loss of Blinding Leads to Possible Bias But Not All Possible Biases Are Actualised

So, blinding throughout a trial is more easily aspired to than achieved. Furthermore,
even if blind is not maintained in a trial, this does not at all automatically entail that
the outcome was subject to bias. Not every possibility for bias will be actualized.
Suppose, for example, that the majority of clinicians involved in a trial have, on the
basis of side-effects and/or marked positive effects, correctly "guessed" the treatment allocation for the majority of participants. But further suppose that – as
happens in properly-conducted trials – the clinicians who assess the outcomes for
each participant are different from those involved in the participants' care during the

---

[14]There are in fact some trickier issues here concerning the case of *apparent* significant clinical
effects. These will be raised, and considered more carefully, *below* pp. 27–29, Sect. 2.3.1.3.

trial, and that, although the clinicians considered as a whole group, are unblinded, those clinicians charged with assessing outcomes can do no better than chance in guessing which treatment each participant was given. If so, then it seems that there can be no question of "observer or reporter bias" having affected the trial's overall outcome.[15]

Or suppose that the outcome measure involved in the trial is "objective" – say, "participant still alive 6 months after treatment" – then, even if outcome assessors have become unblinded, there is no scope for them to introduce "reporter bias" by fudging the outcome for particular participants.[16]

Or suppose, to take a final example, that participants, rather than clinicians, have become unblinded during the course of a trial. This certainly opens up the *possibility* of co-intervention bias: a significant number of those who discover that they are on the placebo arm of the trial may be tempted to seek treatments for their condition outside the trial – an extra drug, let's suppose. But whether or not a participant is taking some particular extra drug can often be tested for – through blood tests for example; if testing reveals that all, or the vast majority, of participants have resisted the temptation to seek additional treatment, then there is no concern about an *actual* co-intervention bias. While these are, perhaps, the clearest-cut cases, similar reassurances *may* be available concerning other potential biases resulting from loss of blinding. If such reassurances are available, then there is no reason to regard the evidential impact of the result of a trial as reduced, on the basis of a "negative" end-of-trial test for blinding.

2.3.1.3 Even If a Trial Has Actually Been Affected by Bias Because of Loss of Blinding, That Trial's Result May Nonetheless Supply Telling Evidence of Effectiveness

Consider a slightly different version of the "toy" example cited earlier. A group of people with heavy colds of recent onset are randomized (at the same time) into experimental and placebo groups. The trial is set to run for 2 hours in total, and the outcome of interest is recovery from the cold. One hour after they are all given a drug (experimental or placebo), roughly half of the whole group exhibit noticeably reduced symptoms; while, at the end of the trial after 2 hours, the participants in the improved half are completely better. The other half have improved barely, if at

---

[15] It seems not to be widely recognised, but this needn't always be the case. 'Blinded' outcome-assessors (in the sense that they have had no contact with the participants while the trial was running) may nonetheless ask questions about side-effects (indeed one would generally want the trial protocol to have them do so), and hence, if there are differences in side-effect profiles, those outcome-assessors may indeed 'guess' treatment allocations before producing their assessments of individual outcomes.

[16] "In general . . . blinding becomes less important to reduce observer bias as the outcomes become less subjective, since objective (hard) outcomes leave little opportunity for bias" (Schulz & Grimes, 2002, 697).

all, and still have pronounced cold symptoms at the end of 1 hour and also at the end of the trial. No doubt the clinicians running the trial will, after the first hour has elapsed, have a good idea of which participants are in which groups; and will therefore have had the opportunity, during the remaining hour of the trial to, for example, give extra attention to those they suppose are taking the experimental treatment. This extra attention may have *some* differential effect, but, given what we know about the natural history of colds and about the (real but) small placebo effect on them, it is surely reasonable, surely "evidence-based" to judge that the effect of the extra attention cannot have been sufficient to have produced an overall result with such a striking effect; and hence not sufficient to invalidate the evidential impact of the result of the trial. That result will still be taken as strong (indeed, in this "toy" case, overwhelming) evidence for the effectiveness of the experimental treatment; and surely correctly so.

Schulz and Grimes (amongst others) talk in this connection of "large clinical effects" as leading to loss of blinding. Strictly speaking, of course, and as noted earlier, what is observed in such cases is an *apparent* clinical effect: a marked difference in the average value of some outcome variable in the two sub-classes of the study population. The "causal" judgement that this difference is the effect of an intervention is always, strictly, a theoretical one. However, so long as the apparent effect size is large, when judged against background knowledge – as it is in my imagined example – then no other explanation of the apparent effect seems remotely plausible.

The medical statistician Stephen Senn has argued even more strongly that, not only is it no problem for blinding to be lost via positive (apparent) clinical effects, it is the *aim* of trials that blinding should be lost in this way. Senn's (2004) was a direct response to the Fergusson et al. (2004) paper cited earlier: "According to Fergusson et al . . ., they [sic] 'consider a trial to be double-blind when the patients, investigators, and outcome assessors are unaware of the patient's assigned treatment throughout the conduct of the trial'. They are quite wrong to do so. The whole point of a successful double-blind trial is that there should be un-blinding through efficacy. That is to say that there should be no incidental reasons, apart from efficacy, as to why the treatments are distinguishable but that the treatments should reveal themselves through efficacy. If the treatments are not distinguishable at all, then the treatments have not been proved different."

It is heartening to hear a medical statistician rejecting the classical statistical hard line that efficacy cannot "reveal itself" during the trial but only after the properly powered and properly planned trial has formally ended and its results have been carefully analysed. Senn clearly goes too far however. First, it is certainly not a necessary condition of a successful clinical trial that it should be "unblinded through efficacy", as Senn suggests it is. There are plenty of very large trials that are generally regarded as positive (null hypothesis refuted at a high level of significance) but which suggest a very small effect of the experimental treatment that would not have been discerned (in any rational way) while the trial was running. Indeed, sadly, very few RCTs have results that are so clear-cut that they would "jump out" at clinicians (as in my imaginary cold example). Moreover, there is an obvious problem with Senn's requirement that "there should be no incidental reasons, aside from efficacy, as to why the treatments are distinguishable". As noted, the loss of blinding undeniably opens up a trial to the possibility of bias. So, there will always be other "incidental" factors that might possibly have an impact on the overall result of the trial. It is only when there is good reason to think that the effect of these factors will be small and the effect size being produced in the trial is large that it is safe to say that there is good evidence that the experimental treatment is superior to the control.[17]

Notice, by the way, that this third consideration points to an important distinction between the two ways in which blinding can be lost during the course of a trial. If the game is given away by unexpected side-effects, in the absence of any apparent large positive effect, then there is always an active worry that the trial may have achieved its positive result as a consequence purely of biases following the loss of blinding. This is especially true because a "positive" result simply means that the null hypothesis of no difference between experimental and control treatments has been "refuted", usually at the 5% level – and this can readily be achieved on the basis of a very small difference in effectiveness between the experimental and control treatments, especially if the trial is a large one. If, on the other hand, blinding is lost because of a large apparent positive effect on one half of the study population then, as we have seen, even if biases have been introduced, the trial outcome may still give good evidential reason to hold that the treatment under trial is indeed effective.

2.3.2 The "Mug's Game" Confusion

So, to summarize Sect. 2.3.1: (a) losing blind may not involve any methodological culpability; (b) losing blind means only that the trial becomes open to the *possibility* of certain kinds of bias, not that it necessarily has actually been affected by any particular bias/es; and (c) even if the outcome of some trial *was* affected by biases, then, at least under some circumstances, it can plausibly be argued that the outcome still gives good empirical support to the theory that the treatment under trial is

---

[17] Jeremy Howick in chapter 6 of his 2011 – see footnote 6 above – claims that 'When failure to successfully double mask a trial results from the dramatic effects of a treatment, the resulting factors arising from [for example] participant and caregiver knowledge are not confounding." But this is open to analogous objections to those just raised against Senn's analysis. First the assertion that the trial outcome was the result of a "dramatically" effective experimental drug is *not* an observational result, it is instead an interpretation of the outcome based on background knowledge. Secondly, the result *will* very likely be confounded by biases produced by the breaking of blind – in the sense that the result will (likely) be produced by a combination of the (supposed) effectiveness of the experimental treatment with biases resulting from the loss of blinding. The result may well have been statistically significant even if those biases had not played a role, but still the result itself was likely affected. Again, judgment based on background knowledge is required to deliver the view that the evidence, in particular the effect size, gives strong support to the theory that the experimental treatment is effective.

effective – so long as there is evidence-based reason to think that the effect of such biases was small, while the effect size "revealed" by the trial was large.

The first of these insights is explicit, and the second and third arguably implicit, in the treatments of Sackett and, particularly, of Schulz and Grimes. Those insights certainly give grounds for rejecting the naïve, but in fact quite widespread view that a "negative" result in an end-of-trial test entails that the result of the trial should be dismissed as of little or no evidential value (a point which Schulz and Grimes themselves emphasise – (2002, 698)). But the claim that we need to be more sophisticated in interpreting end-of-trial test outcomes, particularly if they are "negative", is quite a long way from the conclusion that to carry out such a test is to play a mug's game. We need to look more closely at how Schulz and Grimes and then Sackett arrived at their negative view of the value of end-of-trial tests for blindness.

First, we have talked blithely about blinding being lost as though this was a clear notion; but in fact it is by no means obvious when exactly blinding should be considered to have been lost in a trial. Assuming that the trial was blind initially, in other words that it began with treatment allocation successfully concealed from both clinicians and participants, then, short of illicit access to the randomization code, loss of blinding is not, objectively, a straightforward yes-or-no affair. Participants and clinicians, as noted, may, and often will, be making conjectures as the trial progresses about treatment assignment on the basis of side-effects or apparent positive clinical effects. But in order to say that blinding has been lost: (i) What proportion of participants or clinicians need to have arrived at conjectures about treatment assignments that are correct?; and (ii) How strong does the evidence for those conjectures have to be (and, relatedly one hopes, how sure do participants and clinicians have to be that their conjectures are correct)?

There is no one, objectively correct answer to these questions, but the issue has been decided by social convention. An end-of-trial test (if performed at all) always has the same form. Clinicians are asked to state which arm – experimental treatment or control treatment – they think each of the participants had been assigned to; the proportion of cases in which the clinicians correctly identify treatment arm is noted; and that proportion is compared to the proportion of times they would be expected to be correct if they were simply making random guesses about each participant's allocation. The trial is declared unblinded just in case the clinicians' identification of treatment arm is better to a statistically significant degree than the chance proportion. So, in the most straightforward case where only two arms (experimental and control) are involved, the trial is deemed to have become unblinded so far as clinicians are concerned if the proportion of their correct "guesses" of treatment-arm is statistically significantly different from 50%. Hence, any consideration of how confident the clinicians are in their "guesses", and of whether, and if so how, those "guesses" are based on evidence drops out of the picture (or is assumed dealt with by the requirement that the difference between the actual percentage of correct guesses and the percentage expected "by chance" is statistically significant).

The corresponding test for *participants* (in fact seldom – or, rather, even more rarely – performed) is obviously to ask them individually at the end of the test to state which arm of the trial they believe themselves to have been on – the test being "failed" and the trial regarded as having become unblinded if (and only if) the proportion of the participants who expressed correct views about their own allocation was statistically better than would be expected "on the basis of chance" – that is, in the simplest, two-arm case, where each "guess" had a probability of ½ of being correct. (There is in fact no uniform consensus concerning when to talk of an end-of-trial test as having been "passed" or "failed", having been "positive" or "negative". I will talk of such a test as "failed" whenever the result is deemed to have shown that the trial has become unblinded, even though this will mean that either clinicians or participants (or both) *succeeded* in identifying treatment arm at better than chance rates. To emphasise that this might seem odd (but then so would the opposite convention), I will continue to place "failed" (or "passed") in quotation marks.)

We arrive, then, at the following characterisation of when a trial has become unblinded (reflecting the obvious fact that this cannot apply only when a test for unblinding has actually been carried out): a trial has become unblinded if and only if either the number of clinicians who have correct conjectures about which participants are on which arm (or who would have such conjectures if they were explicitly asked to say which arm each particular participant was on), or the number of participants who have correct conjectures about which arm they themselves are on (or, again, would have such conjectures if they were explicitly asked) is statistically significantly different from what would be expected if they were "merely guessing" – in the simplest case: if they had a probability of 1/2 of being correct each time.

Both Sackett and Schulz and Grimes explicitly adopt this characterisation. Schulz and Grimes, for example, write (2002, 698): "Investigators can theoretically assess the success of blinding by directly asking participants, health-care providers, or outcome assessors which intervention they think was administered . . . In principle, if blinding was successful, these individuals should not be able to do better than chance when guessing the intervention . . .". They immediately add, however: "In practice, however, *blinding might be totally successful* [emphasis supplied], but participants, health-care providers and outcome assessors might nevertheless guess the intervention because of ancillary information . . . If indeed the active (new) intervention materialises as helpful . . . then [the clinicians' and participants'] deductions would be correct more often than chance guesses. Irrespective of their suspicions, end-of-trial tests for blindness might actually be tests of hunches for adverse [side-] effects or efficacy" (Schulz & Grimes, 2002, 698–699).

This is mysterious: first, it is not clear what precise "in principle"/"in practice" distinction is being appealed to, and, more especially, it is completely unclear how blinding can have been "totally successful" if participants and/or clinicians have been able to "guess" treatment assignment predominantly correctly. What exactly is the distinction between testing to see if blinding has been lost and testing to see if adverse side-effects and/or apparent effectiveness have given the treatment assignment away? Short of illicit access to the randomization code, side-effects and apparent positive effects are the (exhaustive, but non-exclusive) ways in which blinding can be lost; and so *of course* in testing for loss of blindness you are at the same time testing for adverse side-effects and/or (apparent) efficacy.

One conjecture might be that, by "successful blinding", Schulz and Grimes mean only that the initial treatment allocation was successfully hidden. But they themselves point out "[while] many medical researchers confuse the term blinding with allocation concealment... [In fact] the term blinding refers to *keeping* trial participants, investigators .... or assessors ... unaware of an assigned intervention, so that they are not influenced by that knowledge" (Schulz & Grimes, 2002, 696). Their statement (2002, 698) that "blinding might be totally successful, but participants, health-care providers and outcome assessors might nevertheless guess [predominantly correctly] the intervention because of ancillary information" seems, then, to be an outright logical contradiction.

Of course, this cannot be what they intended since no one intends to assert a logical contradiction. But further consideration of what Schulz and Grimes might really have thought can be laid aside, because there is surely only one sensible claim here: that blinding may have been completely successful *in the sense that the clinicians organising the trial did everything that they could to maintain blindness*, despite which adverse side-effects and/or clear positive clinical outcomes gave the game away, and hence the trial became unblinded. This hardly helps their argument, however, since this sensible claim clearly does *not* entail their conclusion that end-of-trial tests are of questionable value. To the contrary: independently of how it happened, the fact that blinding has been lost during a trial entails (as they themselves clearly recognize) that identifiable biases *may* have affected its outcome. It then follows that, far from an end-of-trial test being valueless, significant information is in fact provided whatever such a test's result might turn out to be (assuming the result to be genuine, a point which I will take up soon). If the test indicates that blinding was maintained during the trial, then there is no reason to investigate whether certain specifiable "post-randomization" biases may have affected the result of that trial, and so its outcome can carry its full evidential weight concerning the effectiveness or ineffectiveness of the experimental treatment; if, on the other hand, the test indicates that blinding *was* broken, then there are reasons to check for biases – and so, before the result of the trial can be given its full evidential weight, questions need to be answered about whether or not biases did in fact affect that result.

To reiterate what was said above in Sect. 2.3.1.2, if an end-of-trial test provides evidence that blinding was lost during the trial, then this does *not* automatically entail that the trial was in fact biased – only that it might have been. And if the trial was in fact affected by bias, this does not automatically entail that its result cannot provide telling evidence for the effectiveness of the treatment under trial. But these are reasons for interpreting end-of-trial tests and their outcomes sensibly, not for rejecting such tests as without value.

Although, as we shall see, he ends in the same confusion as do Schulz and Grimes, David Sackett's argument for the lack of value of end-of-trial tests is worth separate examination for reasons that will become apparent. Sackett's most sustained and pointed version of the argument is in his 'Clinical Round' paper (2011); and it is there that he explicitly branded performing end-of-trial tests as playing a "mug's game". He presented the argument in that paper alongside reminiscences of his involvement in what came to be known as "The Canadian Aspirin Trial" – a trial now regarded as having been the first to provide evidence that a regular dose of aspirin reduces the risk of strokes and death among people who have suffered from transient ischemic attacks ("mini-strokes").

That trial was a comparative one involving, not just aspirin, but also another potentially effective drug – sulfinpyrazone; and it used a "double-dummy design", which involved participants being randomized into *four* groups – those in the first were given both experimental drugs (aspirin and sulfinpyrazone), those in the second received sulfinpyrazone plus placebo ("aspirin-placebo", a "dummy" pill intended to mimic aspirin), those in the third got aspirin plus (sulfinpyrazone-) placebo and those in the fourth were given both the placebos.

Sackett records (2011, 674) that, as the results emerged, he was "dreaming of a lead article in the *New England Journal of Medicine*" (the world's most prestigious medical journal), largely because the overall outcome was turning out to contradict pre-trial expectations: humble aspirin had not been thought likely to reduce the risk of stroke but the trial had found evidence that it did, whereas sulfinpyrazone was believed, ahead of the trial, to be much the more likely to prove effective, but in fact the trial found no evidence of this.

Sackett recalls that one of the "remaining odds and ends" before a paper could be sent off for publication was the analysis of the results of the end-of-study question-naire that had been used to check that the clinicians involved in the trial had remained blind to which patients were in which groups. This – in line with usual practice, as noted above – had involved asking the clinicians involved to state which group they thought each participant had been in. Sackett and collaborators assumed that, because the trial involved four equal-sized groups (remember the trial had a "double-dummy" design), 25% correct "guesses" would be expected on the basis of chance. Sackett continues the story (2011, 674):

> 'I felt the bullet enter my heart' when our . . . statistician tracked me down on the ward to tell me that our clinicians' correct guesses were . . . statistically significantly different from 25%. Had our triumphant lead article been reduced to an apologetic Letter to the Editor? And why did my [statistician colleague] have a big grin on his face?

The reasons for the grin, apparently, were (a) that the clinicians' predictions about which participants were in which groups were statistically significantly *wrong*; and (b) those clinicians' pre-result predictions, also recorded by the statistician, about which of the two "active" drugs – sulfinpyrazone or aspirin – if either, would prove effective were equally wrong: "[m]ost of them predicted that aspirin would be worthless but sulfinpyrazone would be effective" (Sackett, 2011, 675). So, most clinicians had assumed, in line with their prior conjectures, that those participants whom they identified (notice: identified predominantly correctly!) as faring better in the trial were likely to have been given sulfinpyrazone – an assumption that, it turned out, was at 180 ° to the actual result of the trial.[18]

Sackett spins a fine yarn. But why exactly is its moral supposed to be that in testing for blindness at the end of a trial you are playing a "mug's game"?

First, notice that what Sackett takes to be the crucial aspect of his story – namely that the guesses of the clinicians in his study were statistically significantly *wrong* in terms of which drug, aspirin or sulfinpyrazone, was the effective one – in fact carries no methodological weight whatsoever. The important methodological question is whether those clinicians could, while the trial was still ongoing, distinguish (predominantly correctly) between participants on the treatments that turned out to be effective (whichever they turned out to be – as it happened, aspirin with either placebo or sulfinpyrazone) from those participants on "ineffective" treatments (whichever *they* turned out to be, as it happened, sulfinpyrazone with placebo and placebo with placebo).[19] According to Sackett's story, the test revealed that the clinicians could indeed make that distinction at better than chance rates. It makes no difference from the methodological/evidential point of view if they also (predominantly and now incorrectly) believed that the effective drug would turn out to be sulfinpyrazone rather than aspirin. All that counts is that, since they had, before the end of the trial, (predominantly correctly) identified which participants were on the effective drug, the trial was opened up to all the possibilities of bias we have noted – subsidiary treatment (or "co-intervention") bias, reporter bias and the rest. Although I am happy that he was happy, the grin on the statistician's face was completely misplaced.

Sackett, however, sees the mistaken identification of the likely effective treatment as crucial: "With a 'prior' belief that sulfinpyrazone was effective, when a patient fared well throughout the trial, it was clinically sensible for their neurologist to suspect that they were on it. Similarly, if a patient suffered a stroke during the trial, it was clinically sensible for their neurologist to suspect the double placebo or the aspirin they thought was probably worthless. Thus, our end-of-study test for blindness was exposed as a test for (incorrect [sic]) hunches about efficacy . . . So, the first 'pearl' on offer . . . is that testing for blindness at the end of your trial is a mug's game, because it cannot distinguish the failure of your blinding tactics from their correct [sic] guesses about which treatment was received, based on their experiences of pharmacodynamics, side-effects and trial outcome" (Sackett, 2011, 675).

Taken literally, Sackett here ends in essentially the same inconsistency as Schulz and Grimes did. If clinicians and/or participants are able to guess treatment

---

[18] There is a story to tell about whether the 25% model is correct in this case, but it would only complicate matters unnecessarily to tell it here. The general points about unblinding and end-or-trial tests are independent of that issue.

[19] This is what suggests that the 25% hypothesis was the wrong model. But. to repeat, we shouldn't allow this to spoil the story. Let's assume that Sackett is right that blind was broken, the question at issue is what follows logically from this assumption and this question about logical entailment is, of course, independent of the question of whether or not the assumption is true in the case of some particular trial.

assignment at significantly better than chance rates then, by definition, the trial has become unblinded (despite initial hidden treatment allocation). Hence, except when blind has been broken by illicit access to the randomization code, how *could* an end-of-trial test differentiate the success of their "blinding tactics" from the issue of whether clinicians developed predominantly correct conjectures about which participants were on the effective treatment? If the end-of-trial test finds that the trial has become unblinded, then it would seem that, by definition, the clinicians' blinding tactics have failed (albeit through no fault of their own).

In order to free Sackett's analysis from contradiction, some meaning needs to be found for the phrase "the blinding tactics were a success" that does not entail that blinding need be maintained (and hence an end-of-trial test "passed"). Sackett in fact provides interesting detail on just how painstaking the efforts to keep a trial blind may be in carefully planned trials. He identifies three potential biases (a) contamination of the comparison [control] group with the experimental treatment, (b) cointervention bias and (c) outcome assessment bias. Possible bias (c) was avoided by using a "panel of adjudicators who are blind to treatment" (so they at least would "pass" any end-of-trial test – that is, do no better than chance in identifying treatment allocation). As for (a) and (b) "treatment allocations had been concealed; active drugs and their corresponding placebos were identical in size, color, taste, smell and flotation; we'd told everybody to use acetaminophen for pain; we'd purged uric acid results from all lab reports (sulfinpyrazone is uricosuric); and we'd kept our periodic platelet function test results secret" (Sackett, 2011, 675).

Not all of the considerations that Sackett raises are in fact to do with biases that might be introduced if blind is broken. For example, having clinicians use acetaminophen rather than aspirin to treat any patients in the trial who were in pain was obviously aimed at preventing participants who had been assigned to one of the non-aspirin groups obtaining aspirin as subsidiary treatment alongside the non-aspirin treatment to which they had been assigned. But, for example, not allowing the clinicians access to uric acid results on participants – results which would have given grounds for inferring which participants were being given sulfinpyrazone (which, as Sackett points out, is uricosuric) – *was* an attempt to prevent blind being broken. This is a good illustration of another, subsidiary insight: that if certain side-effects are *expected* from any treatment involved in a trial, then there may be ways in which they can be concealed. If such ways exist, then it is surely good methodological practice to employ them.

Similarly, Sackett makes it clear that sophisticated triallists will pay great attention to making the various treatments similar not only in looks but in other ways that might not initially be considered but which might allow inquisitive clinicians (or participants) to differentiate them – for example, if two types of drug looked the same but smelled differently; or if one type "sank" while the other "swam" when thrown into water (usually in the toilet!).

So, it could be that when claiming that "blinding tactics may be successful" even though blind was broken, Sackett had in mind that all the precautions that were aimed at keeping the trial blind *in so far as possible* were successful. But, even on this reading of his claim, it obviously fails to follow that testing for blindness is a "mug's game". The fact – as we are taking it to be – that, despite all the clever precautions taken in advance, blind was broken in Sackett's Canadian Aspirin trial is important information. It means that a proper methodological analysis of the impact of the result of the trial requires questions to be asked about whether or not the "knowledge" involved in the breaking of blind of which participants were on the effective drug arrived at before the trial had ended had any biasing effect on the eventual outcome. Had the result of the end-of-trial test been "positive" (that is, had the clinicians done no better than chance in discerning treatment assignment) then those questions did not need to be considered. As before, it is certainly important to recognise that "failure" in such a test does not entail that the result was biased, only that it might have been; and, also as before, it is important to recognise that, even if biases did creep in, this does not entail that the outcome can give no solid evidence of effectiveness. But neither of these points means that performing end-of-trial tests is of little or no value.

Sackett's story had a happy ending: his study was published by the *New England Journal of Medicine* (albeit not as the lead article, but certainly as a full article, not the feared Letter to the Editor – (The Canadian Cooperative Study Group, 1978). Sackett records (2011, 675) that "We successfully explained [the situation] to the journal's Editor (if not to one persistently confused referee) and got the trial published." Since what they "explained" to the Editor was that their study "exposed" end-of-trial tests as a test for hunches about side effects and efficacy, and since the assumption that this means there is no need to worry about loss of blinding is itself based on confusion, it seems that it was the Editor that they confused. And it seems likely that the "persistently confused referee" was in fact seeing things clearly: Sackett's colleagues, according to his own story, had become unblinded during the course of the trial and so questions should have been asked about biases, which is not to say, to repeat one final time, that the result *was* biased, only that it might have been.

Sackett develops a secondary argument for downgrading end-of-trial tests for blindness: that revealing the fact that blinding was lost may affect the *perceived* impact of the trial's result. Remember that he "felt the knife go into his heart" when the statistician told him the end-of-trial test had been failed – his reaction being "Had our triumphant lead article been reduced to an apologetic Letter to the Editor?" (Sackett, 2011, 674). Sackett is clearly assuming here that the evidence from a trial known to have become unblinded would have its impact immediately downgraded by the clinical research community in general. And performing tests that are likely to downgrade the impact of your study, *for no good reason*, does indeed seem like a mug's game![20]

---

[20] Indeed, reintroducing the confusion about the allegedly crucial nature of the fact that the prior view of the clinicians was that it would be sulfinpyrazone rather than aspirin that would turn out to be effective, Sackett envisages a "bone-chilling alternative ending [to his Canadian Aspirin story]. What if the neurologists in our ... trial had begun it with the reverse set of hunches, this time thinking that aspirin would probably work and sulfinpyrazone probably wouldn't? Testing for blindness at the end of that trial – forcing us to weaken our conclusions about efficacy and dashing

The qualifier "for no good reason" is, however, clearly crucial here. If trial clinicians were to seek to avoid performing an end-of-trial test knowing that trial had become unblinded and that consequently biases had indeed crept in substantially to affect the result, then this would not count as reasonably refusing to play a mug's game, but rather as evading one's responsibilities to good scientific evidential practice. Presumably Sackett believed that the Canadian Aspirin Trial had been so carefully planned, and the effect-size revealed by the trial large enough (though the latter is arguable – to say the least), that it was legitimate to rule out biases introduced by the acknowledged loss of blinding as an alternative explanation of the observed outcome. But even supposing this is true, to react by in effect recommending that end-of-trial tests are *never* done seems an obvious and significant error: many shoddily performed more recent trials could have hidden (and perhaps have hidden) behind Sackett's "mug's game" claim. The correct reaction is instead to challenge the prevalent community view that loss of blinding immediately and automatically adversely affects the weight of the trial result; and carefully explain why, in your trial at least, provisions were in place to keep the outcome effectively bias-free despite the loss of blinding. It is surely never a good idea to collude in a mistaken view just because it has become socially entrenched. Instead one should aim to re-educate the community and so change its view. But the community's confusion is not to do with whether or not the test separates blinding and side/positive effects (it doesn't because it can't), but instead stems from the assumption that a trial that has become unblinded automatically fails to supply good evidence for the effectiveness of the experimental therapy.

Schulz and Grimes also present a secondary argument, related to, but slightly different from Sackett's. Consideration of this argument will finally lead to a coherent (and dispiriting) concern about the value of end-of-trial tests. Schulz and Grimes write: "Furthermore, individuals [when being quizzed as part of an end-of-trial test] might be reluctant to expose any unblinding efforts by providing accurate responses to the queries – in other words, if they have deciphered group assignments, they might provide responses contrary to their deciphering findings to disguise their actions. That difficulty, along with interpretation difficulties stemming from adverse side-effects and successful clinical outcomes, leads us to question the usefulness of tests of blinding ..." (2002, 699). Laying aside the "interpretation difficulties" as already shown to be confused, Schulz and Grimes are suggesting that clinicians who have in fact discerned which participants are on the experimental treatment have an incentive to hide this fact by not giving true responses to the questions in an end-of-trial test. This is because they fear that if they give accurate responses and hence the end-of-trial test reveals that their trial had become unblinded, then the medical

our hopes of prominent publication – certainly would have been a mug's game!" (Sackett, 2011, 676). This shows how deep the worry that the community will judge on the basis of "unblinded therefore as good as worthless" in forcing them "to weaken [their] conclusions about efficacy". It also, however, aexhibits just how confused Sackett's analysis is: the real story (sulfinpyrazone thought likely to be effective) entails exactly the same concern with lack of blinding as the "blood curdling" alternative.

community is likely to dismiss their trial result as carrying little or no (or, at any rate, greatly reduced) evidential weight. If so, then we cannot rely on the accuracy of end-of-trial tests and this seems – at last – like a good reason to question the utility of such tests.

Schulz and Grimes suggest that clinicians might dissemble in response to an end-of-trial test to cover up the fact that they had correctly discerned treatment allocation (at least to some significant extent). But, again, there are two importantly different reasons that a clinician might have for dissembling. *First* she might know (or, more accurately, have good evidence) that various aspects of the trial protocol made it difficult for bias to affect the result and have good evidence that, if any biases *have* intruded, they would have had only a small effect at most – an effect insufficient to produce a false positive result in the clinical trial. By covering up the loss of blinding, such a clinician, aware of the attitude toward non-blinded trials prevalent in her community, would be trying to avoid an automatic downgrade of the evidential value of their trial's result – a downgrade that she judges (probably implicitly) to be unjustified in this case, and judges it to be unjustified on the basis of evidence. The response to this possibility is the same as the one just given to Sackett's similar but somewhat different concern: namely, that there would be no reason for providing misleading test responses if the clinical trials community were educated into more enlightened evidential practices – into accepting that, in appropriate circumstances, the outcome of a trial that has become unblinded can still provide significant evidence of effectiveness. It would seem, then, that the best advice to such a well-intentioned clinician is to respond truthfully to the end-of-trial questionnaire (as Sackett's colleagues presumably did) and, at the same time, try to re-educate the community.

In this first case, so we are supposing, any bias that affected the trial's outcome as a consequence of unblinding was minimal and so the fact that the end-of-trial blinding test was fudged provides no significant reason to downgrade the evidential value of that outcome. Hence, if we could be sure that all cases of dissembling on end-of-trial tests were like this one, then this possibility of dissembling would form no sort of argument for failing to perform such tests. Not everyone will dissemble, and if they don't, then the test supplies accurate and useful information; while if they do dissemble, then, although the result of the blinding test will be inaccurate, no real evidential harm is done so far as the impact of the outcome of the clinical trial is concerned.[21]

However, there is a *second* reason why a clinician might be motivated to cover up loss of blinding in her trial and hence dissemble when subjected to an end-of-trial test; and this reason is undoubtedly problematic. Clinical triallists generally are under great pressure to produce "positive" results – i.e. outcomes that involve refutations of the null hypothesis of no difference between experimental and control treatments and hence are taken to support (sometimes even to 'establish') the

---

[21] Except, as indicated earlier, that an opportunity has been missed to help eradicate an evidential mistake (that unblinded trials are automatically non-telling).

effectiveness of the experimental treatment. Despite being exposed any number of times as nonsense, the implicit belief that "Not statistically significant means insignificant" – i.e. that "negative results" are of no interest – still seems deeply-embedded in the collective medical psyche. For many years it was next to impossible to get a study with a "negative" result published in medicine. And, although editors have recently wised up somewhat, the problem has by no means entirely disappeared. An even greater worry is the influence of "Big Pharma" which of course has a vested interest in clinical trials yielding "positive results" for any treatment it wants to market. Anyone who thinks this is an exaggerated worry should rush to read (Angell, 2004). Angell was on the editorial staff of *The New England Journal of Medicine* for over 20 years, rising to Editor-in-Chief and writes (2004, xxvi–xxvii): "As I saw industry influence grow, I became increasingly troubled by the possibility that much published research is seriously flawed, leading doctors to believe new drugs are generally more effective and safer than they actually are". So, suppose such an investigator under pressure to produce a "positive" result has discerned treatment allocation during the course of a trial and has used that information to bias the trial outcome – by for example giving special care and attention to those she has (it turns out correctly) conjectured are on the experimental treatment. Such an investigator would have a very good reason to dissemble when completing end-of-trial questionnaires to test for the retention of blinding.

This is certainly worrying. But, although it might provide reason to doubt any "all clear" produced by an end-of-trial test for blinding, it is clearly not at all specific to such tests: it simply adds to the (depressingly long) list of reasons to be suspicious of clinical trial results, especially when financed (and often controlled) by pharmaceutical companies. If the possibility that clinicians will give inaccurate replies to the questions in an end-of-trial test is to be taken as a reason to downgrade the significance of such tests and to leave them out of the guidelines for good practice (as in CONSORT), then, to be consistent, all the usual requirements for good practice should be similarly downgraded. Might clinicians lie when reporting that their trial was properly randomized? Of course they might. Might clinicians lie when reporting that initial treatment allocation was hidden from all clinicians and participants? Of course they might. Might clinicians report outcomes for individual participants inaccurately? Of course they might. But no one, I take it, would infer from these possibilities that imposing requirements such as proper randomization, hidden treatment allocation, accurate reporting of data, etc., is playing a series of mug's games. There is nothing special then, in this regard, about requiring that end-of-trial tests for blinding are performed and their results reported.

Science relies on the honesty of scientists. Of course, self-correcting mechanisms are sometimes in play – principally repetition of the trial by different scientists; but in general it just needs to be taken as the default position that the scientists reporting a result have acted honourably and reported accurately – not allowing the mere possibility of malfeasance to detract from the impact of published results. The fact that clinicians may not respond truthfully to end-of-trial test questionnaires is worrying but is no reason at all for a general downgrade of the methodological value of such tests.

Summarizing, then, my response to the analyses by Schulz and Grimes and by Sackett: the "hunches" issue supplies, on analysis, no coherent reason at all to devalue end-of-trial tests for blinding; and downgrading them is the wrong response to the secondary worry about the reputation, and hence evidential impact, of the trial if such a test is "failed". Instead of holding that performing such end-of-trial tests is a "mug's game", the correct inference from the valid points that we identified earlier in Sect. 2.3.1 as underlying those analyses is surely that it is always a good idea to do end-of-trial tests on trials that begin double-blind but that we should be careful not to over-interpret the results of such tests – in particular, if they turn out to be "negative". More exactly: if the result of the end-of-trial test is that clinicians or participants do not do significantly better than chance in "guessing" treatment assignments there is no issue of post-randomization bias (unless they are dissembling – and we cannot let this mere possibility instil a generalised scepticism). If, however, the result is that clinicians and/or participants do perform at significantly better than chance, then thought needs to be given to the question of whether the biases that breaking of blind makes possible have plausibly been actualized. If there is indeed reason to think that biases have crept in, then the impact of the evidence from the trial's result should be duly downgraded. But if, on the contrary, either (a) the protocol of the trial left unblinded clinicians or participants with little or no scope to use their knowledge of treatment assignment to introduce bias, or (b) the effect size "revealed" by the trial was large and background knowledge endorses the view that biases could, at worst, have contributed only little to that overall effect, then the trial result should still carry its full evidential impact.

2.4   The CONSORT Guidelines Re-revised?

It was always a mistake for the CONSORT guidelines to be as lukewarm about blinding and about tests for the retention of blinding as they were; and the 2010 revision, by omitting mention of end-of-trial tests entirely, made matters worse. The above analysis, if correct, surely mandates a re-revision of these guidelines in this regard.

The current recommendation concerning blinding reads (*www.consort-statement. org*).

Blinding: If done, who was blinded after assignment to interventions (for example, participants, care providers, those assessing outcomes) and how.

An additional clause (in bold below) seems, however, to be required if an accurate assessment is to be made of the weight of evidence provided by a trial's outcome:

Blinding: If done, who was blinded after assignment to interventions (for example, participants, care providers, those assessing outcomes) and how; **if not done, then reasons should be provided for why the biases that lack of blinding makes possible can reasonably be thought, on the basis of the protocol of the**

trial, of background knowledge and of other evidence, to have had at most a minor effect on the trial's outcome.

As for tests for the retention of blinding, the recommendation in the original version of CONSORT was conditional (essentially: "if such tests were done, state who (participants, clinicians, outcome assessors ..) were tested"); while the deliberate omission of any mention of such tests in the 2010 revision surely encouraged the belief that those tests have little or no role in assessing the weight of evidence from the result of the trial, and hence the belief that performing them is barely, if at all, worthwhile. As we have seen, this omission was based on confused thinking, and in fact tests for blinding should ideally always be done (unless the trial is from the outset an unblinded study). This is because those tests always deliver information that is important in the assessment of weight of evidence: if the test is performed and the outcome of the test is "positive" (that is, the trial remained blinded throughout) then the result of the trial carries full evidential weight; if the test is carried out and it is "negative" (that is, blind was likely to have been broken), then issues arise about *possible* biases having affected the result.

As mentioned earlier, the CONSORT team deliberately restricted itself to setting guidelines for how clinical trials should be *reported* – rather quixotically, it might be thought, since how trials are performed is clearly a more important issue than how they are reported (even though the two, of course, interrelate in important ways). However, even restricting attention to the reporting of trials, an entry for 'Tests for Blindness' should surely be reinstated in the guidelines; and the text might read as follows:

Tests for Blindness (At the End of the Trial) If done, then who was tested for retention of blindness (for example, participants, care providers, those assessing outcomes), and with what result; if either not done, or done and the outcome was "negative" (blinding not retained), then state what aspects of the protocol, if any, or of the result provide warrant, given background knowledge, for thinking that the biases that lack of blindness makes possible did not in fact affect the outcome of the trial or affected it only to a negligible degree.

Although it is implicit in this wording, it might be best to make the requirement explicit that trialists should also ideally report, in the case that the end-of- trial test reveals loss of blinding, on whether the people questioned based their discernments of treatment arm on unusual side effects or on large (apparent) positive effect (or on both). This is because, as noted earlier (but invariably ignored in medicine), quite different considerations apply to the two ways in which blind may be broken. If this has happened entirely through noting unusual side-effects, then, there was no clear indication ahead of the loss of blindness that the treatment under trial is effective. Hence the "positive" result in the trial may be entirely the product of the biases that loss of blindness has allowed – especially if the effect size is small and the sample size large. On the other hand, if loss of blinding was brought about by a "large positive effect" of the treatment under trial, then, although strictly only an "apparent" effect, it may well be reasonable to judge on the basis of evidence-based background knowledge, that possible biases introduced in the wake of loss of blinding could not alone have produced an effect of anywhere near that size.

I am sorry that, if implemented, this would make the guidelines more complicated; but accuracy is of course the dominant virtue here. Albert Einstein famously averred: "Physics should be as simple as possible, but not more so". Guidelines too.

References

Angell, M. (2004). *The truth about the drug companies: How they deceive us and what to do about it* (Paperback ed.). Random House.

Benedetti, F. (2014). *Placebo effects* (2nd ed.). Oxford University Press.

Boutron, I., Estellat, C., & Ravaud, P. (2005). A review of blinding in randomized controlled trials found results inconsistent and questionable. *The Journal of Clinical Epidemiology, 58*, 1220–1226.

Ferguson, D., Glass, K. C., Waring, D., & Shapiro, S. (2004). Turning a blind eye: The success of blinding reported in a random sample of placebo controlled trials. *The British Medical Journal, 328*, 432–436.

Howick, J. (2011). *The philosophy of evidence-based medicine*. Wiley-Blackwell (BMJ Books).

Howson, C., & Urbach, P. M. (2004). *Scientific reasoning: The Bayesian approach* (3rd ed.). Open Court.

Hrobjartsson, A., Forfang, E., Haahr, M. T., et al. (2007). Blinded trials taken to the test: An analysis of randomized clinical trials that report tests for the success of blinding. *The International Journal of Epidemiology, 36*, 654–667.

Locke, J. (1689). *An essay concerning human understanding*. Thomas Basset.

Ney, P. G., Collins, C., & Spensor, C. (1986). Double blind: Double talk or are there ways to do better research? *Medical Hypotheses, 21*(2), 119–126.

Noseworthy, J. H., Ebers, G. C., Vandervoort, M. K., Yetisir, E., & Roberts, R. (1994). The impact of blinding on the results of a randomized, placebo-controlled multiple sclerosis clinical trial. *Neurology, 44*, 16–20.

Sackett, D. L. (2004). Why we don't test for blindness at the end of our trials. *The British Medical Journal, 328*, 1136.

Sackett, D. L. (2011). Clinician-trialist rounds: 6. Testing for blindness at the end of your trial is a mug's game. *Clinical Trials, 8*, 674–676.

Schulz, K. F., & Grimes, D. A. (2002). Blinding in randomized trials: Hiding who got what. *The Lancet, 395*, 696–700.

Schulz, K. F., Altman, D. G., Moher, H., & for the CONSORT group. (2010). CONSORT 2010 statement: Updated guidelines for reporting parallel group randomised trials. *The British Medical Journal, 340*, 698–702.

Senn, S. J. (2004). A blinkered view of blinding. *The British Medical Journal, 328*, 1135–1136.

The Canadian Cooperative Study Group. (1978). A randomized trial of aspirin and sulfinpyrazone in threatened stroke. *The New England Journal of Medicine, 299*, 53–59.

Urbach, P. M. (1993). The value of randomization and control in clinical trials. *Statistics in Medicine, 12*(15–16), 1421–1431.

Worrall, J. (1988). The value of a fixed methodology. *The British Journal for the Philosophy of Science, 39*, 263–275.

Worrall, J. (2006). Why randomise? Evidence and ethics in clinical trials. In W. J. Gonzalez & J. Alcolea (Eds.), *Contemporary perspectives in philosophy and methodology of science* (pp. 65–82). Netbiblo.

Worrall, J. (2007a). Why there's no cause to randomize. *The British Journal for the Philosophy of Science, 58*(3), 451–488.

Worrall, J. (2007b). Evidence in medicine and evidence-based medicine. *Philosophy Compass, 2*(6), 981–1022.

Worrall, J. (2010). Evidence: Philosophy of science meets medicine. *Journal of Evaluation in Clinical Practice, 16*(2), 356–362.

Chapter 3
Environmental Decision-Making Under Uncertainty

Joe Roussos, Richard Bradley, and Roman Frigg

Abstract Extreme weather events like hurricanes occur rarely, but when they occur, they cause immense damage. How should decision-makers, both public and private, make decisions about such events? Such decisions face significant and often poorly understood uncertainty. We rework the so-called "confidence approach" to tackle decision-making under severe uncertainty with multiple models, and we illustrate the approach with the case study of insurance pricing using hurricane models. The confidence approach has important consequences for this case and offers a powerful framework for a wide class of problems.

Keywords Uncertainty · Confidence · Insurance pricing · Averaging · Catastrophe model · Hurricane model · Extreme weather event · Climate change

3.1 Hurricane Maria

When Hurricane Maria hit Dominica in September 2017 it devastated the island nation, causing landslides, widespread flooding, and damage to the roofs of almost every home. The prime minister, Roosevelt Skerrit, had to be rescued from his

We thank Tom Philp for numerous discussions about hurricane modelling, and Jan-Willem Romeijn, Sean Gryb, Simon Dietz, and Jonathan Livengood for helpful discussions about different aspects of the paper. Thanks to Wenceslao J. Gonzalez for inviting us to present the paper at the conference *Philosophy of Science Today*, which took place at the Universidade da Coruña, Spain, in March 2020, and to the audience at the conference for stimulating questions.

J. Roussos (✉)
Institute for Futures Studies, Stockholm, Sweden
e-mail: joe.roussos@iffs.se

R. Bradley · R. Frigg
Department of Philosophy, Logic, and Scientific Method, London School of Economics, London, UK

© The Author(s), under exclusive license to Springer Nature Switzerland AG 2022
W. J. Gonzalez (ed.), *Current Trends in Philosophy of Science*, Synthese Library 462, https://doi.org/10.1007/978-3-031-01315-7_3

official residence.[1] The island lost all radio, cell phone and internet services after the storm.

Each year, summer in the Northern Hemisphere brings hurricanes like Maria to the west Atlantic. Also known as "tropical cyclones" or typhoons, these giant, rotating storms wreak havoc in the Caribbean and the southeast of the USA, leading to deaths, evacuations, and billions of dollars of damage each year.

Maria was one of the worst storms on record, and part of the costliest hurricane season on record for the Atlantic, with the final bill for storm damage in 2017 exceeding £228 bn. The root of the damage is the incredibly fast winds that hurricanes generate. The wind itself is strong enough to damage cars, trees, and houses. A "category 1" hurricane, the lowest grade, has winds up to 150 km/h, enough to snap branches off trees and cause flying debris. Category 3, the boundary for a "large" hurricane, has winds strong enough to rip the entire roof from a house. At category 4, not even the wooden walls on American houses withstand the storm. But most hurricane damage comes from water, whipped up by the roaring winds. Meteorologists call this "storm surge": sea water is pushed into fierce waves, metres higher than the usual sea-level. The water slams into the coast and causes flooding for kilometres inland. Coastal buildings are demolished by the frequent pounding of waves during the storm, and flooding seawater erodes beaches and coastal highways, and undermines the foundations of buildings. It is hard to typify the damage caused by category 5 storms like Maria, as this is a catch-all for wind speeds above 252 km/ h. They are catastrophes almost without parallel.

3.2   Extreme Weather Events and Climate Change

Hurricane Maria is an example of an "extreme weather event". An extreme event is one that occurs relatively rarely but has huge impact when it occurs. Droughts, heavy rainfalls, floods, and heat waves, as well as the increased incidence of extremely high sea levels or the more frequent occurrence of particularly hot days are extreme weather events. Not only are these events highly destructive in themselves; they often also have devastating consequences. Heatwaves cause the deaths of vulnerable people; heavy rainfalls cause landslides; and the more frequent occurrence of hot days creates ideal conditions for wildfires.

In its most recent assessment report, the Intergovernmental Panel on Climate Change (IPCC), the United Nations body for assessing the science related to climate change, comes to the conclusion that climate change, whether driven by natural causes or human activities, can result in changes in the likelihoods of the occurrence

---

[1] See, for example, https://www.nytimes.com/2017/09/19/world/americas/hurricane-maria-carib bean.html

or severity of extreme weather events.[2] The IPCC also reports that such changes have indeed been observed since about the 1950s. They report, for instance, that it is likely that the frequency of heat waves has increased in large parts of Europe, that anthropogenic greenhouse gas emissions are likely to be a contributing factor, and that further changes in the future are very likely. They report similar findings for other extreme evens like extremely high sea levels and tropical cyclones.

If we don't want to be hit by these events unprepared, we have to plan. Individuals as well as organisations will have to adapt to the fact that in the future we are more likely to be exposed to extreme weather events than we were in the past, and that these events are likely to be more severe than the ones we have hitherto experienced. Adaption policies put measures into place to help people cope with the effects of climate change, for instance by building flood defences, making buildings hurricane-proof, or increasing the capacity of water storage facilities.

Adapting our infrastructure so that it is able cope with extreme events is expensive, and adaptation requires significant resources. But how much exactly should we invest into adaptation measures? This will depend on what we think the damage that such events cause will cost. On the one hand, we don't want to overspend and invest significantly more into adaptation than any potential damage would be. On the other hand, we don't want to underspend and make ourselves vulnerable to huge losses. To design concrete policies and to secure their implementation, an assessment of the potential damages and their associated costs is indispensable.

Nowhere else is the price of a disaster as "in our face" as in the insurance sector. An insurance company will put a price on the potential damage, assess the likelihood that the damage will occur, and then combine the two into the price of an insurance policy. If they price their policies too low, they will go bust; too high and no one will buy their products. Insurers have to strike the same kind of balance that adaptation policies do. In fact, many of the calculations are the same: how much a state should invest into, say, flood defences will depend on what the expected damage due to flooding is. Looking at insurance pricing therefore gives us insight into how such assessments are made, and the problems that they encounter.

3.3 Hurricane Insurance

For people living in Florida, or on a Caribbean island, the risk of hurricane damage to their home is one of the most serious they face. Naturally, an insurance industry has grown around this risk, offering home-owners protection against the various forms of destruction hurricanes can bring. Residents buy insurance policies that

---

[2]IPCC (2013, 121). For detailed overview of expected changes in extreme weather events of different kinds and their likely causes see IPCC (2013, 110). For a discussion of climate change and the philosophical and methodological question that it raises see Bradley and Steele's (2015), Frigg et al.'s (2015b, c), and Parker's (2018).

guard them against such damage, at frequently high cost: a house insured for £120,000 will cost £2500–6000 *per year*.[3]

The price is so high in part because hurricanes cause significant damage, but also because insurers are so *uncertain* about how risky it is to insure. If you want to sell insurance against something the recipe is simple, with just three ingredients. First, you need the likelihood of the event you're covering (the hurricane). Second, you need an estimate of how damaging these events are when they occur — how much damage, in pounds, does the average hurricane cause in a £120,000 house? Third, you need to obey insurance regulations that tell you how much money you need to have available at any given time. These rules exist to ensure that insurance companies don't go bankrupt and have the money to pay for claims when customers make them.

But the first two ingredients are difficult to work out for hurricanes. Calculating the likelihood of destructive hurricanes requires a detailed understanding of the science of meteorology. Estimating the vulnerability of a building to hurricane damage — in order to determine the monetary value of the damage — requires knowing how it was built, and how the building-materials will withstand the wind and water effects of the storms.

The scientific challenge of predicting hurricanes raises some surprising philosophical challenges. In 2016, we were approached by a team of scientists working for an insurance company who had been reading our previous philosophical work on scientific modelling and decision-making in the face of severe uncertainty. They asked for our help, and thus began a research collaboration on the philosophical challenges of insuring against hurricanes — the first ever, we would bet! In this article, we will share some of what makes hurricane insurance so philosophical interesting. We will look at how such insurance is done today, and why insurers were so unsatisfied that they brought in the philosophers.

3.4   How Do You Price a Hurricane Like Maria?

In most kinds of insurance (health, fire, theft, and so on), insurers calculate the likelihood of the event being insured against by looking at historical statistics. When buying an insurance policy, facts about you (or your house, car, etc.) are used to estimate how likely you are to experience the "event" you are insuring against (fire, theft, etc.). Your car insurance premium is calculated using things like your postcode, age, gender, and even the colour of your vehicle. Insurers looks at the statistics for burglary, accidents, and so on, for people in your area, of your age and gender, or with your colour car. From the insurer's point of view, they do this to try make it more likely that they'll make a profit from your insurance. From your point of view,

---

[3] See https://www.sapling.com/7958883/average-cost-hurricane-insurance

it is important that they do this so that they don't run out of money before they can pay your insurance claim when you make it.

In the case of hurricanes, insurers can't do this, simply because there isn't enough data on destructive hurricanes: even in the USA, which has sophisticated records and experiences hurricanes in most years, there is too little data for actuarial modelling.[4] HURDAT2, the official database for hurricanes striking the Atlantic coast of the USA has ~300 storms to date and only 1/3 of those qualify as "major hurricanes". If the dataset is split by region, the numbers drop precipitously.[5] Compare this to the 6 million car accidents per year in the USA, and you can see why insurers have a much harder time with hurricanes than cars.

But if you live in a hurricane-prone region like Dominica, or south Florida, you *need* hurricane insurance. Insurers also know that, as destructive as Hurricane Maria was, such events are rare. They *should* be able to offer sustainable hurricane insurance, if only they can reliably fill the gap that the missing statistics usually play in their pricing process.

The probability of a hurricane hitting south Florida can be calculated in another way than the statistical approach used for car insurance: using scientific models. These models contain numerical representations of hurricanes: equations from physics and statistics that describe how the storms form, how they grow, and how they move across the Atlantic. In part, these models are based on what we know about the physics of hurricanes. For example, we know that tropical storms get the energy that makes them so ferocious from the sea: the warmer the sea-surface, the more energy is available to "drive" a large storm. But the models also use the statistics we have on where exactly in the Atlantic past storms formed, on periods of higher and lower hurricane activity, and so on.

The insurers treat the models as experts: they take the outputs of their models as an input for insurance pricing, an input that they can't easily check because of the specialist knowledge required to produce it. They have to do this, if they want to avoid hiring their own scientists and building their own models, but it raises some problems.

Insurers know that scientists disagree; not only about some key questions of hurricane science, but also about how to take that science and put it into a computer simulation. The result is that there isn't just one model of hurricane formation in the Atlantic; there are many. When the Florida Commission on Hurricane Loss Projection Methodology, the industry regulator who licenses modelling firms, carried out

---

[4]For a discussion of hurricane modelling for insurance, see Shome et al.'s (2018), and for a discussion of hurricane risk the contributions to Collins and Walsh's (2019).

[5]We're referring to the number of datapoints in HURDAT2 for hurricanes that make landfall on the USA's Atlantic coast. The full database is at http://www.aoml.noaa.gov/hrd/hurdat/All_U.S._Hurricanes.html

their 2007 assessment of the modelling industry, they gathered a collection of 972 models![6,7]

Just knowing about these disagreements makes life hard for insurers, as they have no good way of choosing which model to buy. (Or which modelling company to hire.) The reasons for the disagreements are Greek to them, but any decision they make will boil down to a decision between the different models on offer. But insurers don't like the thought that they're implicitly making a choice on scientific or modelling questions they don't understand.

But if you are selling hurricane insurance, there is another option: you can hire a company like Risk Management Solutions (RMS), a leading modelling firm that uses a collection of thirteen models.[8] RMS's models represent (some of) the different views present in the scientific community. Rather than taking a stand on these disagreements themselves, they try to have a model for each major position. As an insurer, this is an attractive option: it offers you a way to stay out of scientific disputes. But, as we will now show, it doesn't quite solve the problem in the way insurers might hope.

3.5   Growing Dissatisfaction with Hurricane Pricing

RMS uses multiple models because they understand the insurers' discomfort with scientific uncertainty. They also know that insurers need a definitive answer on how to price their products. And so, like many others in their position, RMS combines the outputs from their thirteen models into a just one number. So, any time an insurer asks the RMS software for the probability of a hurricane hitting a particular place, they don't see thirteen different probabilities, they see one: RMS's recommended view, called the Medium-Term Rate.

The Medium-Term Rate is constructed by averaging the answers provided by each of RMS's thirteen models, in what is called a weighted average.[9] In a simple

---

[6]The report is FCHLPM, "Report to the Florida House of Representatives Comparison of Hurricane Loss Projection Models," 2007, https://www.sbafla.com/method/Portals/Methodology/Meetings/2007/20071105_RubioReport.pdf. For a discussion of the 972 models, see Jayanta Guin, "Understanding Uncertainty," AIR Worldwide blog, 2010, http://www.air-worldwide.com/Publications/AIR-Currents/2010/Understanding-Uncertainty/

[7]For discussions of model ensembles and their relation to uncertainty see Knutti's (2010), Marinacci's (2015), Parker's (2010, 2011, 2013), and Stainforth et al.'s (2007a, b). For a discussion of probabilities in relation to ensembles see Frigg et al.'s (2015a) and Smith et al.'s (2014).

[8]Tom Sabbatelli and Jeff Waters, "We're Still All Wondering—Where Have All The Hurricanes Gone?," *The RMS Blog* (blog), October 27, 2015, http://www.rms.com/blog/2015/10/27/were-still-all-wondering-where-have-all-the-hurricanes-gone/

[9]Tom Sabbatelli, "Catastrophe Modeling—Part 2," *The RMS Blog* (blog), September 2, 2017, http://www.rms.com/blog/tag/catastrophe-modeling/page/2/; InsuranceERM, "RMS Responds to AIR's Attack on Hurricane Risk Modelling," Insurance ERM, May 29, 2018, https://www.insuranceerm.com/news-comment/rms-responds-to-airs-attack-on-hurricane-risk-modelling.html

average, the answers are added together the answers and divided by 13. This gives each model equal say, or "weight", in the final answer. In a weighted average, some models can count more than others. RMS decides how much of a say to give to each model by scoring each one on how well it accounts for the past hurricane record.

They measure this by setting each model a "predictive test". The test asks them to calculate the hurricane activity during some historical period, like 1970–1975. When making these calculations, the models are only allowed to use data from before 1970. Each model's outputs for 1970–1975 is compared to the actual data for that period and given a score. The models are tested against many periods, and their overall scores are taken as a sign of each model's skill at predicting hurricanes. These scores then become weights in the average: the better a score, the more that output counts towards the overall answer.

The insurers aren't satisfied with this approach (Philp et al., 2019). That's partly because it is their job to worry about uncertainty, and to constantly strive to gain a better understanding of what they're insuring. But there are specific worries in the case of hurricane insurance, and model averaging in particular, that brought them to us.

The biggest problem is that averaging conceals important information from decision makers about just how uncertain the underlying science is. The averaging process, and the neat software packages that present the averaged results to insurers, focus attention on just one number. To the underwriter using the software, the underlying messy science, and the even messier reality, are swept under the carpet and they tend not to think about the fact that scientists disagree about important questions in hurricane science and modelling.

Although they can't understand the content of these disagreements, the people pricing insurance currently receive no information about them *at all*. But they could, we think, understand and make use of *some* facts about the disagreement. How "spread out" the results from different models are is important information, telling us something about the state of scientific knowledge about a question. The more spread out the results are, the more uncertainty and imprecision there is. This, we say, is valuable information that the decision-maker should use — and, importantly, can *use* even if the details of the disagreement are beyond their grasp. Later we will show how they can use it.

There are also two more technical problems with averaging model results. The first is that the predictive tests that are used to weigh each model's skill use historical data. This data, remember, is too little to use directly for insurance pricing — that is why we needed the simulation models in the first place. But here, the historical data is playing a key role, behind-the-scenes, in determining how the models are evaluated. It remains a weak link.

The second is that there isn't just one way to score the test. For predictive tests like these, there are many different "scoring rules". To see why, let's look at a simpler scenario. You want to know whether it will rain tomorrow, and so you check three different weather services. They say it is 30%, 50% and 80% likely to rain, respectively. Now if it does rain, how do we score those predictions? Some things seem straightforward: the highest prediction did best, because it did rain. But how much better did it do than the 50% prediction? On the one hand, 80% is 30 percentage points above 50%, which should play a role in measuring the difference between them. On the other hand, 50% is barely a prediction at all — it is what you might say if you didn't know anything at all. Shouldn't there be a bonus for "sticking your neck out"?

Statisticians disagree about how to answer these questions, with the result being that there are a great many competing rules.[10] These rules can disagree widely: ranking the predictions in completely different ways, and therefore leading to very different average answers.[11] This puts the insurer in a difficult position: they went to the modelling company because hurricane scientists disagreed, and they couldn't adjudicate that disagreement for themselves. But now it turns out the modellers themselves face a disagreement amongst still other experts: statisticians disagreeing over scoring rules. If insurers don't want their prices to reflect just one view of the science of hurricanes, they can reasonably say they don't want it to reflect just one view of the statistics of scoring predictions, either. But what can they do?

The insurers we work with worry about these problems. They try to compensate for them, "factoring in" their dissatisfaction by, for example, inflating the average probabilities. But while the worry is reasonable, they have no good way of choosing how much to inflate them by.

3.6   Less Precision, More Flexibility

Working with our insurance partners, we have developed a different way to price hurricane insurance.[12] It avoids these problems with averaging and gives insurance decision-makers a more flexible procedure for navigating scientific uncertainty.

Our approach starts with a simple thought: instead of trying to compress the disagreement between the three weather predictions (30%, 50%, 80%) down to one prediction (like the average, 53%), why not simply give a range? "It is 30–80% likely to rain tomorrow."

That's one way of dealing with the disagreement, but it creates its own questions.[13] The simplest question is: how exactly do we form the range? 30–80% includes all the predictions, but often if we consider *every* viewpoint we end up with unhelpfully wide ranges: if the probability is 30% there is no need for an

---

[10] See for example the list of rules at Australian Bureau of Meteorology, "Forecast Verification," 2017, https://web.archive.org/web/20171125111801/https://www.cawcr.gov.au/projects/verification/

[11] For a discussion of this problem in the case of climate models, see Stainforth et al.'s (2007a).

[12] Our approach here reworks a recently developed decision theory called the *confidence approach* to tackle inputs from model ensembles. For a discussion of this approach see Bradley's (2017) and Hill's (2013, 2019).

[13] For a discussion of such "imprecise probabilities", the questions they raise, and how to make decisions with them, see S. Bradley's (2019).

umbrella, but if it is 80% one is practically required. If we consider all the options, we might end up *always* carrying an umbrella, which would be a nuisance.

This raises our second question: how do you make a decision when given a range like 30–80%? If you're told just one number, say 50%, it is simple. You think about your options (take an umbrella, don't) and consider what's likely to happen. If you take the umbrella, there's a 50% chance you carry it for no reason, and a 50% chance it rains and you use the umbrella to stay dry. If you don't take an umbrella, there's a 50% chance you enjoy your day unencumbered, and a 50% chance you get wet in the rain. The classic advice from economists is to "maximise expected utility"; in other words, to pick the option that does best on average. Some numbers can help to see how this works: let's say you really hate getting wet and represent that with the number -10. You don't like carrying an umbrella unnecessarily, but that's only -2. Staying dry using your umbrella is better than that, let's say $+4$. Finally, enjoying a sunny day unencumbered is best, $+6$. If you don't carry an umbrella, then you might enjoy a sunny day ($50\% \times +6 = 3$ units of expected utility), but you might also get wet ($50\% \times -10 = -5$). So, your total expected utility is -2 if you don't take the umbrella. If do you carry the umbrella, there's a chance you do so unnecessarily ($50\% \times -2 = -1$) and a chance it keeps you dry ($50\% \times +4 = 2$), which gives a total of $+1$. You choose the act with the higher expected utility, so you take the umbrella.

But, if you have a range of probabilities for rain (30–80%) and a range for no rain (20–70%), you can't follow this advice. Economists and philosophers have offered various alternative rules for deciding with ranges of probabilities; one example is to be cautious and choose the option that does best in the worst-case scenario.[14] A problem for all of these rules is that they're one-size-fits-all, and you might not want to commit to just one way of making your decision. The economist's rule about maximising expected utility is meant to be universal, and so are the alternatives for ranges. But in practice, people decide differently based on what is at stake. If we're deciding something trivial, like taking an umbrella when we leave the house, we might be happy to pick the middle of the range and decide using it. But if it is a life-and-death decision, we can't afford to ignore worst-case scenarios, and so we'd want to be more cautious and think about the full range.

3.7   The New Method

Our method for making this kind of decision is designed to avoid these problems.

Here is a colourful example to show how it works. Suppose you are deciding whether to place a bet on your favourite contestant, let's call him Kevin, winning a dance contest. To place the bet you pay £50 upfront; if he wins you are paid back your £50 and receive another £50, if he doesn't win you lose your £50. So, you

---

[14] For a review of various alternatives, see Heal and Millner's (2018).

should expect this bet to make you money if the probability of Kevin winning is more than 50%.

But since you don't know the probability of Kevin winning, your situation is a bit like the weather predictions case. So, we will start off by working through various probabilities for him winning, and see which you accept. We start with the widest range: the probability of Kevin winning is between 0% and 100%. You should obviously believe that, but it is no help. We then narrow down the range, in various ways and see which you accept. This is based just on your subjective estimate of his chances, using your extensive experience in dance venues, your knowledge of how judges rule on performances, and your familiarity with Kevin's skills. Let's say that as we suggest various ranges to you, you accept the claim that his chance of winning is between 20–80%, and 30–50%, and so on, down to your best guess for a single number: 42%.

Now if we forced you to do this, it would be natural for you to protest that you aren't very sure at all about the precise number 42%. You're right. The crucial thing to realise is that your protestations hold the key to solving the problem. In fact, you will likely have protested about the 30–50% interval. In fact, every move toward a narrower interval may, and likely will, go hand in hand with decreasing confidence. The method asks you to pay attention to how your unease increases and keep track of your confidence in each of these claims, where "confidence" means how certain you feel that the chance of Kevin winning is in a certain range (or, in the case of 42%, is equal to that number). You should be less confident about the more specific claims; indeed, it would be incoherent to be more confident that the "right" probability is 42% than that it is between 30% and 50% because that range *includes* 42%!

You now have a ranking of your different estimates: most confident 0–100%, next most 30–80%, and so on down to least confident 42%. That tells us how the different ranges compare to one another. But, remembering our example of the umbrella decision and the life-threat decision, we want to get a sense of how confident you are in absolute terms. If your bet was for your life, rather than £50, you would want to be very confident in your probability estimate. Not just *more confident* than in some other guesses, but *confident enough to bet your life on*.

We can do this by putting the different ranges of probability you considered into groups, which we'll call Low, Medium and High confidence. Which category each range ends up in depends on how much *evidence* you used when you decided that the chance of Kevin winning was in that range. The more evidence, the higher the category. Imagine two people making this same bet. Joe is a dance aficionado; he goes to dance shows regularly and knows a lot about these competitions. He knows about each of the competitors, and how they've performed against each other before. Using that experience, Joe judges Kevin's chances to be between 30% and 50%. Roman, on the other hand, is new to dance. He hasn't been to a dance show before and has only just heard of Kevin. He also judges Kevin's chances to be between 30% and 50%, but it is an uninformed guess. Joe's judgement would get High confidence, but Roman's only Low.

Let's say that your experience with dance competitions, and what you know about Kevin, is enough to justify categorising your 30–50% estimate as Medium confidence; your 20–80% estimate as High confidence; and your 42% estimate as Low confidence.

Why do all this? To come to a decision two further issues are crucial: the importance the decision has to you and your attitude to risk. The decision of taking an umbrella is trivial and hence not very important, but if you're shot if you get it wrong, then the decision really matters. How you decide will also depend on whether you value safety or whether you are willing to take risks. An adequate decision algorithm must take these two factors into account, and having various ranges of probabilities ranked by the confidence you attach to them helps you to do so. We now see how.

The "stakes" of a decision reflect how important the decision is to you. We're going to think of stakes as a number on a 0–1 scale, where 0 is totally unimportant and 1 is the most important decision you can imagine. There are different ways we could measure the stakes of a decision, but for now we'll simplify and say that for you the importance of a decision is determined by the worst possible outcome. So in the example of the dance competition the "stakes" are determined by the money you stand to lose: the £50. How important is losing £50 to you? If you are poor, it might be very important! If you're rich, you might not even notice it. For the moment let us say you think this is a moderately important decision and assign it stakes $s = 0.5$.

Second, we want to know how cautious you are. If Joe is very cautious, he might want to be very confident when he makes choices, even if there isn't much at stake. Roman, who is much bolder, is willing to make even important decisions using little evidence. What we want to know for your bet is your answer to the question "how much confidence do you need in order to make moderately important decisions, with stakes around 0.5?" The answer to the question should be one of our confidence levels: Low, Medium, or High. Joe, who is very cautious, will answer High. Roman, who is gung-ho, will say Low.

Let's say that you fall in between them again, and you answer: Medium. This tells you which one of your probability estimates you should use, based on your caution and the stakes (the importance of the decision to you). When we categorised your probability estimates, we put 30–50% into the Medium category, so you should use 30–50% to make your decision.

This is our answer to the first problem for deciding using ranges: which range? The answer is: the range that fits your desire for confidence, which is based on how cautious you are and what is at stake.

You can now use one of the rules that economists and philosophers have suggested for deciding using a range. Let's keep things simple and say you'll use the cautious rule, by choosing the option that you expect to do best if things turn out for the worst, from your perspective. This rule is called "maximin" expected utility. Here's how it works: you start with one of the options (e.g., bet on Kevin), and calculate the expected utility of that option using each of the probabilities in the range (which here is 30–50%). You're looking for the *minimum* expected utility the option has. So, you start by working out the expected utility for 30% (for the moment we'll pretend money and utility are the same): $30\% \times 50 + 70\% \times -50 = -20$. You do the same for all the probabilities in the range, working up to 50% where the expected utility is 0. The minimum of all these values is −20, so that's the only number you need to care about for the "bet" option. You then do the same for the "don't bet" option. This is much simpler as you don't lose or gain anything, so it is always 0. You then compare these minimum values and choose the option with the *highest* worst-case utility (hence, maximin). In this case, that means "don't bet" as 0 is greater than −20.

You would only expect to make money betting on Kevin if the probability of him winning is over 50%. As it gets lower than that, you expect to lose more and more. So because you think the probability of him winning is in the range 30–50%, it is a bad bet! You don't expect to make money based on what you think the probabilities are, and there's only one way you could just break even: if things turned out for the *best,* and the probability was at the very top of the range you think it is in. So, you don't bet.

3.8 Insurance Pricing

That all seems like a lot of effort for betting on a dancer. But in our more complicated insurance context, the machinery of confidence, caution and stakes does important and useful work.

Let's go back to talking about hurricane insurance. Let's imagine an insurer who wants to sell a single insurance contract on house damage due to hurricanes. This is their first contract, and it is for insurance against an event, which we'll label E: "a hurricane strikes Fort Lauderdale in 2025". The contract is for a total value of £100,000 and it is a binary contract: it pays out either £0 if the event does not occur, or £100,000 if it does (there are no intermediate values for partial damage).

Pricing insurance involves working out how much you need to charge to make a profit, in a way that is not too different from deciding whether to place a bet. As in any business, you make a profit in insurance when you take more money in than you pay out. So, the annual price charged for a contract (money in) needs to be larger than the sum of two basic expenses: the expected pay out to customers claiming on their insurance contracts, and the cost of holding money so that it is available to pay those customers if the need arises. Both of these depend on the probability of the hurricane striking Fort Lauderdale. For payments out, that's obvious: you pay out when the hurricane hits. For money held, that's less obvious but it is because the insurance regulator requires insurers to hold capital according to a formula which uses the probability that they will pay out.

Here is an example. If the probability of the hurricane is 1%, then you should expect to pay 1% of the £100,000 insured in any year. That's the first expense: an expected pay-out of £1000. For the cost of capital holdings, we'll simplify and say that the insurer needs to hold the whole £100,000. (This sounds reasonable: after all, if I claim on the contract, I want them to have all £100,000! But in reality, insurers have many customers and they only ever hold a fraction of the money they would need to pay out if everyone claimed at once.) If the cost of capital is 5%, then the second expense, cost of holding capital, is 5% of £100,000 or £5000. So, to make a profit, the insurer must charge more than £6000 per year.

That tells us how insurance is priced if we know the probability of the event. But our original problem was: it is very hard to figure out these probabilities!

Our insurer will estimate this probability by talking to some scientific modellers. Let's imagine that, like Risk Management Solutions, they use a collection of models to advise the insurer. Table 3.1 shows how the information that the insurer gets looks: it's a list with 13 probabilities, each provided by one model constructed by the scientists. The list also contains each model's weight, which is, as we have seen, obtained by scoring each model in a predictive test. In our example Model 1 scored best and gets a weight of 23.7%. Model 10 scored worst and is only going to count for 1.6%.

Remember that in the "standard" approach, insurers will calculate a weighted average of these probabilities. Using those weights, the weighted average probability is 0.0072. We can now price this contract, just as we did above if we take this average to be the probability for the event of a hurricane striking Fort Lauderdale to occur. The first expense, expected pay outs, is $0.0072 \times 100,000 = £720$. The second expense will be the same as before, £5000. So the minimum price is £5720. As we noted above, concerned insurers often inflate these averages for "safety". This might be as crude as doubling the probability from 0.0072 to 0.0144. Going through the calculations with that probability, we get £6440.

3.9 Pricing with Confidence

Pure averaging and averaging with these ad hoc adjustments are unsatisfactory methods. The confidence approach offers an attractive alternative. To apply the confidence approach, we start with forming all the ranges of probability that the decision-maker accepts. The insurer does what you did in our dance example: it gathers all the evidence it can get. This includes all the model results, and what the skill score says about them. But it is not restricted to that: the insurer can take different skill scores into account and see whether they agree; or consider the nature of the different models in the ensemble and how they have been constructed; or weigh up the nature of the scientific disagreements (are they disagreements over principles, or over the application of principles, or over parameter values, or over the use of numerical techniques, or ...); or it can talk to different scientists, including those involved in the construction of the different models, to get the full picture of the state of play in the field. Based on all the evidence gathered, the insurer will then construct various intervals and form an opinion about how confident it can be in these intervals.[15]

---

[15] The details of this are discussed in our (2021).

Table 3.1 "Model outputs" for toy example

Model	m_1	m_2	m_3	m_4	m_5	m_6	m_7	m_8	m_9	m_{10}
$p(E)$	.0070	.0071	.0068	.0074	.0076	.0061	.0083	.0086	.0091	.0092
Weight (%)	23.7	20.7	15.8	11.6	11.5	7.3	3.2	3.0	1.7	1.6

We hope this sounds like common-sense to you. But notice that common-sense is not how things are currently done. Alternative skill scores are not considered when calculating weighted averages, and no other sources of information about either the models, the modelling process, or the state of knowledge in the field is taken into account. Opening up to these issues will allow insurers' scientists to form a much more nuanced picture of the available evidence for and against various scenarios, and come to better founded judgment about the trustworthiness of model outputs, than they could by uncritically relying on mechanically calculated skill scores.

Assume now that the science team of the insurer has gone through this process and, considering all available evidence, has come to the conclusion that it should consider three intervals: [0.007,0.0071] with confidence level Low, [0.0068,0.0076] with confidence level Medium, and [0.0061,0.0091] with confidence level High.

The insurer now has to think about the stakes involved in this decision. This contract is the insurer's first; it will be their whole business and so the risk of going bust is high. Still, no one's life is at stake and there is no impact on anything else (e.g., no other business which might be taken down). So, the insurer concludes that their stakes are moderately high, $s = 0.75$.

Next in line is cautiousness. As insurance against natural catastrophes involves significant uncertainty, this insurer can't be too cautious. So, let us suppose that they only demand High confidence for important decisions of stakes 0.9 or higher. For a decision like ours, with stakes 0.75, they're happy with Medium confidence.

Now we look at the intervals above and see that the relevant interval for that confidence level is [0.0068, 0.0076]. The insurer can now apply the same cautious decision rule we used before, maximin expected utility, and make the choice that turns out best if things go as badly as possible. In insurance, *higher* probabilities are worse, because they make it more likely you need to pay out. So, they care only about the largest probability in this range, 0.0076. The first expense, expected pay out, is $0.0076 \times 100{,}000 = £760$. The holdings are exactly as before. So we end up with a minimum price of £5760.

Pricing method	Price
Averaging	5720
Averaging + "safety" factor	6440
Confidence	5760

It is now interesting to compare the three methods. The "pure" average price is £5720, but as we saw earlier, it has a lot of problems associated with it. The "safety" price, which involves ad hoc adjustments to the average, is much higher, £6440. Our new confidence price is only a little higher than the average at £5760. But to get there, the insurer had to follow a completely different procedure. They used the full set of model outputs. The price depends on how important the decision is, and on how cautious the insurer is. In other decisions, they will get completely different answers. On the other hand, the average and safety prices are always calculated the same way.

3.10   Better Decision-Making

Insurance is meant to put a "price on risk" so that people can pay protect themselves from the unexpected. Insurers would also like to put a price on the kind of uncertainty that we discuss here: not knowing what the probability of some event is. The current options available to insurers are all ad hoc and there is no guarantee for insurance companies that their staff are responding to different risks (hurricane, earthquake, wildfire) in a common and systematic way.

Our approach allows insurers to systematically set this kind of "uncertainty premium". For an insurance company, three of the ingredients discussed above will be a matter of policy. They will need to agree on a way of measuring stakes that allows them to compare the different kinds of decisions they make and decide which are more and less important. Cautiousness can similarly be determined by a high-level decision about how much confidence to demand for decisions of various stakes. Finally, a decision rule will need to be selected; either maximin expected utility or one of its competitors.

With these three elements in place the confidence approach provides a recipe for pricing insurance that is sensitive to all of the evidence available for that decision and which responds naturally (through the cautiousness and stakes) to the different nature of each decision taken. This kind of flexible but systematic treatment of uncertainty is what insurers tell us they have been missing in catastrophe insurance.

The last two paragraphs also demonstrate a second benefit: the confidence approach fits naturally with the kind of distributed decision-making and corporate responsibility found in large insurance companies. The different parts of the recipe are naturally provided by different stakeholders. The cautiousness function is ultimately determined by the shareholders' appetite for uncertainty. The way of determining stakes will be set by senior management in charge of portfolio management and capital allocation. The probability functions themselves, along with their nesting and grouping into confidence levels, come from the science department who cover that particular risk. This is a better fit with how insurance decision-making should work than having underwriters "adjust" scientific estimates of probability individually.

Finally, looking beyond the insurance case, the method we outlined is applicable in all instances of decision-making under uncertainty. This is because nothing depends on the evidence being provided in form of an ensemble of models. The multiple sources of information be experts who hold diverging views, or they can be a mixture of different kinds of sources.[16] The "distributed" nature of the implementation can be an advantage also in other contexts, where different stakeholders will provide the probability functions, and determine the stakes, and set the cautiousness function, which can be a beneficial setup in many social situations.

---

[16]For a discussion of expert elicitation see Cooke's (1991), Morgan's (2014) and the contributions to Martin and Bouman's (2014). For a discussion expert elicitation in the context of climate change adaptation see Thompson et al.'s (2016).

References

Bradley, R. (2017). *Decision theory with a human face*. Cambridge University Press.

Bradley, S. (2019). Imprecise probabilities. In E. N. Zalta (Ed.), *The Stanford encyclopedia of philosophy*. https://plato.stanford.edu/archives/spr2019/entries/imprecise-probabilities/

Bradley, R., & Steele, K. (2015). Making climate decisions. *Philosophy Compass, 10*, 799–810.

Collins, J. M., & Walsh, K. (Eds.). (2019). *Hurricane risk*. Springer.

Cooke, R. M. (1991). *Experts in uncertainty: Opinion and subjective probability in science*. Oxford University Press.

Frigg, R., Smith, L. A., & Stainforth, D. A. (2015a). An assessment of the foundational assumptions in high-resolution climate projections: The case of Ukcp09. *Synthese, 192*, 3979–4008.

Frigg, R., Thompson, E., & Werndl, C. (2015b). Philosophy of climate science part I: Observing climate change. *Philosophy Compass, 10*, 953–964.

Frigg, R., Thompson, E., & Werndl, C. (2015c). Philosophy of climate science part II: Modelling climate change. *Philosophy Compass, 10*, 965–977.

Heal, G., & Millner, A. (2018). Uncertainty and decision-making in environmental economics: Conceptual issues. In P. Dasgputa, S. Pattanayak, & V. K. Smith (Eds.), *Handbook of environmental economics* (Vol. 4, pp. 439–468).

Hill, B. (2013). Confidence and decision. *Games and Economic Behavior, 82*, 675–692.

Hill, B. (2019). Confidence in beliefs and rational decision making. *Economics and Philosophy, 35*, 223–258.

IPCC. (2013). *Climate change 2013: The physical science basis. Contribution of Working Group I to the Fifth Assessment Report of the Intergovernmental Panel on Climate Change* (T. F. Stocker, D. Qin, G.-K. Plattner, M. Tignor, S. K. Allen, J. Boschung, A. Nauels, Y. Xia, V. Bex, & P. M. Midgley, Eds.). Cambridge University Press.

Knutti, R. (2010). The end of model democracy? *Climate Change, 102*, 395–404.

Marinacci, M. (2015). Model uncertainty. *Journal of the European Economic Association, 13*, 1022–1100.

Martini, C., & Boumans, M. (Eds.). (2014). *Experts and consensus in social science*. Springer.

Morgan, M. G. (2014). Use (and abuse) of expert elicitation in support of decision making for public policy. *PNAS, 111*, 7176–7184.

Parker, W. S. (2010). Predicting weather and climate: Uncertainty, ensembles and probability. *Studies in History and Philosophy of Modern Physics, 41*, 263–272.

Parker, W. S. (2011). When climate models agree: The significance of robust model predictions. *Philosophy of Science, 78*, 579–600.

Parker, W. S. (2013). Ensemble modeling, uncertainty and robust predictions. *Wiley Interdisciplinary Reviews: Climate Change, 4*, 213–223.

Parker, W. S. (2018). Climate science. In E. N. Zalta (Ed.), *The Stanford encyclopedia of philosophy*. https://plato.stanford.edu/archives/sum2018/entries/climate-science/.

Philp, T., Sabbatelli, T., Roberston, C., & Wilson, P. (2019). Issues of importance to the (re)-insurance industry: A timescale perspective. In J. Collins & K. Walsh (Eds.), *Hurricane risk* (pp. 1–22). Springer.

Roussos, J., Bradley, R., & Frigg, R. (2021). Making confident decisions with model ensembles. *Philosophy of Science, 88*, 439–460.

Shome, N., Rahnama, M., Jewson, S., & Wilson, P. (2018). Quantifying model uncertainty and risk. In G. Michel (Ed.), *Risk modeling for hazards and disasters* (pp. 3–46). Elsevier.

Smith, L. A., Du, H., Suckling, E. B., & Niehörster, F. (2014). Probabilistic skill in ensemble
    seasonal forecasts. *Quarterly Journal of the Royal Meteorological Society, 140*. https://doi.org/
    10.1002/qj.2403
Stainforth, D. A., Allen, M. R., Tredger, E. R., & Smith, L. A. (2007a). Confidence, uncertainty and
    decision-support relevance in climate predictions. *Philosophical Transaction of the Royal
    Socity A, 365*, 2145–2161.
Stainforth, D. A., Downing, T. E., Washington, R., Lopez, A., & New, M. (2007b). Issues in the
    interpretation of climate model ensembles to inform decisions. *Philosophical Transactions of
    the Royal Society A, 365*, 2163–2177.
Thompson, E., Frigg, R., & Helgeson, C. (2016). Expert judgment for climate change adaptation.
    Philosophy of Science, 83, 1110–1121.

Part II
Philosophy of Artificial Intelligence and the Internet

Part III
Philosophy of Artificial Intelligence
and the Internet

Chapter 4
Artificial Intelligence and Philosophy of Science from the 1990s to 2020

Donald Gillies and Marco Gillies

Abstract The first three sections of the paper are by Donald Gillies. Section 4.1 argues that Artificial Intelligence (AI) is indeed relevant to quite a number of issues in philosophy of science. This paper will focus on one such issue – the problems of induction, confirmation and probability, which are indeed of key importance in philosophy of science. Section 4.2 traces the development of AI from the early 1970s to the 1990s, and Sect. 4.3 examines the implications of this development for philosophy of science as set out in Donald Gillies', 1996 book: *Artificial Intelligence and Scientific Method*. About a quarter of a century has elapsed since that book was published. During that time, there have been enormous advances in AI, and, specifically, in machine learning. The question therefore arises whether conclusions drawn in 1996 from the analysis of machine learning programs, which were state of the art at that time, still hold in the light of the much more powerful machine learning programs of 2020. In Sect. 4.4, Marco Gillies, who uses contemporary machine learning in his research into virtual reality (VR), gives a brief account of some of the advances in machine learning since 1996. He goes on, in Sect. 4.5, to examine how the conclusions stated in Sect. 4.3 need to be modified in the light of advances in AI since 1996.

Keywords Falsification · Induction · Intelligibility · Machine learning · Neural networks

D. Gillies (✉)
Department of Science and Technology Studies, University College London, London, UK
e-mail: donald.gillies@ucl.ac.uk

M. Gillies
Department of Computing, Goldsmiths, University of London, London, UK
e-mail: m.gillies@gold.ac.uk

© The Author(s), under exclusive license to Springer Nature Switzerland AG 2022
W. J. Gonzalez (ed.), *Current Trends in Philosophy of Science*, Synthese Library 462, https://doi.org/10.1007/978-3-031-01315-7_4

4.1   Relevance of Artificial Intelligence (AI) to Philosophy of Science

This paper is concerned with the implications for philosophy of science of the development of Artificial Intelligence (AI). But is AI really relevant to philosophy of science? We will begin by arguing that it is indeed relevant.

In fact new AI results have implications for many questions in the philosophy of science and indeed in general philosophy. However, we will in this paper focus on the importance of AI results for just one group of problems within the philosophy of science, albeit a very important group of problems. These are the problems connected with induction, confirmation and probability. These problems have been central to the philosophy of science since Bacon in the 17th century. We can formulate a few of the questions as follows. Is there a process of induction which can be used to get theories and predictions from empirical data obtained by observation and experiment? If there is such a process, is it concerned with discovery or justification or both? Is there a concept of the confirmation (or disconfirmation) of theories by empirical data? Is this confirmation probability (the Bayesian thesis), or does it differ from probability? Is there such a thing as inductive logic as well as deductive logic?

Now a central technique of AI is *machine learning*, which is a procedure by which a computer obtains theories or predictions from empirical data, that is to say it is a process of induction. It would seem obvious then that the analysis of successful machine learning programs is likely to shed light on the traditional philosophy of science problems about induction, which have just been listed. Indeed simply the fact that machine learning has proved successful has altered many of the earlier discussions of induction. This is illustrated by a debate between Popper and Carnap about whether an inductive logic existed.

In his *Logic of Scientific Discovery* (1934) Popper criticized the claim that there was an inductive logic, and concluded (29): "My own view is that the various difficulties of inductive logic here sketched are insurmountable."

However, Carnap in his 1950 *Logical Foundations of Probability* defended the existence of inductive logic. He writes (192): "Inductive logic is constructed from deductive logic by the adjunction of a definition of *c*."

c here is the confirmation function, so that $c(h,e) = r$ means that the degree of confirmation of h given e is r. Carnap thought that *c* was a probability function to be interpreted in logical terms so that his inductive logic is a form of logical Bayesianism.

Despite this disagreement, Carnap does agree with Popper about one thing, which he describes as follows (1950, 192–193):

> But in one point the present opinions of most philosophers and scientists seem to agree, namely, that the inductive procedure is not, so to speak, a mechanical procedure prescribed by fixed rules. ... This point, the impossibility of an automatic inductive procedure, has been especially emphasized ... by Karl Popper ... The same point has sometimes been formulated by arguing that it is not possible to construct an inductive machine. The latter is

presumably meant as a mechanical contrivance which, when fed an observational report, would furnish a suitable hypothesis, just as a computing machine when supplied with two factors furnishes their product. I am completely in agreement that an inductive machine of *this* kind is not possible.

So in 1950, Carnap, Popper, and, according to Carnap, "most philosophers and scientists" thought that machine learning was not possible. Yet we now know that machine learning exists and has been very successful. Surely this in itself is likely to change views about induction, and specifically the question of whether there can be an inductive logic.

The above considerations led one of the authors of this paper (Donald Gillies) to spend time from the mid-1980s to the end of the 1990s studying AI in order to examine how the new results affected debates in philosophy of science about induction. The results of this investigation were published in his 1996 book: *Artificial Intelligence and Scientific Method*. Chapter 5 of the book continues the debate between Popper and Carnap about inductive logic. Using the new results about machine learning, it argues that there is an inductive logic, but that this has a different form from the one put forward by Carnap.

About a quarter of a century has elapsed since that book was published. During this period, there have been enormous advances in AI, and, specifically, in machine learning. The question therefore arises whether conclusions drawn in 1996 from the analysis of machine learning programs, which were state of the art at that time, still hold in the light of the much more powerful machine learning programs of 2020. This is the subject of this paper. In Sect. 4.3, Donald Gillies summarizes the main conclusions regarding AI and scientific method, which he reached in his 1996 book. Then in Sect. 4.4, Marco Gillies, who uses contemporary machine learning in his research into virtual reality (VR), gives a brief account of some of the advances in machine learning since 1996. He goes on, in Sect. 4.5, to examine how the conclusions stated in Sect. 4.3 need to be modified in the light of advances in AI since 1996. Before going into these questions, it will be helpful to give a brief sketch of the development of AI from the early 1970s to the late 1990s. This will be done in the next section (Sect. 4.2).

4.2 Development of AI from the Early 1970s to the Late 1990s

The first computers in the modern sense appeared in the late 1940s, and by 1950 many people had begun to think of the project of using these new computers to create Artificial Intelligence. Turing sketches out ideas along these lines in his famous 1950 paper: Computing Machinery and Intelligence. Many researchers worked on such plans in the 1950s and 1960s, and formulated ideas, which were later to prove useful. However, it has to be said that, by the beginning of the 1970s, there had been no striking successes in the new field of AI. This led to attacks on the whole AI project.

The most famous was perhaps Hubert Dreyfus' book: *What Computers Can't Do: A Critique of Artificial Reason*, which was published in 1972. Dreyfus expressed considerable doubts about whether computers would ever be able to play chess skilfully, and argues in his book that (197): "further significant progress ... in Artificial Intelligence is extremely unlikely."

Dreyfus himself was beaten at chess by a computer, and showed some signs of irritation at the glee with which the news of his defeat was received by the AI community. Subsequent events showed that he was quite wrong about computer chess. In May 1997, Garry Kasparov, the world chess champion, was defeated by a computer (Deep Blue) in a six game match played under standard conditions. Still, in fairness to Dreyfus, it must be remembered that when he published his book in 1972, research in the AI project had been continuing for more than two decades with little in the way of positive achievements.

However in the early 1970s a breakthrough in AI was produced by the creation of expert systems. The lead here was taken by the Stanford heuristic programming group, particularly Buchanan, Feigenbaum, and Shortliffe. What they discovered was that the key to success was to extract from an expert the knowledge he or she used to carry out a specialised task, and then code this knowledge into the computer. In this way they were able to produce 'expert systems' which performed specific tasks at the level of human experts. The first such system was DENDRAL, which inferred a plausible molecular structure for a compound from its atomic composition and mass spectrogram. Another early expert system (MYCIN) was concerned with the diagnosis of blood infections. It was developed in the 1970s by Edward Shortliffe and his colleagues in collaboration with the infectious diseases group at the Stanford medical school (see Shortliffe & Buchanan, 1975; Davis et al., 1977). The medical knowledge in the area was codified into rules of the form: IF such and such is observed, THEN likely conclusion is such and such. MYCIN's knowledge base comprised over 400 such rules, which were obtained from medical experts.

To test MYCIN's effectiveness a comparison was made in 1979 of its performance with that of nine human doctors. The program's final conclusions on ten real cases were compared with those of the human doctors, including the actual therapy administered. Eight other experts were then asked to rate the ten therapy recommendations and award a mark, without knowing which, if any, came from a computer. They were requested to give 1 for a therapy, which they regarded as acceptable and 0 for an unacceptable therapy. Since there were eight experts and ten cases, the maximum possible mark was 80. The results were as follows (Jackson, 1986, 106):

MYCIN	52	Actual therapy	46
Faculty-1	50	Faculty-4	44
Faculty-2	48	Resident	36
Inf dis fellow	48	Faculty-5	34
Faculty-3	46	Student	24

So MYCIN came first in the exam. This is the first occasion when an Artificial Intelligence program showed itself to be superior to human experts.

However, a problem soon emerged with expert systems. Such systems were based on rules, and so were sometimes known as 'rule-based systems'. Initially the rules were obtained by interviewing experts in the field, but there were two problems with this procedure. First of all, the experts were able to perform a task, but sometimes they did not know the rules they used to do so. Perhaps the knowledge of these rules was unconscious rather than conscious. Secondly the interviewing of experts was a very long and expensive procedure. This difficulty was first pointed out by Feigenbaum in 1977, and came to be known as Feigenbaum's bottleneck.

Machine learning offered a possible solution to these difficulties. Instead of attempting painfully to extract the rules needed for the rule-based system from domain experts, the alternative strategy would be to try to induce the rules using machine learning, from sets of examples provided by the domain experts.

Feigenbaum formulated his bottleneck problem in 1977, and the next few years saw the first attempts to develop machine learning techniques for solving it. One of the first systems, Meta-DENDRAL, was designed to induce rules for use by DENDRAL. It succeeded in this task, and some of the rules were even published in a chemistry journal (see Buchanan & Feigenbaum, 1978). At about the same time, Michalski produced a machine learning system concerned with the diagnosis of soybean disease. The rules were inferred from several hundred correctly diagnosed examples provided by the domain experts (see Michalski & Chilautsky, 1980).

In Gillies (1996, 31–55), two contemporary state-of the art machine learning systems were analysed in detail. They were both logic based. The first, ID3, was concerned with the induction of decision trees. This was developed by Quinlan (see his 1986). The second was a system known as GOLEM, developed by Muggleton and his colleagues. This was applied to the problem of obtaining rules showing how proteins fold up. A protein consists of a linear sequence of amino acid residues, and, depending on the nature of their residues, different proteins fold up in different ways. GOLEM did indeed produce some rules about protein folding, which proved to be accurate (see Muggleton et al., 1992). It is worth noting that the data set used by GOLEM in this problem consisted of about 500 proteins, whose manner of folding had been determined by X-ray crystallography or NMR (nuclear magnetic resonance) techniques.

From this we see that machine learning techniques in AI were initially used in the 1980s and early 1990s to induce the rules used in expert systems, and they tended to be applied to data sets containing a few hundred examples. However, there were already considerable successes in obtaining successful but hitherto unknown rules in scientifically important areas. In the next section we will examine the conclusions for philosophy of science, which Donald Gillies drew in his 1996 book. However, before doing so, it is worth remarking that the developments in AI, which we have just discussed, made considerable use of results from philosophy of science. Shortliffe and Buchanan's, 1975 paper contains 33 references and no less than 14 of these (or over 42%) are to works in the philosophy of science concerned with the confirmation of scientific hypotheses by evidence, and related questions concerned with induction and the interpretation of probability. Among others the following philosophers of science are mentioned: Carnap, de Finetti, Hempel, Popper, Ramsey, and Salmon. Other AI researchers of this period even refer back to Sir Francis Bacon's works published in the 17th century! This shows the possibility of a fruitful interaction between AI and philosophy of science. Ideas from philosophy of science were helpful in the initial development of AI. The results of this development could bring about changes in philosophy of science which in turn might be helpful for further developments in AI and so on.

4.3   AI & Philosophy of Science: The Situation in the 1990s

The 1996 book: *Artificial Intelligence and Scientific Method* contains many quite technical developments in logic and philosophy of science, which were suggested by the new results in AI. For example, as already mentioned, Chap. 5 develops an inductive logic, which is quite different from that of Carnap. In this section, however, we will not discuss these more technical results, but rather give the general conclusions reached regarding philosophy of science. These conclusions were mainly based on the success of machine learning programs. They were four in number.

1. *Induction is real*

Popper wrote (1963, 53): "Induction, i.e. inferences based on many observations, is a myth. It is neither a psychological fact, nor a fact of ordinary life, nor one of scientific procedure."[1]

The successes of machine learning programs show that Popper was wrong on this point. However, it should be added that Popper was not so wrong when he made the remark in 1963. The first striking successes of machine learning did not occur until the late 1970s.

2. *Testing and falsification are used in machine learning*

Testing and falsification are central to Popper's account of scientific methodology, and here machine learning supports rather than contradicts Popper's account. Machine learning programs learn on some training data and the conclusions reached are tested and possibly rejected on some further test data. So we have a sequence of conjectures and refutations. The difference from Popper is that in machine learning the conjectures are generated mechanically by the program rather than intuitively by humans.

---

[1] Donald Gillies heard Popper utter the fateful words: "induction is a myth" when he attended Popper's lectures as a graduate student in the academic year 1966–1967. As far as he can remember, Popper continued: ". . . and those who use the term 'induction', do not know what they are talking about."

3. *The importance of background knowledge*

All the machine learning programs analysed in Gillies (1996) used background knowledge (K) as well as empirical evidence (e). It is concluded (p. 70) that "computer inferences really take the form from K&e infer h rather than from e infer h". For example, in GOLEM's application to the protein folding problem, each of the residues was described in terms of properties which were recognized by scientists in the field to be relevant to how a protein folded. These were properties such as 'being hydrophobic', 'being polar', 'being aromatic' etc. Moreover 9-place predicates were coded into GOLEM and these guided the program to look for laws determining the character of a residue in terms of the 4 residues on each side of it.

4. *Human interaction with the results of machine learning*

The machine learning programs analysed in Gillies (1996) all gave as output rules which were humanly comprehensible. For example Muggleton's GOLEM gave an explicit rule relating to protein folding which is stated in Gillies (1996, 53). This was typical of the period when machine learning was mainly used as a technique for learning the rules of rule-based systems. This situation led to the following analysis of human interaction with the results of machine learning (Gillies, 1996, 54–55):

> There are great advantages in generating rules which are humanly comprehensible, because this allows the following kind of human-machine interaction. Background knowledge supplied by the human scientist is coded into a machine learning program. This generates hitherto unknown, but humanly comprehensible, rules which apply to the domain in question. The human scientist can then examine these rules, and perhaps obtain new insights into the field.

4.4   Machine Learning and AI in 2020

The previous sections have summarised the book *Artificial Intelligence and Scientific Method*, which was published in 1996. Looking back to that time, though the book describes what were then important breakthroughs, it now seems that the great successes of AI hadn't yet started. What in 1996 seemed a niche area of computer science research has now strongly entered the public consciousness with large amounts of (almost certainly excessive) media hype but also many very important practical applications, most often in the work of giant technology companies such as Google and Amazon, but in day-to-day science as well. This section will summarize what has changed in the last quarter century (as well as the surprising amount that has not changed) before the next section discusses the impact of these changes

The 1996 book focused on scientific applications of AI, but also briefly discussed the subject of AI chess playing, making a cautious prediction that chess playing computers might beat a human world champion within a decade. As it turned out, it was only a matter of months between the publication of this prediction and the defeat of Garry Kasparov by IBM Deep Blue, mentioned above. Almost two decades later in 2016, AlphaGo, an AI programme developed by the Google-owned company

DeepMind, defeated world champion Lee Sedol at the game of Go, which is considered far harder to implement on a computer than chess. The distinction between the two victories is instructive. In 1997, machine learning was one technique among many and Deep Blue did not use it, relying on other techniques such as heuristic state space search and some explicitly encoded expert knowledge, analogous to expert systems. However, by 2016, AI and machine learning had become almost synonymous, with most AI problems being tackled using machine learning techniques. AlphaGo used a machine learning method called Deep Reinforcement Learning and most experts would consider it impossible to solve Go playing without machine learning.

The years since 1996 have seen many developments in machine learning. In the mid nineties, there were two major approaches, the use of logic and related techniques exemplified by GOLEM and decision trees (e.g. the ID3 algorithm), and the neural network tradition, which was loosely inspired by the human brain. The late 1990s and early 2000s saw the development of methods based on statistical techniques. This included recasting techniques such as decision trees or neural networks in terms of statistics (e.g. Bishop, 1995) as well as new techniques in the Bayesian tradition such as Bayesian Networks (Jensen, 1996) and Gaussian Processes (Rasmussen & Williams, 2006) and the more frequentist Support Vector Machines and Kernel Method (Cristianini & Shawe-Taylor, 2000).

However, older techniques are still important. While the ID3 algorithm is no longer used in practice, its direct successor, C4.5 (Quinlan, 1993) is still heavily used and is an inspiration for Random Forests (Breiman, 2001), which are still considered a state of the art technique. More importantly the recent successes of machine learning, including AlphaGo, have largely been due to a resurgence of the use of Neural Networks, under the name of Deep Learning (Goodfellow et al., 2016). Given the importance of Neural Networks to machine learning at the time of writing, it is important to briefly summarize how they work.

A neural network consists of a number of simple nodes, called neurons, because they were originally designed as simplified models of the neurons in a human brain. A neuron takes a number of inputs (which are numbers) and produces an output (also a number). In the simplest form of neuron, a perceptron, the output is a sum of the input values after each has been multiplied by a numerical weight. Learning in a perceptron consists of finding optimal values for the weights, so that the inputs in the training data map correctly to the corresponding outputs. Though this type of weighted sum is very simple mathematically, and there are more complex forms of neuron, such as the convolutional neurons used in image processing, simple perceptrons are still the most used form of neuron in state of the art machine learning. The power of modern neural networks does not come from having complex computational units, but by combining together simple units. The output of one neuron can become the input to another neuron, and so on, forming multiple layers of neurons that process data from input to output (for example, from the pixels of a CT scan to a diagnosis).[2] The term Deep Learning, or Deep Neural Network, simply refers to a neural network with many of these layers.

Learning in multi-layer networks happens using an algorithm called back-propagation (Rumelhart et al., 1986). The process begins by assigning random weights to the neurons. The network uses these weights to calculate a predicted output for each of the training examples. It then calculates the error, the difference between the network's output and the correct output, for each training example. This error value is used to make slight changes to the weights of the neurons to bring the predicted output closer to the correct output. The errors are "propagated back" from later layers in the network to previous ones in the opposite direction from the standard calculation.

Neural networks and the back-propagation algorithm were well established in 1996 and are still the most important technique in machine learning today. So in some ways little has changed since the 1996 book. However, the dramatic advances in machine learning seen in recent years have been due to the use of Deep Neural Networks, with many layers of neurons. This has been made possible by two developments. The first was the increase of computing power available, particularly the use of parallel architectures. The deeper the network, the more neurons needed and therefore the more computation involved. The second was a massive increase in the amount of data available to train machine learning algorithms. The GOLEM project described above used about 500 training examples, while the currently popular dataset ImageNet (Deng et al., 2009) has over 14 million examples (and many datasets are even bigger). This increase in the data used makes it possible to train far deeper and more complex models, since the larger number of parameters in a deep model require more data to set them effectively.

The most common interpretation of the function of the many layers of a deep network is that they learn different representations of the data that are useful for the task being performed (Bengio et al., 2013). For example, when training a network to recognise faces in an image, the input to the network will be represented as the pixels of the image. However, the first few layers might learn representations of features in the image, such as edges, or patches of darkness. The later layers might learn more complex and meaningful features representing things like eyes and noses. Early machine learning methods required complex hand written code to detect these types of meaningful features,[3] which were then input to the machine learning. Deep learning makes it possible to learn them directly from the "raw data" such as image pixels.

---

[2] For multi-layer neural networks to work, the output of each neuron must be passed through a non-linear "activation function" before being used as input to the next layer. If this does not happen the result is equivalent to a single, larger weighted sum. These activation functions play an important part in neural networks, but do not particularly affect the argument of this paper.

[3] When describing data used for inputs to machine learning, *Artificial Intelligence and Scientific Method* employed the then standard expressions "attributes of the data", and "properties of the data". These have now been replaced by the expression "features of the data" which will be adopted from now on.

Interestingly the type of scientific problem that machine learning is being applied to is also similar. The GOLEM system (Muggleton et al., 1992), described above, aimed to learn rules to predict protein folding. This is still an important unsolved problem in biology which many researchers are using machine learning to investigate. For example, the team that developed AlphaGo have since applied their work to science and have made important contributions to protein folding (Evans et al., 2018). In general, machine learning is increasingly making important contributions to science, particularly the biological sciences which are characterised by high levels of complexity and potentially large amounts of data. As well as the study of proteins (e.g. Buchan & Jones, 2019), it has been applied to areas as diverse as drug discovery (Williams et al., 2015), modelling neural activity in the Brain (Schultz et al., 1997) and social interaction in autism (Ward et al., 2018).

4.5 AI & Philosophy of Science: The Situation in 2020

Having reviewed the new developments in AI since 1996, we are in a position to see whether the original conclusions of *Artificial Intelligence and Scientific Method* are supported by the new developments. In many ways the methods used and problems tackled are similar to those used in 1996. However, we have seen a number of developments:

- the increasing importance of machine learning relative to other AI techniques
- the increasing use of neural networks
- the use of deeper, more complex, models and larger datasets
- the use of deep neural networks for representation learning

We can now see how these affect the four conclusions:

1. *Induction is real*

This seems to continue to apply. Machine learning is more important than ever and it represents a form of automated induction. In fact, this case is possibly strengthened. One possible counter-argument to machine learning being a form of scientific induction is that the key scientific insight when using machine learning is choosing an appropriate mathematical form for the machine learning model and the automated process is simply setting those parameters. An analogy would be with Kepler's determination of the orbits of the planets. His key insight was the mathematical form of the orbit (an ellipse with one focus at the sun) and determining the parameters of the model (the dimensions of the ellipses) could be automatic. Machine learning models could be seen as being more and more complex mathematical forms. However, as machine learning models become more and more complex and general purpose, for example, deep neural networks, the choice of mathematical form seems less and less scientifically significant. A standard neural network architecture applies to many different problems, and while there are still choices to make, they are likely to be less connected to the science. For Kepler, the choice of ellipse rather than a circle was an immensely scientific insight, for a modern scientist working on protein folding, the choice of using 6 rather than 5 layers of neurons seems to involve very little scientific insight.

However, there is one aspect of the back-propagation algorithm that might be seen to support Popper's view. Back-propagation begins with random weights and uses errors in its output to correct the weights. This use of errors to improve the model could be seen as a form of automated falsification rather than of induction. This does not invalidate the original idea, as many machine learning algorithms are more clearly inductivist, but it does suggest that machine learning could represent an automation of both induction and falsification.

2. *Testing and falsification are used in machine learning*

Testing remains a key part of the practice of machine learning. Machine learning practitioners will divide their data into a training set used as input to the machine learning algorithm and a test set, which is used to test the model once the training has been completed. If the model does not perform accurately on the testing data it is considered to have failed and changes to the training process must be made. This process is primarily to ensure that the model is able to generalise to new data and has not 'overfitted', i.e. it has not learned to recognise irrelevant aspects of the training data. There are a number of specialised training methodologies that have been developed to improve this process. For example, cross validation divides the data into training and test data in multiple ways and performs multiple training steps on these different splits. The results are used to compare different parameters of the learning algorithm. In professional systems it is common to have multiple testing sets. A development set is used to test the model during development. When the model appears to perform well on the development set, a separate test set is used as a final validation (developers often gather new data at this stage). These developments show that testing is a key part of machine learning methodology, supporting the original conclusion.

3. *The importance of background knowledge*

This is rather more problematic. There are certainly those that argue that deep learning requires less background knowledge than previous machine learning approaches.

There are a number of possible ways in which background knowledge can be used. The first is in the choice of the mathematical form of a machine learning model. However, as we have discussed above, this is becoming less scientifically significant. Secondly, the choice of which features of the data to use as input to the machine learning model, this was the major use of background knowledge in the GOLEM system described above. One of the most important aspects of deep learning is that intermediate representations of the data are learned from raw data features such as pixels. This reduces the need for a scientist to choose meaningful features, and thus the need for background knowledge.

There are, however, other uses of background knowledge that remain important. One example is the development of the data sets themselves. This involves curation of examples of data that will be informative to machine learning. The data also need to be labelled, for each instance needs to include the correct output. For example, in the case of protein folding, the input data would be the sequence of amino acids, but the data set would also need to include an output label: the correct 3D folded structure. This label clearly requires very detailed background knowledge.

We can see that the overall trend is for a reduction in the need for background knowledge, but it still remains important in many ways.

4. *Human interaction with the results of machine learning*

The advantages of having rules or models that are humanly comprehensible is an aspect of the scientific process, and should not be changed by advances in machine learning methods. However, as we have seen, neural networks are becoming increasingly important, and they are notorious for being difficult for humans to interpret. In fact, the Muggleton et al.'s (1992) stress on interpretability can be seen as an argument for their method over neural networks. The reason neural networks are hard to understand is that they consist of the interaction of a very large number of very simple units. While this type of model is easy for a computer to process, the number of interacting elements is challenging for humans to keep track of.

The difficulty of interpreting neural networks, and machine learning in general, is increasingly seen as a problem within the machine learning community. For example, O'Neil (2016) has highlighted the social problems that occur when machine learning algorithms make decisions related to health, crime or other social issues without any human being able to check their correctness. Having scientific discoveries that are not humanly understandable seems similarly problematic. This has led to research on machine learning methods that are able to explain their results.

One approach, particularly used in machine learning for medicine, is to develop learning models that are designed to be human intelligible. For example, Wang and Rudin (2015), working in medical diagnosis, sought to develop models that would integrate with medical practice. As such, they used a model that was similar to existing check lists used by doctors. They called these falling rule lists: an ordered list of if-then rules, where each rule is tested in order until a decision can be made. While in current medical practice these lists are human designed, the authors propose an algorithm for learning them.

However, these models are necessarily simplified relative to deep neural networks, and are therefore less powerful in terms of the models they can learn. This can reduce their overall accuracy. It might be possible to limit scientific use of machine learning to only those models that are interpretable, however, this leads to a rather fundamental question: will all the scientific systems we are studying be explainable using only humanly comprehensible rules? If this is not the case, and there seems to be no strong reason to believe it, we are likely to need the assistance of computer models that we do not understand.

One way to combine complex models with human interpretation is to produce local explanations. The model itself remains too complex for humans to understand, but each decision it makes can be explained. For example, Ribeiro et al. (2016) explain individual decisions by fitting a local intelligible model. This work is similar to the work on intelligible models above, but it does not attempt to explain the whole behaviour of a machine learning system through a simpler model. Instead it will create a simplified model (for example, a linear model) for the purpose of explaining the behaviour of a machine learning model on a single data point. The model can very accurately explain the behaviour in the local neighbourhood of that data point, but does not attempt to explain behaviour beyond this.

We can therefore conclude, that the interpretability of machine learning is still a very important issue in science, even if recent developments have made the results of learning less interpretable. This therefore remains an important problem, and research challenge, for machine learning in science.

4.6 Conclusions

This discussion has shown that our four conclusions broadly hold in light of recent advancements in AI, with a few provisos. Machine Learning shows that automated induction is possible, though testing and falsification do also play important parts in machine learning. The use of machine learning in science does not eliminate the need for human experts. It requires background knowledge from humans and it is still important that the results of machine learning are interpretable by humans. Deep Learning, however, has reduced the need for background knowledge and made interpretation more difficult.

These results challenge Popper's assertion that induction is a myth. However, it does not change his idea that induction does not play a part in *human* science. Instead it suggests that AI performs scientific work in a very different way from humans. This is supported by the difficulty of interpreting the results of machine learning: the scientific models produced by machines are very different from those produced by humans. This raises important questions for the future of science.

This paper has, at several points, compared the use of AI in science with the better known applications to game playing, particularly the well publicised defeat of Kasparov by Deep Blue. What is less well-known, is what Kasparov did next. His response to the defeat was to develop a new form of chess competition, called Advanced Chess (also known as Centaur Chess) in which the competitors were humans working together with computers (Kasparov & Greengard, 2017). These combined teams were able to defeat the world's best humans and computers working in isolation. Interestingly, humans and computers tended to take different roles. Computers are superior at the detailed tactics of the game (that champions such as Kasparov excelled at) while the humans were better at the large-scale strategy.

What this teaches us is that computers are unlikely to replace human scientists. Rather computers and humans do science in different ways, and that humans working with computers will combine their different skills to do science in more advanced ways (which is certainly the case today). Understanding these different skills, and creating machine learning systems that humans are able to interact with effectively, is likely to unlock vast possibilities for science. If these endeavours are successful the next few decades should see enormous new advances in science that would not have been possible without the new technologies of AI.

References

Bengio, Y., Courville, A., & Vincent, P. (2013). Representation learning: A review and new perspectives. *IEEE Transactions on Pattern Analysis and Machine Intelligence, 35*(8), 1798–1828.

Bishop, C. M. (Ed.). (1995). *Neural networks for pattern recognition.* Oxford University Press.

Breiman, L. (2001). Random forests. *Machine Learning, 45*(1), 5–32. https://doi.org/10.1023/A:1010933404324, Kluwer.

Buchan, D. W. A., & Jones, D. T. (2019). Learning a functional grammar of protein domains using natural language word embedding techniques. In *Proteins: Structure, function, and bioinformatics.* Wiley, prot. 25842. https://doi.org/10.1002/prot.25842

Buchanan, B. G., & Feigenbaum, E. A. (1978). DENDRAL and meta-DENDRAL: Their applications dimension. *Artificial Intelligence, 11,* 5–24.

Carnap, R. (1950). *Logical foundations of probability* (2nd ed., p. 1963). University of Chicago Press.

Cristianini, N., & Shawe-Taylor, J. (2000). *An introduction to support vector machines: And other kernel-based learning methods.* Cambridge University Press.

Davis, R., Buchanan, B. G., & Shortliffe, E. H. (1977). Production systems as a representation for a knowledge-based consultation program. *Artificial Intelligence, 8,* 15–45.

Deng, J., Dong, W., Socher, R., Li, L.-J., Li, K., & Fei-Fei, L. (2009). ImageNet: A large-scale hierarchical image database. In *IEEE conference on computer vision and pattern recognition (CVPR).* Available on ResearchGate.

Dreyfus, H. L. (1972). *What computers can't do: A critique of artificial reason.* Harper & Row.

Evans, R., Jumper, J., Kirkpatrick, J., Sifre, L., Green, T. F. G., Qin, C., Zidek, A., Nelson, A., Bridgland, A., Penedones, H., Petersen, S., Simonyan, K., Crossan, S., Jones, D. T., Silver, D., Kavukcuoglu, K., Hassabis, D., & Senior, A. W. (2018). De novo structure prediction with deep-learning based scoring. In *Thirteenth critical assessment of techniques for protein structure prediction (Abstracts)* (pp. 1–4).

Gillies, D. (1996). *Artificial Intelligence and scientific method.* Oxford University Press.

Goodfellow, I., Bengio, Y., & Courville, A. (2016). *Deep learning.* The MIT Press.

Jackson, P. (1986). *Introduction to expert systems.* Addison-Wesley.

Jensen, F. V. (Ed.). (1996). *An introduction to Bayesian networks.* UCL Press.

Kasparov, G. K., & Greengard, M. (2017). *Deep thinking: Where machine intelligence ends and human creativity begins.* John Murray.

Michalski, R. S., & Chilautsky, R. L. (1980). Learning by being told and learning from examples: An experimental comparison of the two methods of knowledge acquisition in the context of developing an expert system for soybean disease diagnosis. *Journal of Policy Analysis and Information Systems, 4,* 125–161.

Muggleton, S., King, R. D., & Sternberg, M. J. E. (1992). Protein secondary structure prediction using logic-based machine learning. *Protein Engineering, 5*(7), 647–657.

O'Neil, C. (2016). *Weapons of math destruction: How big data increases inequality and threatens democracy.* Crown/Archetype.

Popper, K. R. (1934). *The logic of scientific discovery.* 6th Revised Impression of the 1959 English Translation, Hutchinson, 1972.

Popper, K. R. (1963). *Conjectures and refutations: The growth of scientific knowledge.* Routledge & Kegan Paul.

Quinlan, J. R. (1986). Induction of decision trees. *Machine Learning, 1,* 81–106.

Quinlan, J. R. (1993). *C4.5: Programs for machine learning*. Morgan Kaufmann Publishers.

Rasmussen, C. E., & Williams, C. K. I. (2006). *Gaussian processes for machine learning*. The MIT Press.

Ribeiro, M. T., Singh, S., & Guestrin, C. (2016). "Why should I trust you?": Explaining the predictions of any classifier. In M. Gillies & R. Fiebrink (Eds.), *ACM SIGCHI workshop on human-centered machine learning*. ACM.

Rumelhart, D. E., Hinton, G. E., & Williams, R. J. (1986). Learning representations by back-propagating errors. *Nature, 323*, 533–536.

Schultz, W., Dayan, P., & Montague, P. R. (1997). A neural substrate of prediction and reward. *Science, 275*(5306), 1593–1599. https://doi.org/10.1126/science.275.5306.1593

Shortliffe, E. H., & Buchanan, B. G. (1975). A model of inexact reasoning in medicine. *Mathematical Biosciences, 33*, 351–379.

Turing, A. (1950). Computing machinery and intelligence. *Mind, 59*(236), 433–460.

Wang, F., & Rudin, C. (2015). *Causal falling rule lists*. Available at: http://arxiv.org/abs/1510.051 89

Ward, J. A., Richardson, D., Orgs, G., Hunter, K., & Hamilton, A. (2018). Sensing interpersonal synchrony between actors and autistic children in theatre using wrist-worn accelerometers. In *Proceedings of the 2018 ACM international symposium on wearable computers – ISWC '18* (pp. 148–155). ACM Press. https://doi.org/10.1145/3267242.3267263

Williams, K., Bilsland, E., Sparkes, A., Aubrey, W., Young, M., Soldatova, L. N., De Grave, K., Ramon, J., de Clare, M., Sirawaraporn, W., Oliver, S. G., & King, R. D. (2015). Cheaper faster drug development validated by the repositioning of drugs against neglected tropical diseases. *Journal of The Royal Society Interface, The Royal Society, 12*(104), 20141289. https://doi.org/10.1098/rsif.2014.1289

Chapter 5
Whatever Happened to the Logic of Discovery? From Transparent Logic to Alien Reasoning

Thomas Nickles

Abstract Can recent developments in deep, artificial neural networks (ANNs), machine speed, and Big Data revolutionize scientific discovery across many fields? Sections 5.1 and 5.2 investigate the claim that deep learning fueled by Big Data is providing a methodological revolution across the sciences, one that overturns traditional methodologies. Sections 5.3 and 5.4 address the challenges of alien reasoning and the black-box problem for these novel computational methods. Section 5.5 briefly considers several responses to the problem of achieving "explainable AI" (XAI). Section 5.6 summarizes my present position. If we are at a major turning point in scientific research, it is not the one initially advertised. The alleged revolution is also self-limiting in important ways.

Keywords Deep learning · Artificial neural networks · Big Data · Scientific revolution · Alien reasoning · Black-box problem · Explainable AI

5.1 Deep Learning: A Methodological Revolution?

Recent developments in machine learning by deep, artificial neural networks (ANNs) provide an opportunity to consider the future of methodology of scientific discovery, understood here as scientific and technological innovation in a broad sense.[1] Developed by Geoffrey Hinton, Yann LeCun, Yoshua Bengio, and many others, this approach — in some ways a "turbo" version of the connectionist program of the 1980s and 1990s (Buckner & Garson, 2018) — has achieved notable success,

---

[1] In this paper the term 'discovery' names a traditional topic area in philosophy of science. It is not, for me, an achievement term meaning the disclosure of a final truth about the universe. I prefer the generic term 'innovation' as noncommittal to strong scientific realism. (I am a selective realist/nonrealist.) It is also a term that recognizes that there are many varieties of techno-scientific innovation besides empirical discoveries and the invention of new theories.

T. Nickles (✉)
Foundation Professor of Philosophy Emeritus, University of Nevada at Reno, Reno, NV, USA
e-mail: nickles@unr.edu

© The Author(s), under exclusive license to Springer Nature Switzerland AG 2022
W. J. Gonzalez (ed.), *Current Trends in Philosophy of Science*, Synthese Library
462, https://doi.org/10.1007/978-3-031-01315-7_5

81

especially since 2012, with highly visible victories over the best chess and go players and impressive leaps forward in practical applications such as language translation, classifiers, and recommendation systems. Given their ability to detect subtle statistical patterns when millions or billions of parameters are in play, various deep learning algorithms provide marvelous ways of processing the huge databases ("Big Data") now available in several sciences and elsewhere. That ANNs largely program themselves is a major step toward the grand AI goal of giving a machine a problem to solve without having to tell it how to solve it.[2] The absence of explicit human programming can make such machines more exploratory, more potentially innovative.[3]

Its strongest advocates immediately proclaimed that the ability to process Big Data vaulted today's machine learning to the status of a scientific revolution, a *general, methodological* revolution (rather than a revolution in one specific scientific specialty), one that would consign traditional scientific research to the past.

> [T]he reinvention of discovery is one of the great changes of our time. To historians looking back a hundred years from now, there will be two eras of science: pre-network science, and networked science. We are living in the time of transition to the second era of science (Nielsen, 2012, 10).

So wrote quantum computation expert Michael Nielsen, in his book *Reinventing Discovery: The New Era of Networked Science.*

Already in 2008, Chris Anderson, then editor-in-chief of *Wired* magazine, provocatively announced "The End of Theory: The Data Deluge Makes the Scientific Method Obsolete."

> The new availability of huge amounts of data, along with the statistical tools to crunch these numbers, offers a whole new way of understanding the world. Correlation supersedes causation, and science can advance even without coherent models, unified theories, or really any mechanistic explanation at all (Anderson, 2008).

Anderson's central claim was that models are defective approximations, owing to our data-poor past; but now we are getting so much data that "the data can speak for

---

[2]See, e.g., Koza (1992, ch. 1). Although the successes of machines such as AlphaZero are remarkable in this regard, we must still be alert for hype. It took a tremendous amount of work to configure and tune its architecture in order to achieve the dramatic results.

[3]Think of evolutionary computation, say, in the form of genetic algorithms (Koza, 1992 and successor volumes), or, more recently, the use of genetic algorithms to evolve scientific models from big data, partly to overcome human cognitive limitations, e.g., the "Genetically Evolving Models in Science" (GEMS) project at the London School of Economics (Gobet, 2020). Better, think of the "endless forms most beautiful" (Darwin, 1859) that biological evolution, as a counter-intuitive, exploratory process, has produced. (I don't mean this as an exact analogy. The rough idea is that the instances of a biological species genetically incorporate the phylogenetic wisdom of long, multi-generational evolutionary processes, while a machine trained on many examples, whether historical human examples or self-generated ones, comes, ontogenetically, to embody much wisdom, e.g., as how to play chess or to translate French into Chinese.) An open, exploratory process may not be very efficient; but, in the long run, it is more likely to produce genuinely novel results than one that is humanly goal-directed.

themselves." No need for the old method of hypothesis and test. Here we have shades of Francis Galton and Karl Pearson!

Reaction against such views was swift, and, by now, many application biases and other flaws in particular ANNs have come to light. Some of the bloom is off the rose. As Judea Pearl puts it: "I am an outspoken skeptic of this [data only] trend, because I know how profoundly dumb data are about causes and effects" (Pearl & Mackenzie, 2018, 16). Raw data don't interpret themselves. The anti-causal statements simply recapitulate the mistake made by some of the founders of modern statistics. While the ANN boosters speak of an *anti*-causal revolution, Pearl speaks of "the causal revolution." The irony here is that, according to critics such as Pearl, transparent causal modeling is precisely what is needed to complete the current computational revolution.

Still, to ask where the sciences may be headed in the age of ANNs remains an important philosophical question. At the very least, ANNs are research accelerators — tireless research assistants quickly processing the terabytes of data that are pouring in from the sensor revolution and social media. The question is how better to characterize this scientific progress.

5.2 Deep Learning Success and Hype

Today's ANNs constitute one approach to machine learning. 'Deep' refers to the number of layers of artificial neurons (units) rather than to intellectual profundity or underlying causality. Today's ANNs have several layers of hidden units — sometimes dozens of layers with millions of units — arranged in complex structures, and hence millions or billions of parameters, compared with the two- or three-layer toys of first- and second-generation connectionism.

With some overlap, today's ANNs divide into three main types of learning. In *supervised* learning, the machine is fed a large training set for which the correct answers are prelabeled and from which it learns to extract features and form an implicit hypothesis or set of rules or model that it can then apply to classify items in future data beyond the training set. For example, the training set may involve many thousands of pictures of dogs of various breeds in various contexts. A successful machine/algorithm will then be able reliably to classify by breed the dogs spotted in new pictures beyond the training set. (The pictures are presented to the machine as large, structured arrays of numbers.) The manual labeling is an arduous, expensive task. This limited degree of human instruction is still enough to bias the machine's learning in such a way as to blunt its innovative potential.

By contrast, in *unsupervised* learning, the data is not labeled, hence the machine is more likely to find previously unknown structure in the data by keying on features that it learns. Since many big databases in the sciences will be unlabeled, this approach is badly needed. Importantly, both kinds of machines program themselves in the sense that, as they respond to the huge training sets and to new data, they learn by changing their activation values and some hyperparameter values, thereby forming implicit models for the domain.[4]

Third, there is *deep reinforcement learning*, in which ANNs learn by a process somewhat reminiscent of Skinnerian reinforcement learning, the machine acting as an *agent* in an environment with feedback, including reward.[5] By agent, I mean that the machine acts on an environment (often artificial or simulated) and senses some of the consequences of that action. It does not just sit there quietly filtering and classifying the items in big databases the way the other two types of machines do. There is no explicit training set. The machine effectively generates its own training set by internalizing information gained as a product of its action on the environment, information relevant to its reward signal as represented by a cost function in the algorithm. Here it is important to find a good balance between exploration and exploitation. The 2015 success of the DeepMind company with Atari video games advertised the potential of this approach (Mnih et al., 2015). DeepMind then graduated to the classic games of chess and go. Today's most famous machines are *generative adversarial networks* that engage in "self-play." That is, they in effect divide themselves into two opposing players that play zillions of games against each other, constantly improving their play. This in contrast to being explicitly hand-programmed with thousands of lines of symbolic code, including human chess heuristics, as was IBM's Deep Blue when it defeated chess champion Garry Kasparov in 1997.

The self-programming capability of today's machines is remarkable. Since deep learners extract their own features from the data, in deep learning there is no longer the old, machine-learning need for experts to hand-program in thousands of lines of code, drawing on their *feature-engineering* expertise about dog breeds, human faces, financial markets, chest X-rays, linguistics, game strategies, and such.[6] Success in deep learning calls for a different kind of expertise. Today's platforms of tool libraries such as TensorFlow and Keras, plus computer power and the availability

---

[4]Hyperparameters are those values that set the structure of a given ANN before training, that classify the type of machine or "model" it is, e.g., the number of layers of which kinds (input, convolutional, output, etc.), the parameters that govern a specific convolutional configuration, unit biases, number of training epochs, and so on. This as opposed to the momentary values of the activation functions of individual units or input arrays (for example).

[5]Recall that, in Skinnerian learning theory, behavior is shaped on the basis of the reward signal alone (along with information about the new state of the environment as a result of the behavior), without explicit instruction, as an ontogenetic extension of biological evolution. Biological evolution is also exploratory in this respect, as Skinner (1981) recognized. Ontogenetic learning is an evolutionary shortcut—and more flexibly adaptive. Sejnowski (2018, 149) points out the difference that a ANN has a many reward signals, one for each output unit, whereas there is just one signal to the animal brain. The brain proceeds to solve the credit-assignment problem.

[6]Roman Frigg (in discussion) raises the interesting question who or what should get credit for a new discovery or problem solution made principally by a machine. My answer, to a first approximation: Presumably, the machine and everyone associated with designing and building it, as in today's movie credits, when even the person who brings the ham sandwiches for lunch gets a mention. But that does not answer the question of order: first author, last author, etc.

of huge, open-source databases, make today's work in some ways easier than before (Krohn, 2020, 11ff). That's the good news. The bad news is that it is often difficult to know which features the ANNs have selected and how they are weighted.

Despite the remarkable technical and commercial successes, even the founders are now acknowledging the shortcomings of current deep learning. These machines are not Einsteins. In fact, current machines have zero understanding about what they are doing and are unable to explain themselves to us humans. Their opacity is a serious problem. They are black-boxes, albeit, far more powerful than older, white-box versions of AI.[7] Even the experts have little idea how they arrive at their conclusions, or how to make the algorithms flexibly adaptive to small changes, or how to debug them when things go wrong. Despite their remarkable utility in some domains of application, many of them suffer from a *long tail* of failures, often ludicrous and morally insensitive mistakes. In practical applications, many biases have been disclosed, although the biases often originate in the training data from which the machines learn.[8]

The most hyped early successes have been in the field of closed, full-information, fixed-rule games such as chess and go. As we now realize, these are very special cases as far as general, adaptive intelligence goes. The usual problems with blank-slate, statistical-correlational empiricism soon surfaced for real-world applications. Deep learning and Big Data alone provide narrow, shallow, brittle results that are often difficult to integrate with previously achieved theoretical understanding. ANNs alone cannot handle abstract knowledge representation and hence not even logical reasoning. Nor does success in one domain transfer to neighboring domains.[9]

In some (but not other) respects, today's successes are the result of scaling up the connectionist approach of the 1970s and 1980s — more layers, greater speed and memory, more data, etc. Surprisingly, then, the present approach does not scale well, in part for the same reasons. The nets keep growing bigger, but the bigger nets are too expensive and power hungry for all but the largest companies or institutes to

---

[7] White-box models such as linear regression, decision trees, and traditional production systems lack the predictive power of the new ANNs, nor do they scale well to more complex problems. 'Black box' and 'white box' are, of course, two extreme points on a continuum.

[8] For discussion of biases and other limitations, see the website AINow.org, Marcus and Davis (2019), Brooks (2019), Nickles (2018a), Broussard (2018), Eubanks (2018), Noble (2018), Knight (2017, 2020), Somers (2017), Sweeney (2017), Wachter-Boettcher (2017), Weinberger (2014, 2017), Zenil (2017), Arbesman (2016), Lynch (2016), O'Neil (2016), Nguyen et al. (2015), Pasquale (2015), and Sjegedy et al. (2014). Worse, insofar as opaque machines shape our understanding of the phenomena or systems to which they are applied, the latter themselves become opaque (Pasquale, 2015, Ippoliti & Chen, 2017).

[9] Both *generalization* within a task domain, beyond the training set, and *transfer of training* to other domains are challenges. These are not the same. To be fair, there has been some recent progress on both fronts. For *transfer learning* see Krohn (2020, 188f, 251f). See also the Gillies paper (Chap. 4, this volume) on the failure of current deep learning to handle background knowledge.

afford them.[10] They lack intuitive physics and common sense, and proposed solutions to the common sense problem do not scale well in real-time response.[11]

The learning process remains extremely tedious, often requiring training sets of many epochs, each containing millions of items. By contrast, children (supposedly) learn quickly, both from explicit instruction and from a few examples of dogs and kitty cats. And today's machines are easily tricked in ways that would not fool a three-year-old child.

The original hype suggested that deep learning was a big step toward artificial *general* intelligence (AGI), as well as revolutionizing scientific research. Unfortunately, there are a potential infinity of common life situations. Imagine trying to generate and label the huge data sets to cover each of these situations, necessary to train an AGI via supervised learning! Moreover, today's ANNs are Cartesian in separating computer mind from robot-effector body. But implementing embodiment, so that the machine or robot learns from "real life" experience, faces major problems as well.

In sum, the old adage that what computers do well, we humans do badly—and vice versa—still holds for deep learning machines.[12]

All of that said, the striking successes of deep learning do raise fascinating philosophical issues, questions relevant to the future of philosophy of science and to innovative research in particular. Even though AGI remains a distant goal, today's

---

[10] The generation of large databases, including training sets, can be very expensive. And the dollar and energy cost of deep learning machines is *increasing* according to a Moore-like exponential law. As in many domains, increasing performance only a few percent (e.g., in error reduction of translation programs) may require an exponentially large increase in computational power and expense. Here's a recent statement by Jerome Presenti, head of the AI division of Facebook.

> When you scale deep learning, it tends to behave better and to be able to solve a broader task in a better way. So, there's an advantage to scaling. But clearly the rate of progress is not sustainable. If you look at top experiments, each year the cost is going up 10-fold. Right now, an experiment might be in seven figures, but it's not going to go to nine or ten figures, it's not possible, nobody can afford that.
>
> It means that at some point we're going to hit the wall. In many ways we already have. Not every area has reached the limit of scaling, but in most places, we're getting to a point where we really need to think in terms of optimization, in terms of cost benefit, and we really need to look at how we get most out of the compute [sic] we have. This is the world we are going into (Knight 2019).

See also Knight (2020) and Thompson et al. (2020) on the computational limits of deep learning.

[11] Human beings apply common sense in real time, employing intuitive physics as well as local knowledge, whereas serious attempts to program common sense, such as Douglas Lenat's Cyc project, based on traditional symbolic entries, make processing ever slower as more verities are added. While Cyc is an old approach, hand-crafted in the traditional way, the Paul Allen Institute in Seattle is now taking a new, crowd-sourced, machine-learning approach to common sense.

[12] Of course, we hardly understand how we do it either. This has been called "Polanyi's paradox": many of the things we do most easily, in both thought and action, are the product of "tacit" skills that we are unable to articulate. As several experts have pointed out, computers can best adults on intelligence tests and games of intelligence such as chess but cannot compete with the perception, flexibility, and learning capabilities of a one-year old. See Walsh (2018, 136).

special purpose machines are methodologically interesting in themselves. Anyway, general intelligence of a *human* sort would not appreciably change the nature of scientific research (although it might accelerate it some), since we already know how to produce more humans.

A most intriguing feature of today's ANNs is that, they exhibit surprising, some say *alien*, forms of reasoning. This strangeness is fascinating for two reasons. First, it casts human forms of intelligence into perspective. We have long recognized that our reasoning abilities, our cognitive capacities, are bounded; but could our rationality be bounded in ways that we have not suspected? Second, the new types or patterns of reasoning might further accelerate scientific innovation, sometimes leading to new breakthroughs. Although correlation is not causation, the correlations can provide heuristic guidance to new research, including causal modeling.

But there is a tradeoff here. The alien reasoning candidates exacerbate the black-box problem, for these machines do not disclose the method of their apparent madness. In the rest of the paper, my focus will be on the topics of alien intelligence and the black-box problem.

5.3   Alien Intelligence

Famous for his work on bounded, ecological rationality, Herbert Simon's most important contribution to the discovery debate was to point out that discovery involves problem solving, problem solving involves search, efficient search through large problem spaces requires heuristic guidance — and that computer scientists were already formalizing different kinds of search. Hence, they already had much of cognitive-epistemological interest to say about the discovery process. Simon's early, pathbreaking AI work with Allen Newell adopted the strategy of constructing problem-solving computer programs that simulated human thinking at the level of the conscious protocols of groups of ordinary subjects given sets of problems to solve (Newell & Simon, 1972; Langley et al., 1987).[13] The 1972 programs employed only symbolic logic and domain-general heuristics such as hill climbing and backward chaining. That is, they included no specific empirical scientific information. The Newell and Simon approach modeled AI on human intelligence and on a historically contingent conception of human reasoning (Dick, 2011; Erickson et al., 2013; Daston, 2015). Some years later came the more successful "expert systems" or "knowledge engineering" approach to AI that attempted to

---

[13] At the time, this was a reasonable starting point for attempting to mechanize problem solving as the core process of scientific discovery. I am not claiming that Newell and Simon thought human intelligence was the gold standard. Recall that Simon's Nobel Prize was for his economics work on our "bounded rationality." For discussion of ways in which current work surpasses Simon and other developments, see Gillies and Gillies (Chap. 4, this volume) and Gonzalez (2017).

strengthen AI by programming in the domain-specific knowledge and heuristics used by human experts in a given specialty area.[14]

Both of these highly original approaches were human-centered, while deep learning is not so much; and that is the point I want to emphasize. "The most important thing about making machines that can think is that they will think differently," writes Kevin Kelly, the founding editor of *Wired Magazine*. "[A]s we continue to build synthetic minds, we'll come to realize that human thinking isn't general at all but only one species of thinking..." (in Brockman, 2015). Alan Turing's "imitation game" employed a human standard of intelligence, but he already had asked: "May not machines carry out something which ought to be described as thinking but which is very different from what a man does?" (Turing, 1950, 435). In fact, some early AI experts did aim to discover *non*-human ways of thinking, not merely to automate, speed up, and otherwise enhance human-type cognitive processes. Stephanie Dick (2015) points out that the mechanized proof school called "automated reasoning" or "automated deduction" separated itself from early AI by explicitly aiming to go beyond human modes of reasoning and rejecting the idea that all cogent reasoning could be reduced to a simple, linear, step-by-step process. She highlights J. A. Robinson's resolution principle.[15] Simon had a leg in both camps.

In short, we must not simply assume that human competence is the very paradigm of general intelligence.

One of the most astute observers of the digital transformation now underway is David Weinberger.

> We are increasingly relying on machines that derive conclusions from models that they themselves have created, models that are often beyond human comprehension, models that "think" about the world differently than we do. But this comes with a price. This infusion of *alien intelligence* [my emphasis — TN] is bringing into question the assumptions embedded in our long Western tradition. We thought knowledge was about finding the order hidden in the chaos. We thought it was about simplifying the world. It looks like we were wrong. Knowing the world may require giving up on understanding it. (Weinberger, 2017; cf. Weinberger, 2014)

Consider some popular examples. In 2017, four-hours of autonomous learning by self-play was sufficient for a chess version of AlphaZero to win or draw every single game of a 100-match contest against the previous world champion, Stockfish.8. Grandmaster Peter Heine Nielsen (advisor to Magnus Carlsen, the current human chess champion) told the BBC: "I always wondered how it would be if a superior species landed on earth and showed us how they played chess. Now I know."[16] It's

---

[14] Today's ANNs are pretty domain-specific in their architectures and in the tuning of hyperparameters but not in the sense of containing much substantive domain-specific knowledge. Sometimes little or none is needed. More generally, it remains a problem of how ANNs can build on extant domain knowledge. Again, see Gillies and Gillies (Chap. 4, this volume) on background knowledge.

[15] See Gillies (1996) on the resolution principle and Prolog.

[16] Dorigo (2017). Thanks to Mat Trachok for the reference. In a later, 1000-game match, a later version of AlphaZero again crushed (the then-latest version of) Stockfish.

an arresting thought that we're surely more likely to encounter alien intelligence through artificial intelligence than from a visit from extraterrestrial aliens (Russell, 2019, ch. 1).

Other experts speak of "superhuman performance." Here is the abstract of a technical summary paper published in *Science* by the DeepMind group, the people who developed the Alpha series of winning chess and go algorithms.

The game of chess is the most widely-studied domain in the history of artificial intelligence. The strongest programs are based on a combination of sophisticated search techniques, domain-specific adaptations, and handcrafted evaluation functions that have been refined by human experts over several decades. In contrast, the AlphaGo Zero program recently achieved superhuman performance in the game of Go, by tabula rasa reinforcement learning from games of self-play. (Silver et al., 2017)

They proceed to describe AlphaGo Zero's generalization to AlphaZero, which can play championship chess and shogi (Japanese chess) as well as go. AlphaZero easily defeated earlier Alpha models, by means of far more efficient hardware and software.

Some of the moves of the Alpha machines have been so unorthodox as to look like stupid blunders when first made, e.g., an early queen sacrifice or moving the king to the center of the board in the middle game. Korean go champion Lee Sedol was so flummoxed by an unprecedented move by AlphaGo that he had to leave the room for a while. But what first appeared to many analysts to be a bug turned out to be a feature. Sedol lost that game. Today's players are trying to learn the bold, new strategies, as if they were playing against intelligent aliens from the planet Zork.

Part of the problem of alien intelligence and/or alien reasoning is that we don't have a crystal- clear understanding what intelligence is, or even what reasoning is, let alone alien reasoning as a negative category (Buckner, 2019). Writes Toby Walsh:

No one knows the outer boundaries of what alien intelligence could be. Human history, anthropological studies, animal and plant research, science fiction, and current learning machines are all expanding our imagination. One salient thought is that we should not anthropomorphize intelligence. Whatever it is, human intelligence is the product of a highly contingent evolutionary path on a single planet. (Walsh, 2018, 181)

A few comments will have to suffice here. We do know that the meanings of 'reason' and 'intelligence' have shifted historically, even recently.[17, 18] The vagueness of our concepts of intelligence and alien reasoning make it difficult to say precisely what evidence of alien reasoning we may already have — or could have in the future. One obvious place to look for evidence is the behavioral performance itself. The

---

[17] On historical changes in the concept of reasoning, see Dick (2011, 2015), Erickson et al. (2013), and Mercoier and Sperber (2017). On differences between AI and human reasoning and intelligence, see Gonzalez (2017) and the references therein.

[18] Does Davidson (1974) on conceptual schemes deny that we can meaningfully attribute radically alien intelligence to other entities (intelligence without intelligibility to us)? I join those who reject this idea as anthropocentric, and as containing a whiff of verificationism.

crazy-looking behavior of ANNs (including the patterns of *mis*classification made by many ANNs today) are what first inspired the label 'alien'.[19]

Another sort of evidence is *generative* rather than *consequential*, in that the physical basis or mode of generation of the successful behavior is so unorthodox.[20] It already bothers some people to think that anything nonconscious or "mechanical" could be intelligent. Many today still find biological evolution alien in the sense of being a "strange inversion of reasoning," radically counterintuitive to what Dan Dennett calls the traditional "mind first" approach.[21] In the biological world, ant colonies, jellyfish, octopuses, and even slime molds can learn and act intelligently, in some sense, despite not having a central, controlling brain. In fact, nonhuman animals and plants behave cleverly in their niches without having a developed language. And today's AI experts well enough understand how ANNs themselves operate to be amazed that devices of this sort can be so effective. This surprise is related to the black-box problem, since it signals that not even the experts understand precisely how it is possible that the machines provide such excellent results. Thus, their surprising success and the existence of a black-box problem can be taken as indicators of alien intelligence.[22]

The "black-box problem" will be the topic of the next section. Meanwhile, we need to keep in mind an important question. If we suppose that human modes of thinking are limited, then how far can *human* science go in the "alien" direction? For our science must remain human-centered in some significant respects if it is to

---

[19] Dennett applies "a strange inversion of reasoning" to both Darwinian evolution and Turing machines. Animals operate on a "need to know" basis, guided by "free-floating rationales" programmed in transgenerationally by evolution. On this view, *rational* behavior need not be backed by explicit reasoning at any level of description, nor by the possession of language. Nonhuman animals cannot explain to us why they behave as they do. Rejecting "mind first," Dennett (1995, chs. 1, 3, 2017) implies that we also must reject "method first." Successful methods are inductive outcomes of research, not available to us a priori. Methods are cranes, not skyhooks. Dennett reworks the above-mentioned material in Dennett (2017). See also Mercoier and Sperber (2017) and Buckner (2019).

[20] Expert behavior produced by deep learning machines could scarcely have been imagined twenty years ago. The AlphaZero machines do not start with any human expert or domain knowledge at all, yet they achieve championship level through self-play in only three days.

[21] We must be careful not to spread the extension of 'alien' so widely that most of our own behavior is alien to us (see Wilson 2004). Although the human brain is also a highly distributed system that we understand only dimly, we do have the aforementioned evidence concerning our huge differences from machines regarding general intelligence as a basis for claiming that they operate in a way much different from us. When it comes to understanding each other, however, especially within the same culture, we rely on our high-level, belief-desire model of behavior (say, in the form of Dennett's intentional systems model), a "theory of mind" that we can fairly reliably attribute to others. The result is that we can black box micro-biological implementation levels with much cognitive, real-time gain. Still, that many of our ordinary human epistemic activities are alien to ourselves has always been a main source of "the discovery problem." It generalizes to the problem of understanding expert scientific practice.

[22] On the other hand, some AI experts are attempting to make progress by reverse-engineering the human visual system, which, like other animal sensory systems, is far more computationally intensive than automated logical reasoning is.

remain *our* science, something that we can understand well enough to use at least instrumentally—and to trust.

5.4   The Black-Box Problem

"As generations of algorithms get smarter, they're also becoming more incomprehensible...," observed computer scientist Jon Kleinberg and economist Sendhil Mullainathan (in Brockman, 2015). *The black-box problem*, also called *the opacity* or *transparency problem* or *explainable AI (XAI) problem* for deep learning, is the problem of understanding how or why the device arrived at its conclusions. Unlike Skinner, we want to know something about what is happening inside the black box. In fact, there are multiple black-box problems, depending on the questions being asked.[23] For present purposes, however, I shall usually lump them together as "the black-box problem."

Complex interaction/adaptive networks quickly challenge the limits of human cognition, and ANNs are no exception. Yet ANN technology can help us deal with extant networks (such as real neural networks and protein-gene interaction networks in biological cells) and in designing complex technologies (such as the coordination of large-scale satellite systems and automated transportation systems).

We don't need to know everything that is going on inside the black boxes, for there is much that is not relevant to our questions about how to explain their behavior. And we do know *something* about what is inside the boxes, since human experts have designed the ANN architectures and set or tuned their hyperparameters. Unfortunately, what we do know does not help us much with the black-box problem. Today's sophisticated machines employ a massively *distributed* (rather than a *localist*) architecture in the sense that the individual units have no definite conceptual meaning. Even a feed-forward machine can take billions of parallel-processing steps involving subtle statistical correlations to identify a Golden Retriever in a picture, with probability p. Typically, the mapping of input to output is highly nonlinear; and, once we add the recurrent or feedback loops that many machines require, the processing becomes even more opaque to us in terms of why they behaved in the way they did. Even the experts describe the construction and tuning of ANNs as a mystery-shrouded art. Typical machines possess zillions of hyperparameters that may require setting or tuning, and they all involve performance trade-offs.

---

[23] Explainability or interpretability or transparency in the use of machine models is itself a complex topic that I must gloss over here. There are several different kinds of opacity, some harder to handle than others, some more relevant than others to the epistemic or moral justification questions we need to ask. (The issue becomes broadly "rhetorical" in the sense that we seek an appropriate match between what the machine can "tell" us and what questions we, the audience, need answered.) For discussion, see, e.g., Lipton (2018), Humphreys (2004, 2009), Winsberg (2010), Sullivan (2019), and Creel (2020).

Consequently, opaqueness presents serious challenges to scientific research as well as to wider social applications. It is not simply a question of correlation without causation. Opaqueness means that we have no explicit models, let alone causal models, that we can work with in traditional ways, wiggling this parameter and that in order to see how the system responds.[24] Thus, much traditional heuristic guidance is lost. While some limited generalization technologies exist (e.g., dropout, which permits better generalization to examples not in the training set), generalization remains a problem, especially generalizations that build on expert background knowledge. Likewise, there are abstraction technologies, e.g., convolution and pooling for visual processing (Krohn, 2020; Buckner, 2018); but abstraction in the wider sense employed by research scientists is missing. Today's deep learning cannot synthesize or integrate hardly any sort of knowledge, since it cannot adequately *represent* knowledge claims. In addition, much scientific work requires precise physical and other practical skills that today's AI robotics is nowhere close to achieving. There are also serious *reproducibility* problems, given that the amount of trial-and-error fiddling is often so great (and typically unpublished) that it is far from clear how the successes were achieved. And how can such machines be applied reliably in novel research contexts where we do not yet know what the good results are? Besides all that, many of today's algorithms are proprietary, for commercial reasons, or those of general security or privacy, state secrets, or even attempts to *resist* government regulative intervention (Burrell 2016; Barber, 2019; Tian et al., 2019; Mitchell et al., 2019).

Finally, much future work at scientific and translational frontiers will remain data poor. We should not forget that, historically, around 1800, the problem of sparse data in novel theoretical contexts was a major motivation for the revival of the method of hypotheses, as a departure from data-driven empiricism (Laudan, 1981).[25]

What about quantum computing? This would surely make the black-box problem even worse. Digital computing largely displaced analog computing for two reasons. Digital circuity is easier for humans to understand in terms of Boolean logic and causal narratives; and, since 1950, digital hardware has advanced far more rapidly (Fritchman, 2019). Although quantum computing would be a huge technological advance, it would be even harder for humans to understand, even more sensitive to noise than analog is, and perhaps too expensive for ordinary research labs.

---

[24] I am here talking about substantive, domain-specific models such as the Hodgkin-Huxley model of neuron action-potential transmission. In a different sense, deep learning experts often describe the specific architectures of their machines as models.

[25] In some cases, it is possible to simulate much of the data used to train a machine.

5.5   What Can Be Done to Obtain Explainable AI (XAI)?

Explainable AI (XAI) is now an increasing concern of the ANN inventors as well as the big AI corporations, DARPA[26] (the U.S. Defense Advanced Research Projects Agency), and similar entities in other countries. I only have space for a few hints.

One group — let's call them *the evaders* — attempts to evade the problem rather than to solve it, saying that we should not use today's ANNs in high-risk or morally sensitive situations. The European Union has introduced a set of General Data Protection Regulations governing the use of AI, including the right to an explanation for decisions reached (EU GDPR, 2016, Recital 71). This measure does address the problem that the opaqueness is sometimes deliberate, for proprietary reasons. However, evasion remains unpopular if treated as a final resolution, for it largely ignores the potential of deep learning. Given the complexity of social problems, should we not try to expand the reach of research beyond human limitations?

Next are the *mitigators*. Their response to opacity is to mitigate the problem by employing extensive *consequential testing* of machine outputs (cf. Humphreys, 2004, 150). It is consequential testing that has taught us most about the strengths and weaknesses of the ANNs we have today, by deliberate trial-and-error experimentation as well as the episodic detection of errors, including biases; and testing is already a mainstay of both scientific research and engineering design. Like the evaders, the mitigators do not aim to *solve* the transparency problem, at least not deeply; but they are surely correct that testing can help to justify the practical application of ANNs by exploring the limits of their reliable performance. After all, we need testing to inspire trust even when we *do* understand the working of a new machine. Testing precedes trusting!

Third are the optimistic *deniers*, who claim that there is no problem, because (they say) we already have enough transparency, or at least as much as we have with ourselves. One argument is that scientific explanation and engineering design routinely black-box the implementation of lower-level processes as irrelevant to the task at hand.[27] Typically, it is irrelevant to know how a computer program computes averages, orders lists, or computes a sigmoid function, let alone how the underlying electronics works. Dennnett (1971, 1995, 2017) points out that, when dealing with the normal behavior of "intentional systems," we do not need to know the design and physical stances. (Indeed, in the course of normal behavior, we do not *want* to know, as that would likely interfere with smooth behavioral flow.) A second denier argument, at which I hinted above, is that we don't know the neural-level basis of our own cognition, so we should not worry about the opacity of ANNs. The deniers are right up to a point, on both counts; but they miss the main target. The XAI problem for today's ANNs is precisely that we don't know what the learning

---

[26] See DARPA and Gunning (2016).

[27] See, e.g., Polanyi (1966), Wilson (2004), and Sullivan (2019) for versions of these ideas, also Strevens (2017) on levels of explanation. On minimal model explanations and universality, see Batterman (2002) and Batterman and Rice (2014).

machine or robot is doing even at the highest relevant level, in terms of "reasons" for deciding or otherwise acting as it does. In the case of humans, we do possess a "theory of mind," fallible though it is, that we utterly lack with today's ANNs.

Now we come to the *confronters*, those who do try to solve the opacity problem head-on. One approach is to treat the black boxes themselves as domains of the universe about which we construct and test our own, humanly constructed models (cf. Lipton, 2018; Creel, 2020). This is problematic insofar as the "real" models used by the machine are alien to us — although we may still be able to produce useful, simplified (e.g., linearized) models, as we often do in the sciences — and in the everyday, human "theory of mind."[28] For deep reinforcement learners, learning from their actions on an environment, we can imagine coming to an appreciation of their Dennett-style "free-floating" rationales by doing the same sort of experimental modeling we do on biological organisms, e.g., game and decision theoretic modeling based on a cost function.

Fifth, the *integrators* are confronters who aim to integrate today's sorts of ANNs with symbolic forms of computation. The human brain seems to be a hybrid of several different kinds of processing in various centers. ANN critics such as Gary Marcus and Ernest Davis (2019) recommend such an approach, and, increasingly, so do founders such as Bengio (2019). Pedro Domingos (2015) takes this idea to the limit in his project to integrate all five major approaches to AI in a "master algorithm."

Some of the most promising work is that of Judea Pearl on "the causal revolution," namely, the use of causal models as represented by causal diagrams, in conjunction with Bayesian networks. Philosophers of science at Carnegie-Mellon, led by Clark Glymour, have strongly influenced Pearl. Bengio at Montreal is currently trying to employ Pearl's ideas.[29] Naturally, there are serious challenges in scaling up this approach to complex systems of zillions of variables and feedback loops.

Finally, we have the *mergers* (those who merge items) as distinct from the integrators. A popular example is Elon Musk's rather secretive Neuralink Corporation, which aims to develop prosthetic, human brain "neural lace" implants,

---

[28] See, e.g., Tishby and Zaslavsky (2015) and Kindermans et al. (2017). While achieving a general, abstract-theoretical understanding would be a major breakthrough, there remains the question how this could be applied to particular cases such as helping Maria understand why, specifically, her application to medical school was denied. (This question again underscores that there are distinct black-box issues.) There are still other approaches not mentioned above. Consider the idea of reverse-engineering generative image classifiers, for example, to determine which features are most significant in the machine's categorization decisions (as in the code-audit approach). For example, when asking it to generate pictures of dogs, the length and shape of the ears is particularly important in today's machines' ability to discriminate dog breeds. One can also look for patterns in adversarial examples in the hope of better understanding the major processing pathways, much as we study visual illusions to determine what heuristic shortcuts our visual system is making.

[29] See Glymour and Cooper (1999), Spirtes et al. (2000); Pearl (2000); Woodward (2003), Pearl et al. (2016); Pearl and Mackenzie (2018). All approaches have their critics, of course, including Pearl's. For example, see Gillies and Sudbury (2013).

motivated by the cyborgian idea that we shall come to understand new modes of thinking/calculating better by merging with them, i.e., by combining human and machine intelligence. The long run goal is some form of transhumanism.[30] Ray Kurzweil has long touted a positive form of transhumanism compared to Musk's more defensive posture.

An opposite sort of merging is to place more emphasis on robotics, i.e., on embodied cognition and agency, giving up the Cartesian separation of mind and body and what John Dewey (1929) called a "spectator theory of knowledge." The way to have advanced systems learn with a grounded, causal understanding of the world is to have them interact with the world in many ways. This would require a fairly general intelligence plus a far more rapid (and safer!) learning method, else the robot's parts would wear out before it learned very much. So, a pure *tabula rasa* approach, learning from data alone, would not work. Perhaps intuitive physics could be programmed in as a kind of genetic endowment; and much training could employ interactive gaming platforms, virtual reality devices, and even movies (e.g., Ullman et al., 2017).[31]

Right now, no one can know how things will turn out, but here are four reasons for pessimism concerning XAI. First, there has been little progress so far; but, to be fair, it is early in the game. Second, and more importantly, although we may make progress, it is likely to be outrun by further advances in AI that will be even more foreign to human ways of thinking (Arbesman 2016, 82). Third, future technology will open the way to research on real-world systems of increased complexity and weirdness, the domains into which frontier science is already expanding. We cannot simply assume that the universe is fully intelligible to us. In these domains, deep learning is likely to add to the problem of nature's own possible intelligibility to us. Fourth, solving the black-box problem in a humanly satisfying way would reduce

---

[30] For references and a critique of transhumanist approaches as challenging our humanity, see Schneider (2019).

[31] I mention also the use of fast, simple statistical algorithms, from Paul Meehl (1954) to Gerd Gigerenzer today. Success here, e.g., in rapid medical diagnosis, has led philosophers such as J. D. Trout to reject the quest for understanding insofar as it is subjective and cannot serve as a proxy for reliability. Others such as Henk de Regt defend the importance of understanding as a scientific goal (Trout, 2002; Bishop & Trout, 2005; de Regt et al., 2009; de Regt, 2017). The "fast and frugal" approach goes in the opposite direction from today's ANNs, with their thousands of variables, instead employing just two or three principal factors. For example, Gigerenzer and Todd (1999) include several examples of such algorithms. Their aim is to contribute to "ecological rationality," based on the sets of heuristics that animals, including people, can use in real-time behavioral decisions. In several applications, these quick applications produce better results than more complex ones and better than human experts. One application of this sort of approach to deep learning might be a "code audit," near-final stage that reports only the few variables that played major roles in the final classification. While the examples provided by Gigerenzer's team are impressive, they offer critics little hope that they can replace anything close to the full range of black-box machines. Moreover, although their simplicity makes them easier to understand and use, these approaches themselves *exhibit* the black-box problem rather than solving it. For we will still lack the intuitive *causal* understanding that scientists have traditionally sought, as well as heuristic guidance as to how to further develop our models (but see Nickles, 2018a, b).

AI back within the limits of human intelligence. And the more alien the black box, the more limiting would be this human constraint.

Better stated, this would not constitute a genuine solution to the black-box problem at all. Strange to say, in this respect *solving the opacity problem would be disappointing insofar as it tamed alien reasoning*. So, there is a whiff of optimism, after all, to a resistant black-box problem, insofar as it signals that alien reasoning may be advancing the scope and depth of research.

We seem to be caught in a dilemma. Insofar as we succeed in solving the black-box problem, we have not escaped a human-centered conception of cogent reasoning, and we have dimmed the exciting potential of alien intelligence. But, insofar as we don't solve it, its successful use in the sciences will force us to reassess some time-honored goals and standards of scientific research. As Google's Cassie Kozyrkov (2018) writes:

> Some tasks are so complicated you cannot automate them by giving explicit instructions. . . .
>   Now that we're automating things where we couldn't have handcrafted that model (recipe/instructions) — because it's way too complicated — are we seriously expecting to read it and fully grasp how it works?

5.6 Conclusion

There is no question that the advent of powerful ANNs is a turning point that adds powerful new tools for innovative research. Although they fall well short of the old dream of a general logic or method of discovery, their existence, alongside the automated analytic, modeling, simulation, computational and other tools of today's research labs, brings joy to traditional logic of discovery advocates. And it should cause the logical empiricist deniers of a philosophically relevant topic of discovery to think again.

Nonetheless, on the specific question whether "alien" ANNs can be made sufficiently transparent, I am among the pessimists, albeit with the aforementioned bonus of alien reasoning. Hence, I reject the claim that anything today's ANNs are producing a general, revolutionary *overturning* of traditional scientific methods and goals. I see deep learning as greatly enhancing rather than replacing the traditional toolkit.

Here is my current position:

1. Achieving artificial general intelligence (AGI) remains a very remote goal.
2. Not even *human* AGI would give us a general method of discovery, for we already have human general intelligence without a general method of discovery.
3. Achieving ordinary *human* AGI would add little to the scientific enterprise (only a linear scaling-up of effort), since we already have an easy way to produce humans.
4. If today's ANNs are not a large step toward *human* AGI, that is thus of only modest consequence for the future of scientific discovery. *Non*human intelligence, whether AGI or not, is a different story.

5. Today's highly *dedicated* machines (i.e., specific-purpose machines of limited generality) are already interesting for the hints they provide of nonhuman, "alien" intelligence. That is their most exciting use in the sciences and elsewhere.

6. Alien intelligence is not a new idea in the sciences, since it has long been documented in the biological world and, to some extent, in earlier computer science.

7. Both biological and artificial forms of alien reasoning confront us with the black-box problem.

8. The black-box problem looks unsolvable for today's kinds of deep neural networks.

9. Worse, incremental progress in explainable AI probably will be swamped by a rapid deepening and complexification of future machines and their application to complex systems in the natural and social worlds. The problem is general. It arises for any form of alien intelligence.

10. There is thus a *black-box tradeoff* in the use of alien intelligence to advance *human* scientific innovation. The more alien the machine, the more potentially path-breaking it is—and also the more unintelligible. But the more we tame such machines, the more we are left with scientific inquiry as usual. An appropriate slogan for this tradeoff might be: *The more we learn, the less we understand.*

11. Despite today's limitations, our ANNs provide valuable additional tools for discovery. These are tools among many others. They do not replace the explanation and understanding that derives from traditional, causal-model-based reasoning. The new tools may revolutionize some data-rich domains of research, but they do not constitute the sort of general scientific revolution touted in the original hype.

12. The original hype about ANNs revolutionizing science missed the target. Insofar as causal-theoretical understanding is eliminated, it is not because it is no longer needed; on the contrary, it is because it is no longer possible in today's strict ANN regime of science.

13. Despite the pessimism, mine is not a du Bois-Reymond-style *Ignoramibus* (Finkelstein, 2013). We may, in the future, design machines with radically different architectures or an integrated combination of architectures that will diminish opacity and the other limitations. But how to do so without retreating to human limitations on reasoning? *Quis scit?* Who knows?

At its most esoteric, scientific investigation has always skirted the edges of human intelligibility — from Newton on instantaneous action-at-a-distance to quantum physics to the molecular biology of cells and brains to the interactive complex systems of economics. These cases are perhaps harbingers of what is to come. Deep-learning-based science constitutes a leap in level of opacity — and now, as just noted, not only for the "deepest," most esoteric sciences such as fundamental physics but for *all* areas of application of such machines, including the "special sciences" (biology, psychology, anthropology, sociology, etc.) and the many wider social and commercial uses. Insofar as alien reasoning came to dominate research, the traditional goal of scientific understanding would have to be abandoned. In the balance of scientific goals, there would be a shift away from explanation and toward prediction. Several current varieties of scientific realism would become increasingly harder to defend as science advanced.

No one today knows how far deep learning — *or any other mode of alien reasoning, modeling, simulation* — will advance, but it could be that we shall face the advertised major trade-off. While alien intelligence would provide exciting advances, this tradeoff would draw us increasingly into the strange epistemological situation in which *the more we know* (in some sense of know), *the less we understand*. Obviously, it would therefore also raise difficult issues concerning trust, trust in both the technical computational processes and the experts who certify them, including in their application to human affairs, perhaps widening the current public distrust of science. To some extent it already does.[32]

Insofar as future AI does achieve the sort of contextual, domain sensitivity and integration with background knowledge that it now lacks — yet without becoming more explainable — we would approach a dystopian extremum imagined by mathematician Steven Strogatz (2018). He envisions "a day, perhaps in the not too distant future, when AlphaZero has evolved into a more general problem-solving algorithm; call it AlphaInfinity," one that could address serious scientific and social problems yet greatly surpass our ability to comprehend it.

> Suppose that deeper patterns exist to be discovered — in the ways genes are regulated or cancer progresses; in the orchestration of the immune system; in the dance of subatomic particles. And suppose that these patterns can be predicted, but only by an intelligence far superior to ours. If AlphaInfinity could identify and understand them, it would seem to us like an oracle.

> We would sit at its feet and listen intently. We would not understand why the oracle was always right, but we could check its calculations and predictions against experiments and observations, and confirm its revelations. Science, that signal human endeavor, would reduce our role to that of spectators, gaping in wonder and confusion.

Acknowledgements I am honored to contribute to these 25th anniversary *Jornadas*. My thanks to Wenceslao J. Gonzalez and his staff for organizing the conference under difficult (Covid-19 pandemic) conditions. I am also grateful for helpful comments from my fellow participants, Roman Frigg and (especially) Donald and Marco Gillies. See their chapters in this volume.

References

Anderson, C. (2008). The end of theory: The data deluge makes the scientific method obsolete. *Wired, 16*, 106–129.
Arbesman, S. (2016). *Overcomplicated: Technology at the limits of comprehension*. Penguin/Random House.

---

[32] Descartes once observed that, the more science teaches us about how human sensory systems work, the more skeptical we become about the knowledge of the world they seem to impart to us.

Barber, G. (2019). Artificial Intelligence confronts a 'reproducibility' crisis. *Wired*, online, September 16. Accessed 16 Sept 2019.

Batterman, R. (2002). *The Devil in the details: Asymptotic reasoning in explanation, reduction, and emergence*. Oxford University Press.

Batterman, R., & Rice, C. (2014). Minimal model explanations. *Philosophy of Science, 81*(2), 349–376.

Bengio, Y., et al. (2019). A meta-transfer objective for learning to disentangle causal mechanisms. *arXiv*:1901.109112v2 [cs.LG], 4 February.

Bishop, M., & Trout, J. D. (2005). *Epistemology and the psychology of human judgment*. Oxford University Press.

Brockman, J. (Ed.). (2015). *What to think about machines that think*. Harper.

Brooks, R. (2019). *Forai-steps-toward-super-intelligence*. https://rodneybrooks.com/2018/07/. Accessed 30 Sept 2019.

Broussard, M. (2018). *Artificial unintelligence: How computers misunderstand the world*. MIT Press.

Buckner, C. (2018). Empiricism without magic: Transformational abstraction in deep convolutional neural networks. *Synthese (online) September, 24*, 2018.

Buckner, C. (2019). Rational inference: The lowest bounds. *Philosophy and Phenomenological Research, 98*(3), 697–724.

Buckner, C., & Garson, J. (2018). Connectionism and post-connectionist models. In M. Sprevak & M. Colombo (Eds.), *The Routledge handbook of the computational mind* (pp. 76–90). Routledge.

Burrell, J. (2016). How the machine 'thinks': Understanding opacity in machine learning algorithms. *Big Data & Society*, January–June, 1–12.

Creel, K. (2020). Transparency in complex computational systems. *Philosophy of Science*. Online 30 April 2020. https://doi.org/10.1086/709729

DARPA, & Gunning, D. (2016). *Explainable Artificial Intelligence (XAI)*. https://www.darpa.mil/attachments/XAIIndustryDay_Final.pptx. Accessed 8 Dec 2019.

Daston, L. (2015). Simon and the sirens: A Commentary. *Isis, 106*(3), 669–676.

Davidson, D. (1974). On the very idea of a conceptual scheme. *Proceedings and Addresses of the American Philosophical Association, 47*, 5–20. Reprinted in Davidson, D., *Inquiries into truth and interpretation* (pp. 183–198). Oxford University Press.

de Regt, H. (2017). *Understanding scientific understanding*. Oxford University Press.

de Regt, H., Leonelli, S., & Eigner, K. (Eds.). (2009). *Scientific understanding: Philosophical perspectives*. University of Pittsburgh Press.

Dennett, D. (1971). Intentional systems. *Journal of Philosophy, 68*(4), 87–106.

Dennett, D. (1995). *Darwin's dangerous idea*. Simon & Schuster.

Dennett, D. (2017). *From bacteria to Bach and back: The evolution of minds*. Norton.

Dewey, J. (1929). *The quest for certainty*. Putnam.

Dick, S. (2011). AfterMath: The work of proof in the age of human-machine collaboration. *Isis, 102*(3), 494–505.

Dick, S. (2015). Of models and machines. *Isis, 106*(3), 623–634.

Domingos, P. (2015). *The master algorithm: How the search for the ultimate learning machine will remake our world*. Basic Books.

Dorigo, T. (2017). Alpha Zero teaches itself chess 4 hours, then beats dad. *Science 2.0*. https://www.science20.com/tommaso_dorigo/alpha_zero_teaches_itself_chess_4_hours_then_beats_dad-229007. Accessed 25 Aug 2019.

Erickson, P., et al. (2013). *How reason almost lost its mind*. University of Chicago Press.

EU GDPR. (2016). *Recital 71*. http://www.privacy-regulation.eu/en/recital-71-GDPR.htm. Accessed 10 Dec 2019.

Eubanks, V. (2018). *Automating inequality: How high-tech tools profile, police, and punish the poor*. St. Martin's Press.

Finkelstein, G. (2013). *Emil du Bois-Reymond: Neuroscience, self, and society in nineteenth-century Germany*. MIT Press.

Fritchman, D. (2019). *Why analog lost*. http://Medium.com. Accessed 30 Nov 2019.

Gigerenzer, G., & Todd, P. (Eds.). (1999). *Simple heuristics that make us smart*. Oxford University Press.

Gillies, D. (1996). *Artificial Intelligence and scientific method*. Oxford University Press.

Gillies, D., & Sudbury, A. (2013). Should causal models always be Markovian? The case of multi-causal forks in medicine. *European Journal for Philosophy of Science, 3*(3), 275–308.

Glymour, C., & Cooper, G. (Eds.). (1999). *Computation, causation, & discovery*. MIT Press.

Gobet, F. (2020). Genetically evolving models in science (GEMS). *Research program at the London School of Economics*. http://www.lse.ac.uk/cpnss/research/genetically-evolving-models-in-science. Accessed 26 July 2020.

Gonzalez, W. J. (2017). From intelligence to rationality of minds and machines in contemporary society: The sciences of design and the role of information. *Minds & Machines, 27*(3), 397–424.

Humphreys, P. (2004). *Extending ourselves: Computational Science, empiricism, and scientific method*. Oxford University Press.

Humphreys, P. (2009). The philosophical novelty of computer simulation methods. *Synthese, 169*(3), 615–626.

Ippoliti, E., & Chen, P. (Eds.). (2017). *Methods and finance: A unifying view on finance, mathematics and philosophy*. Springer.

Kindermans, P.-J., et al. (2017). Learning how to explain neural networks: PatternNet and PatternAttribution. *ArXiv*:1705.05598v2. 24 October. Accessed 28 Dec 2019.

Knight, W. (2017). The dark secret at the heart of AI: No one really knows how the most advanced algorithms do what they do. *MIT Technology Review*, 11 April 2017, 55–63.

Knight, W. (2019). Facebook's head of AI says the field will soon 'hit the wall'. *Wired*, December 4. Accessed 10 Dec 2019.

Knight, W. (2020). Prepare for Artificial Intelligence to produce less wizardry. *Wired*, July 11. Accessed 26 July 2020.

Koza, J. (1992). *Genetic programming: On the programming of computers by means of natural selection* [the first volume of a series]. The MIT Press.

Kozyrkov, C. (2018). Explainable AI won't deliver. Here's why. Hackernoon.com, September 16. https://hackernoon.com/explainable-ai-wont-deliver-here-s-why-6738f54216be. Accessed 19 Nov 2019.

Krohn, J. (with Beyleveld, G., & Bassens, A.). (2020). *Deep learning illustrated*. Addison-Wesley.

Langley, P., Simon, H., Bradshaw, G., & Zytkow, J. (1987). *Scientific discovery: Computational explorations of the creative process*. The MIT Press.

Laudan, L. (1981). *Science and hypothesis*. Reidel.

Lipton, Z. (2018). The mythos of model interpretability. *Queue* (May–June), 1–27. https://dl.acm.org/doi/pdf/10.1145/3236386.3241340. Accessed 2 Aug 2020. Original, 2016 version in Archive.org > cs > arXiv:1606.03490.

Lynch, M. (2016). *The Internet of us: Knowing more and understanding less in the age of big data*. Liveright/W. W. Norton.

Marcus, G., & Davis, E. (2019). *Rebooting AI: Building artificial intelligence we can trust*. Pantheon.

Meehl, P. (1954). *Clinical versus statistical prediction: A theoretical analysis and a review of the evidence*. University of Minnesota Press.

Mercoier, H., & Sperber, D. (2017). *The Enigma of Reason*. Harvard University Press.

Mitchell, M., Wu, S., Zaldivar, A., Barnes, P., Vasserman, L., Hutchinson, B., Spitzer, E., Raji, I. D., & Gebru, T. (2019). Models cards for model reporting. In *FAT* '19: Proceedings of the Conference on Fairness, Accountability, and Transparency, 29 January 2019* (pp. 220–229). https://doi.org/10.1145/3287560.3287596

Mnih, V., et al. (2015). Human-level control through deep reinforcement learning. *Nature, 518*, 529–533.

Newell, A., & Simon, H. (1972). *Human problem solving*. Prentice-Hall.

Nguyen, A., Yosinski, J., & Clune, J. (2015). Deep neural networks are easily fooled: High confidence predictions for unrecognizable images. In *2015 IEEE conference on computer vision and pattern recognition* (CVPR), pp. 427–436.

Nickles, T. (2018a). Alien reasoning: Is a major change in scientific research underway? *Topoi.* Published online, 22 March. https://doi.org/10.1007/s11245-018-9557-1

Nickles, T. (2018b). TTT: A fast heuristic to new theories? In D. Danks & E. Ippoliti (Eds.), *Building theories: Hypotheses and heuristics in science* (pp. 169–189). Springer.

Nielsen, M. (2012). *Reinventing discovery: The new era of networked science*. Princeton University Press.

Noble, S. (2018). *Algorithms of oppression: How search engines reinforce racism*. NYU Press.

O'Neil, C. (2016). *Weapons of math destruction: How big data increases inequality and threatens democracy*. Crown.

Pasquale, F. (2015). *The black box society: The secret algorithms that control money and information*. Harvard University Press.

Pearl, J. (2000). *Causality: Models, reasoning, and inference*. Cambridge University Press.

Pearl, J., & Mackenzie, D. (2018). *The book of why: The new science of cause and effect*. Basic Books.

Pearl, J., Glymour, M., & Nicholas, J. (2016). *Causal inference in statistics: A primer*. Wiley.

Polanyi, M. (1966). *The tacit dimension*. Doubleday.

Russell, S. (2019). *Human compatible: Artificial Intelligence and the problem of control*. Viking.

Schneider, S. (2019). *Artificial you: AI and the future of your mind*. Princeton University Press.

Sejnowski, T. (2018). *The deep learning revolution*. The MIT Press.

Silver, D., Hassabis, D., et al. (2017). Mastering chess and shogi by self-play with a general reinforcement learning algorithm. arxiv.org/pdf/1712.01815.pdf. Accessed 28 Aug 2019.

Sjegedy, C., et al. (2014). *Intriguing pro-erties of neural networks*. arXiv.org/pdf/1213.6199.pdf. Accessed 5 February 2018.

Skinner, B. F. (1981). Selection by consequences. *Science, 213*, 501–504.

Somers, J. (2017). Is AI riding a one-trick pony? *MIT Technology Review*, September 29.

Spirtes, P., Glymour, C., & Scheines, R. (2000). *Causation, prediction, and search* (2nd ed.). The MIT Press.

Strevens, M. (2017). The whole story: Explanatory autonomy and convergent evolution. In D. Kaplan (Ed.), *Explanation and integration in mind and brain science*. Oxford University Press.

Strogatz, S. (2018). One giant step for a chess-playing machine: The stunning success of AlphaZero: A deep-learning algorithm, heralds a new age of insight — One that, for humans, may not last long. *New York Times*, December 26.

Sullivan, E. (2019). Understanding from machine learning models. *British Journal for the Philosophy of Science*, online. https://doi.org/10.1093/bjps/axz035. Accessed 21 Sept 2019.

Sweeney, P. (2017). *Deep learning, alioen knowledge and other UFOs*. http://medium.com/inventing-intelligent-machines/machine-learning-alien-knowledge-and-other-ufos-1a44c66508d1. Accessed 18 Nov 2017.

Thompson, N., Greenwald, K., Lee, K., & Manso, G. (2020). The computational limits of deep learning. arxiv.org/pdf/2007.05558.pdf. Accessed 16 July 2020.

Tian, Y., et al. (2019). ELF OpenGo: An analysis and open reimplementation of AlphaZero. arxiv.org/pdf/1902.04522.pdf

Tishby, N., & Zaslavsky, N. (2015). Deep learning and the information bottleneck principle. arxiv.org/pdf/1503.02406.pdf. Accessed 5 Feb 2018.

Trout, J. D. (2002). Scientific explanation and the sense of understanding. *Philosophy of Science, 69*(2), 212–233.

Turing, A. (1950). Computing machinery and intelligence. *Mind, 59*(236), 433–460.

Ullman, T., Spelke, E., Battaglia, P., & Tenenbaum, J. (2017). Mind games: Game engines as an architecture for intuitive physics. *Trends in Cognitive Sciences, 21*(9), 649–665.

Wachter-Boettcher, S. (2017). *Technically wrong: Sexist apps, biased algorithms, and other threats of toxic tech*. Norton.

Walsh, T. (2018). *Machines that think: The future of artificial intelligence*. Prometheus Books.

Weinberger, D. (2014). *Too big to know: Rethinking knowledge*. Basic Books.

Weinberger, D. (2017). Alien knowledge: When machines justify knowledge. *Wired*, April 18. Accessed 17 Mar 2018.

Winsberg, E. (2010). *Science in the age of computer simulation*. University of Chicago Press.

Woodward, J. (2003). *Making things happen: A theory of causal explanation*. Oxford University Press.

Wilson, T. (2004). *Strangers to ourselves: Discovering the adaptive unconscious*. Harvard University Press.

Zenil, H., et al. (2017). *What are the main criticism and limitations of deep learning?* https://www.quora.com/What-are-the-main-criticsms-and-limitations-of-deep-learning. Accessed 5 Feb 2018.

Chapter 6
Scientific Side of the Future of the Internet as a Complex System. The Role of Prediction and Prescription of Applied Sciences

Wenceslao J. Gonzalez

Abstract The future of the Internet as a network of networks has many faces. The focus here is on the scientific side, analyzing the various steps relevant to its possible future as a complex system. (I) The ontological framework is a reality articulated at three major layers: the technological infrastructure (Internet *sensu stricto*); the Web; and cloud computing, practical applications (apps) and the "mobile Internet." These are based on designs within artificial environments whose configuration can above all be analyzed in terms of dualities of a complex system, such as structure and dynamics, internal and external perspectives, epistemological and ontological factors. (II) The philosophico-methodological analysis of the Internet — in a broad sense — includes the scientific side, the technological facet and the social dimension. The central aspects of the scientific side are applied science and application of science. This involves the sciences of the Internet and other sciences related to the network of networks. (III) The problems of epistemological and ontological complexity of the Internet as a whole require both an internal and external perspective. They affect the structural complexity, the dynamic complexity and the social dimension of the network of networks. (IV) The role of prediction and prescription concerns above all applied science that deals with complexity. Thus, the task of prediction in the face of the complexity of scientific activity and the social dimension is analyzed together with the task of prescription when facing the complexity of the Internet from the scientific side. This leads to the question of evaluation and meta-evaluation of future studies. (V) The coda offers some final reflections on the question of how to deal scientifically with the future of the Internet in a broad sense.

Keywords Scientific side · Future · Internet · Complex system · Prediction · Prescription · Applied Sciences

W. J. Gonzalez (✉)
Center for Research in Philosophy of Science and Technology (CIFCYT) of the University of A Coruña, Faculty of Humanities and Information Science, Ferrol (A Coruña), Spain
e-mail: wenceslao.gonzalez@udc.es

© The Author(s), under exclusive license to Springer Nature Switzerland AG 2022
W. J. Gonzalez (ed.), *Current Trends in Philosophy of Science*, Synthese Library 462, https://doi.org/10.1007/978-3-031-01315-7_6

6.1   The Many Faces of the Issue of the Future
of the Internet in the Broad Sense

A particularly important issue is the future of the Internet — in the broad sense — in structural and dynamic terms,[1] due to the enormous relevance that this complex system has for science, technology and society as a whole.[2] From a philosophico-methodological point of view, the analysis of this issue — how its future might be — has many faces. It can be approached from the scientific side, the technological facet or the social dimension (Gonzalez, 2018a). Thus, in order to deal with the problem of the future of the Internet as network of networks, a comprehensive philosophical analysis of the complex system needs to consider these three main components: scientific, technological and social. In this regard, the focus here will be on the scientific side.

6.1.1   The Focus on the Scientific Side

The scientific side has received proportionately less attention than the technological facet and the social dimension.[3] The scientific side also has many faces, as it is at least threefold: (1) sciences directly oriented to the development of the network of networks, which is diversified in favor of its different ontological layers (the Internet itself, the Web, and cloud computing, practical applications [apps] and "mobile Internet"); (2) sciences that use this network of networks for their own development as disciplines (economics, communication sciences, information science, etc.); and (3) sciences that deal with the emerging properties of the network of networks, where many data that appear are new and often they are big data (the focus of data science, but also of interest for the Network science).[4]

---

[1] It is striking that, in some important books on the Internet, there is no chapter specifically devoted to the future of the Internet or that it is, in fact, focused primarily on the role of prediction in this network of networks. This is the case with Dutton (2014). This extensive book, which has 26 chapters and 607 pages, has only a brief mention of "Internet Studies and future studies of" in the thematic index (related to page 13) and does not have a specific entry for prediction, which appears in passing on the occasion of "Internet and predictive failures" (connect to page 41).

[2] Since 2019 more than half of the world's population uses the network of networks, cf. Meeker (2019, 334 pages). See, in this regard, The Economist (2019). Some companies like Alphabet (owner of YouTube, Google Search, Gmail, Android, Chrome, Maps, Play Store, Drive and Photos) are estimated to cover the entire spectrum of users and reach 4 billion people that use their products or services. See The Economist (2020).

[3] This can be seen in publications directly focused on this topic, such as the book of Winter and Ono (2015a).

[4] The contents that come from the emerging properties are also of interest to network science, which deals with the relationships that are established with data available in that layer. On the different types of science related to the network of networks, see Gonzalez (2018b, 155–168, especially, 158–161).

In turn, scientific research also has at least three faces: (a) basic science, which seeks preferably to expand available knowledge and focuses on explaining and predicting; (b) applied science, which deals with solving specific problems through prediction and prescription; and (c) application of science, which deals with the use of applied research in the various contexts of the natural, social or artificial world. Within these contexts, which can vary in a very appreciable way, the application of science has to single out the solutions to the specific problems posed.[5]

Here the analysis will address the issue of the future of the Internet — in the broad sense — from the perspective of applied science, analyzing this topic in a setting of complexity, with some remarks related to the other two types of scientific research. This preference for the applied science, instead of basic science and application of science, is because the applied research seeks the solution to specific problems in a practical sphere (Niiniluoto, 1993, 1995),[6] which is the case of the Internet as a whole. In this regard, the task to be done on the future (in the short, middle or long run) includes *prediction* on this complex system — in each of its layers as a network of networks — and *prescription* on what patterns to follow to address the specific problems posed at the various layers of its complex reality.[7]

Looking at the future of the Internet as a whole from the scientific side requires starting with what this reality looks like now (i.e., its actual *being* as a complex reality, mainly artificial, articulated at several ontological layers) and consider the problem with respect to the future in each of these three main layers (the Internet itself, the Web and the cloud computing, practical applications [apps] and the "mobile Internet"). That future, depending on the degree of control of the variables, is presented with completely different degrees of certainty, which in epistemological and methodological terms lead to the distinction between foresight, prediction in the strict sense and forecast (Gonzalez, 2015a, 68–72).

Prima facie, it seems clear that the scientific models on how that future *might be* and how it *ought to be* ("descriptive" and "prescriptive" models) are quite different for epistemological, methodological and ontological criteria. The task of applied science is first to consider the *possible future*, which may be in the short, medium or long run, but also in the imminent term and in the very long term. Then applied science has to provide guidelines to steer it (e.g., in order to face tomorrow's threats in terms of security).

Although the future of the network of networks has many faces, *prima facie* the possible outcome of this complex system in the future might be located between the two potential poles: (i) the *continuation improvement* based on an efficient

---

[5]The characteristics of basic science, applied science and application of science and their differences are analyzed in Gonzalez (2015a, 4, 18, 32–40, 70–71, 151n, 321 and 325).

[6]A problem in the realm of applied research — or even in the field of application of science — can give rise to research at a more abstract level, to the point of giving rise to basic science. This is the case of Leonhard Euler, who found inspiration for his mathematical theory of graphs from the Könisberg bridge problem. Tiropanis, T., *Personal communication*, 21.10.2020.

[7]There are a number of similarities with the case of economics as an applied science. See Gonzalez (1998).

collaboration between the three main sources (scientific research, technological innovation and social contribution), with the consequent optimization of the Internet in the broad sense, and (ii) the possibility of *disruption*, either from a negative internal dynamic to the network of networks or from external dynamics that disrupt its normal operation (mainly due to an economic or political motivation).[8]

6.1.2   Steps to Face the Possible Future

To face that possible future — which presumably can be situated somewhere between these two poles — and so be able to guide it requires analyzing its present. This first step in the paper includes making explicit the *being of the network of networks*, where its structural and dynamic configuration in terms of layers stands out. These ontological layers are mainly three: (I) Internet *sensu stricto*, which is the technological infrastructure that supports the rest of the complex system; (II) the Web, which relies on this infrastructure to perform a plethora of tasks;[9] and (III) cloud computing along with practical applications (apps), which also includes the "mobile Internet" (this version on the move is common in smart phones). This involves characterizing all three layers as such and their orientation towards the possible future.

These three layers or tiers are conceived here as successive in dynamic terms, but not necessarily as dependent in terms of their structure. This involves that there is no intrinsic dependence — technological or otherwise — of the third level on the second level. Rather, the articulation is of this type: (a) the Web is developed on top of the network (including mobile Internet) in order to have navigation and content; (b) the cloud is built on top of the network (including mobile Internet) to provide resources for the storage of information and computing; and (c) the cloud is later than the Web in terms of its composition and has interdependencies with it, but is not necessarily built on top of it.

The second step in the analysis is to emphasize that this complex system of the network of networks is a polyhedral reality. Consequently, it has to be approached from different angles. Thus, it has to take into account the contributions of the

---

[8] At least since 2007, the possibility of disruption has been raised, cf. Feldmann (2007). See also Schönwälder et al. (2009, 27). The dystopian or negative future has also been expressly considered as one of the future possibilities for the case of the Web, cf. Hendler (forthcoming).

[9] One of the most interesting questions about the Web is to analyze how its dynamics, which has had internal and external elements, has led to the current situation. In this dynamic, one of the outstanding factors is how the semantic web and web 2 were proposed. An analysis of these options was carried out at the time in Floridi (2009). At the present time, with respect to the future, it seems possible that "web3 will not dislodge web2. Instead, the future may belong to a mix of the two, with web3 occupying certain niches." The Economist (2022, 54).

scientific side, the technological facet and the social dimension, because they are three main sources in its configuration as a complex system. Within science, the philosophico-methodological analysis should preferably attend to *applied science*, which is commonly followed by the *application of science* in a bidirectional relation.[10] At the same time, the analysis should consider that there are sciences directly related to the Internet as a whole (like Network science, Web science, and Internet science) [Tiropanis et al., 2015; Barabási, 2013; see also Wright, 2011], which have taken on special importance in recent years for its development.[11]

Thereafter, the third step here is to deal with complexity, since the analysis regarding the future of this network of networks is oriented to go deeper into complexity (structural and dynamic, but also pragmatic).[12] This needs to address both the *internal* perspective of structural and dynamic complexity as well as the *external* perspective of complexity due to the social dimension. Epistemological and ontological factors intervene here too, all of which influence the short, medium and long-term future of the network of networks as well as at its micro, meso and macro levels.

After that the fourth step is on prediction and prescription in this setting with the focus on the area where applied science deals with complexity. The analysis has to look in depth at two complementary ways: (i) in the role of *prediction*, considering its triple configuration philosophico-methodological (prediction as ontological, epistemological and heuristic),[13] and (ii) in the role of *prescription*, which depends on the values assumed in the internal governance of the network of networks and in the values accepted mostly in the social dimension.[14] This distinction between prediction and prescription can be seen very clearly in disciplines that are goal-oriented, such as Internet science and Web science, which are interdisciplinary in their configuration and are aiming to look after the Internet and the Web respectively.

---

[10] On the differences between applied science and application of science in the context of complexity, see Gonzalez (2020c, 251–257, 259, 262–267, 270, 273, 275–276, and 279n-280).

[11] In addition to the development of the sciences directly connected to the Internet, other disciplines, such as communication sciences, contribute to enhance the network of networks within an environment of complexity. See Gonzalez and Arrojo (2019).

[12] For pragmatic complexity see Gonzalez (forthcoming-a). Pragmatic complexity is better appreciated in the application of science, as has been seen in the vaccination processes against Covid-19 (Gonzalez, forthcoming-b). But it can also be detected when in the network of networks it is necessary to single out solutions for specific groups or companies or customize solutions for individuals with special needs.

[13] On prediction as ontological, epistemological and heuristic, see Gonzalez (2015a, 108–112).

[14] The case of economics, which is a dual science (a design science, within the sciences of the artificial, and a social science) is representative in this respect, cf. Gonzalez (1998).

6.2 An Ontological Framework: A Reality Articulated in Three Major Layers and Its Philosophical Analysis

Ontologically, the network of networks that configure the Internet in a broad sense began with the design of the technological infrastructure and its services (Internet in the strict sense),[15] which was primarily aimed at communicating machines, even though it also provided the basis that enable communication between people (as in the case of email). Later, it expanded with the design of the Web (cf. Berners-Lee, 1999; see also Berners-Lee & O'Hara, 2013; Shadbolt et al. 2013), which sought to directly communicate people and allowed users to contribute to content, especially with Web 2.0, and already had that network as a support for its structure and dynamics.[16] Thereafter, the development of this complex system reached the design of a set of elements composed of cloud computing, practical applications (apps)[17] and the "mobile Internet."[18] These three major layers are particularly relevant for the future of this complex system.

6.2.1 Three Main Designs in an Artificial Environment

Each of these progressive designs is within a primarily artificial environment, which configures a layer or ontological level of this complex system of network of networks. De facto, "the Internet was designed in a layered fashion — the goal was that the applications running on top of the packet forwarding service should not be able to affect it" (Clark, 2018, 37). In this design of the whole Internet through successive layers, three of which — already mentioned — are particularly important. The organizations related to each layer depend on internal factors (scientific and technological), but are externally conditioned by social variables (institutional, legal, etc.). Both combined lead to a remarkable variation in terms of decentralization and concentration:

(1) Internet in *sensu stricto* is at the bottom, where "are all the protocols that allow different sorts of networks and devices to exchange information, or 'internetwork' (hence internet). At that level, it is still largely decentralised: no single company

---

[15] The main actors in the process of developing the network have made a detailed synthesis mainly from the internal perspective: Leiner et al. (1997). An extended version has a different title: Leiner et al. (2009). A history made primarily from the external perspective, see Greenstein (2015).

[16] Mobile Internet has led to much more interaction on the Web between people than previous designs.

[17] There is a difference between apps and the Web, insofar as apps can enable "islands" where one only accesses the resources available via the practical application (app) and not on the Web. De facto, many apps do not have a Web interface. It happens that this tendency to have "islands" was criticized by T. Berners-Lee a few years ago. Tiropanis, T., *Personal communication*, 21.10.2020.

[18] On "mobile Internet" and prediction, see Yap et al. (2020).

controls these protocols" (The Economist, 2018, 5). (2) "The next layer up—everything happens on top of the internet itself — has become much more concentrated. This is particularly true of the web and other internet applications, which include many consumer services, from online search to social networking" (The Economist, 2018, 5). (3) All the extensions that the network has spawned are in another layer, where centralization is rampant: "most people use one of two smartphone operating systems: Apple's iOS or Google's Android. Cloud computing is a three-horse race among Amazon, Google and Microsoft" (The Economist, 2018, 5).

6.2.2   The Configuration and the Analysis in Terms of Dualities

Although the trajectory followed by the three major layers has taken several decades, it is necessary to emphasize the kind of *configuration* of the structure and dynamics of the Internet from the beginning, which also affects its future. In this regard, the design with the goal of *generality* is one of the reasons why the Internet has been successful. "In fact, there are two important aspects of generality represented in the Internet: generality with respect to the applications that run over it and generality with respect to the sorts of communications technologies out of which it can be built" (Clark, 2018, 42). Another important idea which shaped the dynamics of the Internet — and is relevant to its future — was *open-architecture networking*. This feature was there almost from the beginning and was introduced around 1972 (cf. Leiner et al., 2009, 24).

This configuration of the Internet as a whole can be thought of philosophically in terms of dualities of a complex system, such as structure and dynamics, internal and external perspectives, epistemological and ontological factors. This set of distinctions is relevant for the future. Thus, the layered structure and dynamics of this network of networks — with its three major layers — can be analyzed from the internal and external perspectives, where the main factors are epistemological and ontological.

In connection with the *internal* perspective, the scientific side and the technological facet are crucial. According to the level of performance, each layer has a structure with characteristic features and a dynamic that follows a temporality that is in line with the level where it moves. From the epistemological and ontological point of view, this complex system involves many levels of aggregation, organization and interaction. Meanwhile, when a complex system based on layers is seen from the *external* perspective of social dimension, which is especially relevant for the future, then the complexity of the Internet is based on the fact that it "is a uniquely vast emergent system — a platform for untold forms of economic, social and personal interaction" (Schultze & Whitt, 2016, 63). The criteria of social epistemology and social ontology can be used to analyze the novelty brought about by this complex system.

6.3   The Philosophico-Methodological Analysis of the Future of the Internet

When the future of the Internet is analyzed in philosophico-methodological terms, in addition to the novelty of the field of study itself, there are new questions to be addressed, many of which have artificial status, that is, they are oriented towards specific goals. There are elements of these designs that are often associated with infosphere content (e.g., when it comes to promoting information science through the network of networks). But they are not analyzed here from the perspective of the philosophy of information but from the philosophy and methodology of science. It also happens that this approach is not primarily oriented towards basic science, as has been done preferentially in other philosophico-methodological fields of study. Here the interest in the analysis of applied science prevails, followed by attention to the application of science.

For this approach to really produce fruitful results in philosophico-methodological terms, a rather holistic view is needed as a starting point. Thus, when the focus is on the scientific side of the three major layers of the Internet, then it is worth highlighting the relevance of this network and the Web for the scientific progress in many ways. This includes two *structural* complementary features of scientific activity: (a) the Internet as a whole is a new field for scientific research,[19] and (b) the Internet in the broad sense is a key instrument for doing science now and for doing it, moreover, in several directions.

As a new field for scientific research, the future of this network of networks will involve epistemological, methodological and ontological aspects, which will be played out in the short, middle, and long run as well as at the micro, meso, and macro levels. In addition to the scientific side, there are also the technological facet and the social dimension of the Internet as a complex system. They are interconnected in the three main layers mentioned, both within the set and within each one of them. This is clearer in the case of the Web. Furthermore, the intertwining is seen when the Web science was founded as new scientific discipline: "a policy question arises as to how to control access to data resources being shared on the Web. This latter question has implications both with respect to underlying technologies that could provide greater protections, and to the issues of ownership

---

[19] "In the last 25 years, over 300.000 academic papers have been written that explicitly mention some aspect of the World Wide Web or the Internet in the title, abstract or keywords of the publication. The overall pattern of this growth is one of nearly continuous linear upward growth since 1990" (Meyer et al., 2016, 1160).

in, for example, scientific data-sharing and grid computing" (Berners-Lee et al., 2006, 770).

Even when the origin of the Web science is presented, the explanation given is a science that underlies the Web from a "socio-technical perspective." Thus, "ten years ago, the field of *Web Science* was created to explore the science underlying the Web from a socio-technical perspective including its mathematical properties, engineering principles, and social impacts" (Hall et al., 2016, 1). In this regard, the philosophico-methodological analysis of this scientific activity — and, in general, what is under the heading of "sciences of the Internet" — leads one to maintain that they are primarily *sciences of the artificial*, which are goal oriented.[20]

Furthermore, the sciences of the Internet perform tasks of applied sciences in the two-way relationship with the application of these sciences to the network, the Web or cloud computing, apps and the "mobile Internet." *Applied science* seeks the solution to specific problems, such as security (of the network itself, which is constantly expanding and with new vulnerabilities, the Web or the "mobile Internet"), scalability, connectivity, privacy, semantic web, etc. Meanwhile, the *application of science* uses scientific knowledge in diverse contexts. This knowledge, which is the result of the application in specific cases, has repercussions in order to have adequate guidelines for the solution of problems within the part of applied science. In addition, there is here a nexus between scientific creativity and technological innovation.[21]

Besides the structure of the scientific side in an actual threefold undertaking (scientific, technological and social), there is a *dynamic* of a complex process embedded in *historicity*, which can be found on the trajectory of the Internet as a whole.[22] If the focus is on the *scientific side*, then we can find historicity in several ways: (i) Historicity of science as a human activity, which changes over time in ends and means. This has been the case in the sciences of the Internet: new sciences appeared and novel developments in terms of aims, processes and results have been made. (ii) Historicity in the diverse kinds of sciences (natural, social and artificial) used by the sciences of the Internet as interdisciplinary enterprises, which, in turn, should take into account the technological facet and the social dimension. (iii) Historicity of the agents that do or use the sciences of the Internet. In this regard the research task includes three kinds of historicity: (a) in the social context where the research is made, because there are changes in the historical setting and,

---

[20] This being goal-oriented feature of the sciences of design was emphasized by Simon (1996, xi).

[21] An analysis of the distinction between "applied science" and "application of science" in connection with scientific creativity and technological innovation is available in Gonzalez (2013a).

[22] Both in the history of the Internet narrated by its main actors (previously cited in the 1997 and 2009 versions) and in other later reconstructions of the network's trajectory — such as in Greenstein's book on *How the Internet Became Commercial* — the existence of successive turning points is highlighted as a result of decisions on design or from interaction with other agents, which also completely changed the initial plan: from a public service network, used for very defined uses of an informative and communicative nature, to a network of networks with strong commercial content and a powerful social dimension for citizens.

sometimes, in completely unexpected ways (as has happened with the pandemic caused by Covid-19); (b) in the relations with the institutions — public or private — funding the research; and (c) historicity in their interactions with other colleagues, which may or may not allow the continuity of research teams.[23] It seems that this motley historicity may continue in the future in these three main directions.[24]

Based on this motley historicity of scientific activity, it seems likely that the future of the Internet will bring about changes that might lead to new structural and dynamic problems of complexity. They might not be purely "internal" to the network, insofar as they can also be connected to the "external" terrain of the social dimension of the Internet.[25] This paper addresses some of these aspects of the prospects of the network in scientific terms and examines the role of prediction and prescription. Epistemologically, it considers the three main layers on the basis of the present knowledge to anticipate the possible future and the need for patterns of action to solve the problems prospected. Methodologically, this scientific side of the future of the Internet requires the bidirectionality mainly between the facets of applied science and application of science.

6.3.1 The Scientific Side, the Technological Facet and the Social Dimension

Dealing with the scientific side of the future of the Internet — where the scientific, the technological, and the social components are interconnected — requires an initial framework. (I) The future of the network depends largely on the *creativity* of the scientific activity, insofar as it solves specific problems and contributes to the development of the technological platform. (II) The prospects of the Internet are also in need of the *innovations* in the technology available, which presumably will lead to new objectives, procedures and results.[26] This requires taking into account that the technological platform of the Internet is at least twofold regarding scientific activity: (a) the network *supports* scientific research, making it possible in this sphere, and (b) it serves as a *conditional medium* for the scientific activity made through this network.[27] (III) The future of the Internet will also involve the *social*

---

[23] An analysis of these three facets of historicity can be found in Gonzalez (2011b).

[24] On the methods to study the history of the Web, see Priestley et al. (2020).

[25] These kinds of "external" changes have repercussions for the organization of the structure in the network of networks. This means that a number of external factors, including national or international security reasons, can lead eventually to internal changes in the structure organizational of the layers.

[26] Technology is initially "external" to the sciences of the Internet, but it may become "internal," insofar as the viability of the use of the designs of the network — aims, process, and results — need technology to have effectiveness or to be efficient.

[27] In addition, "the structure of the Internet topology highly impacts the traffic distribution" (Uhlig, 2010, 167).

dimension (i.e., the use of the network by individual agents, groups, organizations, "multi-agent systems,"[28] etc.), insofar as the social dimension interacts with the scientific activity and the technological platform of this network.[29]

All three of the above include epistemological and methodological aspects regarding the short, middle or long run and ontological elements in the future at the micro, meso or macro level (such as the local apps, some social networks of academic content and new versions of the Internet Protocol, IP). Frequently, the emphasis on these aspects and elements has been on the technological platform (Internet in the strict sense) and the social dimension.[30] Here the focus is on scientific activity, which in the internal perspective includes multiple interconnections with future technological tasks and in the external perspective has the constant input of the social dimension.[31]

Placed in this threefold setting, the analysis will be mainly on the "internal" aspects of the Internet indicated, rather than on the "external" factors, which have been studied already.[32] Accordingly, the analysis of the scientific activity is here primarily focused on "Internet activity" (i.e., the activity itself of this network of networks as human endeavor) and, secondarily, it pays attention to "Internet as activity" amongst others, which is when the activity of the Internet is seen as being interrelated with other human activities (social, cultural, political, legal, economic, ecological, etc.).

6.3.2   Key Aspects of the Scientific Side: Applied Science and Application of Science

Due to the practical environment of the three major layers of the Internet in the broad sense, scientific activity appears to be at least twofold: applied science and application of science. In this regard, the future of the Internet will require a scientific activity in two main directions: (I) scientific contributions for specific *problem-solving* in the practical realm of the network, the Web and cloud computing, apps and the mobile Internet. Applied science should seek solutions to specific problems, such as bandwidth and storage capacity, connectivity, security, scalability, accessibility, etc. (II) The new applications of the scientific knowledge to be used in order to enlarge the capabilities of this complex system. This application requires considering

---

[28] On "multi-agent systems" see Floridi (2014, ix and 180–189).

[29] This social dimension includes legal issues, some of them very important, such as the right to privacy.

[30] See, for example, the papers published in Winter and Ono (2015a).

[31] Among the social repercussions are the economic ones, cf. Brynjolfsson and McAfee (2011).

[32] On the contextual features related to the Internet, see the set of elements discussed in Floridi (2014, 1–24 and 167–216). See also Gonzalez (2020a).

variations in the contexts the use of the network of networks, circumstances related to individual agents, groups, organizations, nations, international institutions, etc.

These two directions of scientific activity are complementary and can have feedback between them. Methodologically, this two-way or bi-directional relation between applied science and application of science includes several aspects: (a) it allows us to understand issues of the *origin* of the sciences of the Internet (as a process of "scientification" of the patterns used by professionals for solving specific problems) [cf. Gonzalez & Arrojo, 2019]; (b) it contributes to the advancements regarding *procedures and methods* — quantitative and qualitative — of the sciences of the Internet based on in the use of the Internet, the Web, etc. (cf. Ackland, 2013); (c) it has many repercussions in terms of *results* for the network of networks, such as the need to improve "predictive models for understanding the impacts of information use across the Internet" (Hendler & Hall, 2016, 704). Thus, there is an increasing interest in the relations between research and professional practice in order to achieve improvements, which will continue in the future.[33]

Both directions include the need for prediction and prescription:[34] the problem-solving of specific problems in the practical domain (applied science)[35] and the use of scientific knowledge in a variety of new contexts (application of science).[36] Scientific activity in *applied science* requires us to consider prediction of the possible future to anticipate the problems in the short, middle or long run. It also includes prescribing the possible course of action to solve the problems detected at the micro, meso or macro level. Meanwhile, the *application of science* involves a diversity of possibilities in the use of the scientific knowledge available (a variety of contexts, which can be very wide-ranging). Thus, the relation with the social dimension of the Internet is clear.[37]

Bi-direction between them appears, because the application of science has to deal with the "external" demands of the users of the network (individuals, groups, organizations, governments, etc.). But these demands can generate "internal" problems for applied science, which should be solved scientifically in order to enlarge or improve the Internet available at each moment. In the sciences of design, the "internal" and "external" aspects are commonly interrelated and applied science as

---

[33] Along with the internal dynamic is the external dynamic, where the predictions made should take into account the changes that occur in social activity that affect the bi-directional relationship indicated, so that the predictions should consider internal as well as external aspects.

[34] Herbert Simon was aware of the importance of prediction and prescription in systems modeling, but he did not develop the philosophico-methodological distinction between applied science and application of science. See Simon ([1990], 1997b).

[35] The general characteristics of the applied sciences can be found in Niiniluoto (1993). Additional details are available in Niiniluoto (1995).

[36] Niiniluoto also points out the existence of application of science as different from applied science. The application of science involves the use of scientific knowledge in the diverse circumstances of the social milieu, cf. Niiniluoto (1993, 1–21, especially, 9 and 19).

[37] In the case of the Internet of Things, the feature of the application of scientific solutions is more noticeable for the users than in the case of the network itself.

well as application of science are needed. This is also the case in the sciences of the Internet, such as Network science,[38] Web science and Internet science.

Artificial Intelligence — another science of the artificial — (Gonzalez, 2017a; Simon, 1995) also has a role in the future of this network of networks.[39] Some developments of AI will have an impact on the Internet.[40] The repercussions may take several forms, many of them are located at the layer of the Web, such as those related to machine learning for better translations, the use of algorithms for data economy, or Artificial Intelligence as the basis for new designs concerning audio-visual communication on the Internet (cf. Gonzalez, 2017b). In addition, AI has a role with respect to the processing of the cloud computing data.[41] This has contrib-uted to that, in recent years, big technological companies, such as Microsoft (Azure), Amazon (AWS) and Alphabet (Google cloud), have invested in cloud computing, with very profitable results.[42]

Versatility is the main feature of this repercussion of Artificial Intelligence on the Internet as a whole, including the technological infrastructure. AI is used for a motley variety of scientific purposes (communicative, informational, economic, educational, etc.). This use of AI is present in those sciences that have a direct relation to the Internet, such as Network science, Web science (cf. Hendler & Berners-Lee, 2010), or Internet science, ... Again, the development of AI can be conceived in terms of applied science or as an application of science. De facto, research teams of important firms related to the network of networks, such as Alphabet, highlight the use AI in new ways and at the layers of the Internet in the broad sense.[43]

---

[38] In addition to the network and the Web, Network science also investigates issues related to the third layer (cloud computing, apps and the "mobile Internet"). Thus, "Network Science explores phenomena that include, but are not limited to, the Web or the Internet" (Tiropanis et al., 2015, 82). This means that Network science has two different tasks, depending on the layer in which it works: on the one hand, it contributes to the configuration of the structure of the network and the Web; and, on the other hand, it develops a task with the contents available in cloud computing or on the mobile Internet.

[39] "There is no doubt that the prosperity of AI is inseparable with the development of the Internet. However, there has been little attention to the link between AI and the Internet" (Liu et al., 2017, 377).

[40] The impact of the future Artificial Intelligence will be also on society and business firms, cf. Makridakis (2017).

[41] "The cloud computing environment offers development, installation and implementation of software and data applications 'as a service.' Three multi-layered infrastructures namely, platform as a service (PaaS), software as a service (SaaS), and infrastructure as a service (IaaS), exists" (Vasuki et al., 2018, 3842).

[42] The increase is very noticeable in "Google's cloud business, which includes G Suite, its package of professional online services, is growing at more than 50% a year. Revenues are expected to reach $13bn this year, contributing 8% to Alphabet total" (The Economist, 2020, 17).

[43] In addition to its main and best-known companies for users (Google, YouTube, Gmail, Android, Chrome, etc.), Alphabet has "other bets," which are "its non-core businesses, now number 11, each with its own capital structure. These include Access (offering fibre-optic broadband), GV (which invests in startups), Verily (a health-care firm), Waymo (a developer of autonomous cars) and X (a secretive skunk works engaged in all manner of moonshots). Commercially these ventures seem

Through Google, Alphabet has invested in research on Artificial Intelligence focused on the application of science, as a support to increase the quality and the possibilities of human knowledge using deep learning (see, for example, Dean et al., 2012). This enlargement of human possibilities can also be seen in the provision of services to specific customers through the cloud and the big data accumulated in it. In this regard, "the cost of storage has considerably reduced with the advent of cloud-based solutions. In addition, the 'pay-as-you-go' model and the concept of commodity hardware allow effective and timely processing of large data, giving rise to the concept of 'big data as a service.' An example of one such platform is Google BigQuery, which provides real-time insights from big data in the cloud environment" (Vasuki et al., 2018, 3842).

6.3.3   The Sciences of the Internet and Other Sciences Related to the Network of Networks

Concerning the scientific activity in this realm of the network of networks, there are at least three possibilities: (1) the "sciences of the Internet,"[44] which *work directly* in this field. These are the disciplines focused on the problems of the network itself,[45] the Web or cloud computing, apps or the mobile Internet, which include in any case network science, Web science, and Internet science (on specific aspects of this ambit) [Tiropanis et al.; 2015]. (2) The disciplines that search for novelty supported by the network of networks. They used this complex system to develop new aspects of traditional sciences, such as economics (Gonzalez, 2019, 2020b), information science, education or communication (Gonzalez & Arrojo, 2019). This search for some new goals (e.g., in financial economics, the information retrieval, the education on line, audiovisual communication via Internet, etc.) covers a large number of options.[46] (3) The new directions in sciences — with approaches that are usually transdisciplinary or interdisciplinary — that study the *emergent properties* on the

---

only loosely connected to the core. What links them to the main business is information processing — and specifically these days Artificial Intelligence (AI), which powers everything from search to Waymo's self-driving cars" (The Economist, 2020, 15).

[44] The complementary approach from the internal perspective is through the study of the features of the technological platform of the Internet. On its design, see Hanseth and Lyytinen (2010).

[45] Certainly, the scientific and the technological components are not "disconnected" or in a "parallel" situation, insofar as they are in active interaction due to the relations between scientific creativity and technological innovation are bidirectional. Cf. Gonzalez (2013a).

[46] From a scientific point of view, the use of the Internet for communication sciences can have an impact on internal aspects (new concepts, novel contents, etc.) as well as on external elements (the users, media organization and policies, media policy and regulations, etc.). Furthermore, from a technological perspective, the use of the network can have consequences in terms of innovation for the technological platform. See, in this regard, Küng et al. (2008).

Internet, due to the intensive and extensive use of the network,[47] such as data science or Network science. The existence of these properties — mainly all sorts of data that are now available through the network — opens the door to disciplines directly focused on what depends on the use of this artificial platform.

(1) Due to novelty in the design of the three main layers, the "sciences of the Internet" — the first group — have made the central contributions to this network of networks. Furthermore, they have been developed in recent decades.[48] Certainly, their scientific activity addresses problem-solving in the realm of the Internet itself (such as connectivity, scalability or security), but they are also useful for the scientific community as a whole (cf. Berners-Lee et al., 2006, 770). In this regard, scientific activity shows creativity in order to reach novelty within the infosphere.

Epistemologically and methodologically, these "sciences of the Internet," although they have a clear interdisciplinary content, belong primarily to the sphere the sciences of the artificial,[49] because they work on designs oriented toward solving issues raised by enlarging or improving of the Internet itself as well as the "Internet of Things."[50] Thus, "sciences of the Internet" are in an artificial domain, which seeks to enlarge or improve human possibilities through new aims and novel processes in order to reach new results.[51] Even though they can have analytic elements, they are predominantly synthetic (cf. Berners-Lee et al., 2006, 769).

(2) There is more variety in the sciences of the second group, which use the network as a tool for specific purposes, because they work on different kinds of problems. In this case, the problems regard issues like financial economics, information retrieval, education on line or audiovisual communication. However, the first and second group share common ground: they are applied sciences, as they seek the solution to specific problems by means of finite number of steps. In addition, they are also sciences of the artificial, conceived as sciences of design, and, therefore, goal-oriented (Simon, 1996, xi).

---

[47] In addition to these emergent properties through the use of the Internet, there are emergent properties of the network itself, i.e., as a technological platform (with an open, share, heterogenous and evolving dynamic). See, in this regard, Hanseth and Lyytinen (2010, 1–19; especially, 10).

[48] A clear case is the Web science. The World Wide Web was designed by Tim Berners-Lee around 12 March 1989. It was on 30 April 1993 when the CERN opened it to the public and at no cost to users, cf. Floridi (2014, 18). But its "official" beginning as the scientific field of the Web science is associated with 2006 and the publication of the paper Berners-Lee et al. (2006).

[49] The pioneer in this field was Herbert Simon, especially relevant with his book *The Sciences of the Artificial*, whose third and final version was published by The MIT Press, Cambridge, MA, 1996. The first edition was in 1969 and the second in 1981. This field has different aims, processes, results, and consequences from the social sciences.

[50] On the "Internet of Things" see Ornes (2016) and Park (2019). IoT involves new legal problems (Zhang & Zhang, 2017).

[51] They have some similarities with economics, which is also a science of design (within the sciences of the artificial) in addition to being a social science. Cf. Gonzalez (2008a).

Consequently, insofar as both groups related to the network of networks are applied sciences, they use prediction and prescription. Thus, these sciences look for the problems in the future related to a practical domain, and this requires the use of predictions to anticipate possible events (positive and negative, within a temporal scale and a degree of reliability). This guide to the future given by predictions opens the door to the role of prescription, which is when the patterns of solutions of the problem are proposed (Simon [1990], 1997b), as happens in economics (Gonzalez, 2008a, 2015a, ch. 12, 317–341).

(3) Data science is the main case of this third kind of scientific activity related to the network of networks. In this regard, "much of the interest in data science today arises because of the vast stores of information, both structured and unstructured, that have become available through the Web. The growth of available information is also leading to increasing the use of data analytics in many fields" (Hall et al., 2016, 2). Similarly, the Internet is contributing data and driving requirements for analysis either via supporting the Web or through the Internet of Things.[52] Thus, data science has grown intensely and in many ways in recent years.

From a philosophico-methodological stance, data science is sometimes seen as a "transdisciplinary" field, which builds upon a number of relevant disciplines rather than an "interdisciplinary" area in a genuine sense (cf. Cao, 2017a, 59–68; especially, 59; and Cao, 2017b). It is a new scientific activity that deals with "big data,"[53] whose relevance has increased especially as support for scientific research in quite different fields, although it raises epistemological and ethical problems,[54] in addition to methodological difficulties.

When these three groups of disciplines related to the network of networks are analyzed with respect to the future and in connection with applied science in a context of complexity, then we need to clarify the epistemological, methodological and ontological characteristics that are important for prediction and prescription. (a) The epistemological scope of the knowledge of the possible future can be related to the immediate, short, middle, long run or very long run;[55] (b) methodologically the focus of the research can be on structural aspects or dynamic ones; and (c) the ontological status depends on the level of reality (micro, meso or macro level) where the problem in the future might be located (such as aspects of the Internet service provider or the local apps, some social networks or certain aspects of the mobile Internet, or the backbone infrastructure of the network). These are the aspects to

---

[52] In the case of the Internet of Things, see Siow et al. (2018).

[53] In addition to the complexity of the network of networks and each of its layers, there is a complexity in the very content they carry or transmit, whether that content is structured ("information") or unstructured ("data"). Cf. Vasuki, Rajeswari and Prabakaran (2018, 3841–3844; especially, 3841).

[54] See, in this regard, Floridi (2014, 14–17).

[55] These five options, in the case of economics, are considered in Gonzalez (2015a, 66, 192, 219 and 251).

consider so that the prediction can be a guide for the prescription of the patterns for problem-solving and can deal with complexity — structural and dynamic — in the Internet as a whole.

A new area of problems is in the third group of sciences, where *new properties emerged* because of the use of the network, the Web or cloud computing. The use of Internet data can be useful for many fields such as "open science," insofar as it offers relevant information accessible for any user of the Internet that can lead to genuine knowledge.[56] In addition, we have scientific fields related to information and its management in specific domains. This is the case of bioinformatics, which seeks to combine big data and complex analysis of workflows in data driven science (cf. Allmer, 2019). This type of research has been called the "Internet of Science" (IoS). In this regard, "the main aim of the internet of science is to re-establish collaboration as the first principle of science. The IoS enables collaborative work on developing and testing software, developing and testing data analysis workflows, and joint reporting. It binds together all stakeholders while enabling the tracking of individual and joint efforts via DOIs" (Allmer, 2019, 5).

Commonly, structured data and information of certain kinds available in the Internet — in the broad sense — can be particularly relevant for social sciences (and the social aspect of the health sciences). This is the case insofar as the data gathered through this network of networks represent to a large extent a part of everyday life (cf. Askitas & Zimmermann, 2015). Thus, they are *new social aspects* that cannot be measured otherwise, even when they are social phenomena (economic, cultural, etc.) already present in the society before the Internet (e.g., tourism or international migration).

Actually, from an epistemological point of view, Internet data can be used for predictions related to economic phenomena (such as unemployment, consumption goods or tourism), for detecting medical problems (such as distribution of flu, spreading of Covid-19,[57] malaise and ill-being during the economic crisis), for measuring social complex problems where the traditional data is incomplete (such as international migration or collective bargaining agreements in developing countries), etc. Also, it can gather useful information in terms of predictions in the immediate run or "nowcasting," which gives relevant information faster than the traditional procedures.

Methodologically, there are still major problems in data analysis and issues on data reliability (cf. Askitas & Zimmermann, 2015, 2), as can be seen in the ongoing Covid-19 pandemic, which depends on the reliability of primary sources of information. This has repercussions not only on the formation of applied science (the knowledge for the elaboration of vaccines) and the way of carrying out the application of science (the appropriate therapy for each type of patient or, even, for single patients with atypical medical conditions) but also on the decisions of public

---

[56] On the distinction between data, information, and knowledge, see Rescher (1999).

[57] To predict the spreading of Covid-19, mathematical models can use data from the "mobile Internet," which can complement data from institutional sources or specialized organizations.

authorities for the possible future in the social milieu (measures on lockdown or restriction of movements of people).[58]

6.4 The Internet and the Epistemological and Ontological Problems of Complexity: Internal and External Perspectives

Complexity is a feature of the Internet as a whole, largely due to the interaction between the scientific side, the technological facet and the social dimension. If the focus is on *scientific activity*, then two kinds of philosophico-methodological problems related to complexity may appear in the foreseeable future of the main layers of the Internet. These layers share the presence of artificiality, which opens new doors for complexity.[59] Following the internal perspective, we have two fronts. On the one hand, there are issues related to the structural complexity of the Internet, which are mainly epistemological and ontological; and, on the other, there are matters surrounding the dynamic complexity of this network of networks, which require dealing with several forms of changes over time that have also epistemological and ontological components. Thereafter, the external perspective comes with the social dimension of the Internet in the broad sense.

Looking forward, the analysis requires us to consider that scientific creativity has an *internal* perspective, which can be found in the whole set of constitutive elements of science (language, structure, knowledge, method, activity, ends, and values),[60] and an *external* perspective (the relations with the milieu: social, cultural, political, etc.). Both have consequences for technological innovations. In this regard, "the future Internet is an ever evolving composite of devices, services, and usage models. (...) Since we have no access to an accurate prediction of future network services, any proposed future Internet architecture must be able to evolve over time. Its adaptability and flexibility are crucial to deal with changing requirements, incorporate new functions, and accommodate continuous innovations" (Yin et al., 2014, 14).

---

[58] This decision-making concerns various types of institutions, both national and international. See, in this regard, Gonzalez (2021).

[59] See, in this regard, the analysis made for the case of economics: Schredelseker and Hauser (2008).

[60] These constitutive elements of science also are the key components for its conceptual distinction from technology.

6.4.1 Structural Complexity

Central elements of structural complexity are epistemic modes and ontological modes. These are general pieces of any complex system.[61] They are available in this network of networks. The three main layers — the Internet itself, the Web and cloud computing, apps and the mobile Internet — show complexity within them and in their internal articulation as a system. In this regard, there is a frequent tension between centralization and decentralization, leading to an emphasis on hierarchical configuration and a fully distributed operational framework. This affects the present and the future of this complex system as such and the way governance is carried out, which generates another polarity between a universal network in each of the layers and a thematic fragmentation (economic, educational, industrial, informational, entertainment, etc.). In addition, this also leads to the problem of a two-tier network: one with high speed and good performance as opposed to a slower network with fewer services.

Hence, the internal and the external perspectives have a place in the structural complexity of the Internet as a whole. By appearing in the structural alternatives mentioned, it seems likely that present issues on the epistemic and ontological modes will continue in the future at least in the short and middle run. In addition to the articulation of the structure as such, there are problems about the topology, which are clear in the case of the Web, where the rules of connectivity can be associated with those of scalability, as has happened in the transition from the early Web to later versions (Berners-Lee et al., 2006, 770). Thus, when the analysis goes to the *epistemic modes* of structural complexity, these can be of three kinds, according to the descriptive, generative and computational factors:

(1) Descriptive complexity is where length of account for an adequate description increases. This happens commonly with the new innovations in the three main layers of the Internet. Thus, an adequate description of the present state of the knowledge on the Internet requires more details than the original ARPANET, and this will increase in the future. (2) Generative complexity is where the set of instructions to produce something new increases. This in the case with a new kind of specialized network or a new design for sophisticated webpages. Furthermore, the generative complexity, as happens with the Web science, involves the intertwining of the scientific activity, the technological angle, and the social dimension.[62] (3) Computational complexity is where the amount of time and effort involved in solving a

---

[61] This analysis is based on the general modes of complexity proposed by Nicholas Rescher (1998, 1–19, especially, 9).

[62] "Web science is about more than modeling the current Web. It is about engineering new infrastructure protocols and understanding the society that uses them, and it is about the creation of beneficial new systems" (Berners-Lee et al., 2006, 770–771).

problem requires more collaborative creativity.[63] This is the case in the use of big data and machine learning to improve the quality of the services through this network of networks.[64]

Ontological modes of the structural complexity can be compositional, strictly structural, and functional.[65] On these three counts the Internet as a whole — an artificial world with empirical and virtual elements — is now more complex than the original precedents, as happens with ARPANET and the present network, which includes a developed version of the technological infrastructure and the services. At present we have the following, which is expected to continue at least in the short and middle term:

(i) Compositional complexity is related either to the number of components involved (constitutional complexity) or the heterogeneity of the components (taxonomical complexity). This can be seen in the contrast between the initial website, designed by Berners-Lee, and the current one, which shows a much greater complexity in terms of the number of components and their typological variety. (ii) Strictly structural complexity deals with the different ways of arranging the components (organizational complexity) or with the relationships in the modes of inclusion and subsumption (hierarchical complexity).[66] In the first case we have the protocols of the Internet or the Web, which play a very important organizational role for arranging the components,[67] whereas the second case can be seen on institutional websites. (iii) Functional complexity is diversified into the variety of modes of operation (operational complexity)[68] or the kind of rules governing the phenomena

---

[63] According to Walter Isaacson, "collaborative creativity" is a common feature of the innovations related to the scientific activities that support the development of the information and communication technologies. See Isaacson (2014, 1).

[64] This is one of the reasons for the success of the services currently provided by the main technology companies in this sector (such as Google with its program translate, which has improved over the years, especially in some languages).

[65] Among the functional cases are the operational ones. In this regard, an interesting aspect to be discussed is whether increasing the level of freedom of a system makes it more complex from an operational viewpoint. Cases for study might be the big communicative corporations developed in recent decades.

[66] Rescher maintains that, in hierarchical complexity, the higher-order units are "always more complex than the lower-order ones" (Rescher, 1998, 9). Within the sphere of the communication sciences, when the support for broadcasting is the network, this phenomenon can be seen in the case of some corporations with multi-media platforms and presence in multiple countries.

[67] The Internet itself as technological network is organized with protocols such as Transmission Control Protocol, TCP, Internet Protocol address, IP, and the Domain Name System, DNS. Meanwhile, the Web has among its main protocols HyperText Transfer Protocol, HTTP, and File Transfer Protocol, FTP.

[68] This kind of variety can appear with the combination of social and artificial operations. This possibility can be seen in the use of some social networks of the Internet for communicative purposes, where the performance of the public and the elaboration of designs (e.g., for over-the-top television) end up in operational complexity.

at issue (nomic complexity).[69] The mobile Internet, which has connectivity with two different types of wireless architectures (homogeneous and heterogenous networks), exemplify this kind of functional complexity.

6.4.2 Dynamic Complexity

When thinking of the future of the Internet as a complex system articulated in layers, the *dynamic complexity* of the network of networks is particularly relevant.[70] Each layer generates a dynamic of change that encourages complexity. In this respect, the focus is usually first on the technological facet,[71] thereafter on the social dimension and then — to a lesser extent — on the scientific side. A philosophical analysis shows that threefold dynamics (scientific, technological and social) include some features — epistemological, methodological, ontological and axiological — that are pertinent for the future of the three main layers and, consequently, for society as a whole.

Below the intersection of the scientific side, the technological facet and the social dimension there are the specific issues of the dynamics of each one of them. The kind of problems is characteristic of the sciences of the Internet — mainly, the Web science — and includes cases leading to the use of the tools of complex dynamic systems analysis.[72] These problems of complexity can appear in the other two groups of sciences pointed out here, insofar as they are connected with the layer of the network and depend on it for their aims, processes and results.

(I) Epistemologically, the dynamic complexity is connected to novelty (in contents, functions, applications, etc.) of this network as a multivariate platform, which was initially focused on information and communication (Graham, 1999). (II) Methodologically, there is a combination of processes in each of the main layers (cf. Clark, 2018, 37; and The Economist, 2018, 5), which depend on factors of structural complexity (mainly, epistemological and ontological).[73] These procedures and methods can be understood in terms of methodological pluralism (cf. Gonzalez,

---

[69] This can be the case of the questions "how many links are possible among how many nodes" (Floridi, 2014, 23).

[70] A specific analysis of dynamic complexity on the Internet as an information and communication platform from the perspective of the sciences of design and the role of prediction can be found in Gonzalez (2018b).

[71] "What causes network requirements to change over time? In my view, it is the interplay among three important drives: new developments in network and computer technology, new approaches to application design, and changing requirements in the larger context in which the Internet sits" (Clark, 2018, 301).

[72] The analysis of the development of the Web has shown that the Web has "areas that have largely been studied by physicists and mathematicians using the tools of complex dynamics system analysis" (Berners-Lee et al., 2006, 770).

[73] Besides Rescher's general analysis of complexity, see also another angle: Bishop (2007).

2020d). De facto, there is an increasing sophistication of aims of the research (most of them genuinely new, insofar as they were not available in the original design of the Internet).[74] (III) Ontologically, there are several levels in the network (micro, meso and macro), which are related to different kinds of designs and their changes over time. (IV) Axiologically, the values related to ends and means change based on new needs and novel options. In this dynamic of values, there is a genuine interaction over the time between an internal and an external side of the network,[75] where initially external values — of social or environmental origin — can become internal.

On the one hand, some experts in Web science "argue that the development of the Web has followed an evolutionary path, suggesting a view of the Web in ecological terms" (Berners-Lee et al., 2006, 770). Furthermore, there is the idea that complexity increases in evolution seem to underlie not only dynamic complexity but also structural complexity. If so, then the question is how complexity increases in evolution. In this regard, within evolutionary biology, there are three general lines to characterize the mechanisms for increasing complexity in biological systems: (a) internalist mechanisms, (b) externalist mechanisms, and (c) undriven mechanisms (i.e., theories invoking no driving force at all) [cf. McShea [1991], 1998, 626].

But, on the other hand, it happens as a matter of fact that the approaches based on "mechanism," both biological and philosophical, do not cover the terrain of dynamics complexity of the sciences of the Internet, in general, or the other two groups of sciences (those which use the network of networks for new developments in their disciplines or the new disciplines that deal with the emerging properties of this complex system). They cannot grasp either the depth of historicity that scientific creativity and technological innovations involve or the trait of the bi-directional or two-way relations between internal and external factors of complexity, which are crucial in the sphere of the sciences of the Internet and their consequences for economics or communication sciences, which are artificial as well as social.

Prima facie, these changes in the Internet — in the broad sense — may be characterized according to the idea of new processes, in terms of the evolution of some aspects (i.e., adaptations and mutations), which to some extent were already available,[76] or through the concept of historicity, with different types of changes over time, which can include horizontal as well as vertical novelty, and which are open to the idea of revolutions (either from an "internal" source or from an "external" cause).

---

[74]Generality is one of the reasons of the success of the design of the Internet. This includes generality regarding the applications that run over it as well as generality concerning the information and communication technologies out of which new can be constructed. Cf. Clark (2018, 42).

[75]This can be seen in Tim Berners-Lee's history of the Web (Berners-Lee, 1999).

[76]In the case of evolution, when is related to biological phenomena, Daniel McShea asks that "the reader do not equate complexity with progress" (McShea [1991], 1998, 626). But McShea recognizes that evolutionary trends are frequently related to an increasing adaptability and a growing control by organisms over their environment (cf. Simpson, 1949).

(i) "Process" is a very broad term,[77] which is even more basic than "evolution."[78] But it seems clear that not all processes are related to changes over time (e.g., in machines). (ii) "Evolution" is a different concept, which has been changing in content. This concept has been analyzed in detail and has had a wide projection in both science and philosophy (cf. Bowler [1983], 2009; Gonzalez, 2008b). It can describe some of the changes undergone by the Internet as a whole. In addition, there are some authors that describe the future of the Internet in terms of "evolvability." But they offer a description of intensive and extensive changes that goes beyond what is a proper evolution, insofar as there are changing requirements, new functions and continuous innovations (cf. Yin et al., 2014, 14). (iii) "Historicity" includes evolution and revolution, gradual changes and strong shifts, those able to get something really revolutionary (conceptually, practically or heuristically).[79] Thus, the changes over time of the network of networks can be explained in terms of historicity, which includes revolutionary shifts (changes can be seen retrospectively[80] or considered prospectively [cf. Varghese & Buyya, 2018]). This characteristic of historicity makes the social impact of the Internet more understandable.[81]

Looking backwards, the trajectory of the Internet, the Web or cloud computing, apps and the mobile Internet is more a revolution — scientific, technological[82] and social[83] — than an evolution. What started as a technological infrastructure — a somehow closed and limited enterprise — has reached out far beyond expectations with all the ramifications of the successive versions of the Web or the developments of cloud computing. The Internet as such was initially focused on information and

---

[77] On the characterization of "process," see Rescher (1995).

[78] In addition to "process," Rescher has also developed a set of ideas regarding evolution in epistemological terms. Nevertheless, "process" seems a more basic notion in his approach, insofar as he discusses the "Varieties of Evolutionary Process," cf. Rescher (1990, 5–12).

[79] These strong shifts have not been uncommon in recent decades, such as the introduction of audiovisuals on the Web, making it expressly interactive, or the design of smart phones (such as Apple iPhone) for mobile Internet, which has turned them into de facto laptops for many functions.

[80] This was soon detected by Simon, who saw its repercussions for the organizations, cf. Simon ([1986], 1997a).

[81] On the distinction between "process," "evolution," and "historicity," cf. Gonzalez (2013b).

[82] If we follow the changes operated in the network of networks according to layers, then the revolution in the Internet in the strict sense would be predominately technological, while the revolution in the Web would be primarily scientific. In the case of the cloud computing there is a combination of both. On the concept of "scientific revolution," I have developed an alternative to Thomas Kuhn (especially with respect to the initial stage) and with important nuances regarding Paul Thagard's conceptual revolutions. Cf. Gonzalez (2011a, b).

[83] This social revolution can be seen with the multiple impact of the Covid-19 pandemic, especially in countries that have had strict lockdown. Face-to-face universities were transformed in just a few days into online universities, something that no one had foreseen and it has been like this for months, and many public and private companies became telework centers. See Gonzalez (2020e). These social changes affect individual, group and organizational activities, both public and private. It can also be argued that the constant use of screens also affects cognitive skills.

communication, but now as a network of networks reaches all spheres of contemporary society (cf. Floridi, 2014, ix, 90–94 and 219).

Historicity and complexity can be linked to two kinds of *novelty* in the trajectory of the Internet as a whole, which might continue in the future: (a) changes of a horizontal or longitudinal type came from creativity by enlargement of something already available;[84] and (b) changes of a vertical or transversal type came from the genuine creativity of an intense novelty or a somehow revolutionary modification (email, YouTube, electronic banking, etc.). Both kinds of novelty — horizontal or longitudinal and vertical or transversal — can be used for the reconstruction of the history of the Internet. They are also elements for predicting its future, insofar as these variables can continue into the possible future.

6.4.3   The Social Dimension and Complexity

Epistemological and ontological problems are also present in the social dimension, which is a key factor of complexity of the network of networks both in structural terms and in dynamic ones. The social dimension, which is the main focus of the external perspective, is present in the three main layers of this complex system, not only the Web and the cloud along with apps and mobile Internet, but also the background technological infrastructure and the providers of technological services through specialized companies (cf. Gonzalez, 2020a). In addition, the social dimension is interconnected to the scientific side and the technological facet, which involve the intentionality of human actions in a social environment.

While the scientific activity of the Internet — in the broad sense — is mostly "internal," the social dimension of this network is predominantly "external." But the Internet works as an open system, where the "internal" and "external" aspects interact both structurally and dynamically. *De facto*, the social dimension depends on creativity at the three ontological levels (micro, meso and macro),[85] where the creativity of individuals, groups and organizations increases the structural complexity and the dynamic complexity of the three of them.[86]

Besides the role of the users (individuals, groups, organizations, etc.) in the undeniable relevance of the social dimension of the Internet as a whole,[87] there are

---

[84] Some changes, such as the design of apps, can be seen as an "evolution" rather than a "revolution," insofar as apps "rely on the same web architectures" as web browsing, cf. Hendler and Hall (2016, 704).

[85] This involves the assumption that there is a social ontology, which is the case of the Internet when its users interact with the artificial designs made scientifically.

[86] The organizations themselves can be configurated in terms of complexity, cf. Simon (2001). On complexity and its study, see Strevens (2003).

[87] "The Internet is inextricably intertwined with almost every sector of society, increasing its complexity and bringing forth numerous opportunities and challenges. It has been only 50 years from its earliest conception in the early 1960s, to its present state as a vast, interconnected network

increasing legal discussions (e.g., the legislation in place in the European Union since May 25th, 2018) [cf. GDPR, 2016], which is a new factor of complexity, because it affects contents as well as new possible realities (e.g., in Web 3.0).[88] Moreover, the social dimension can increase both intensively and extensively in the near future, mainly for two reasons: (a) there are new possibilities that will be added to the network of networks in short, middle and long run, and (b) there are still many millions of people in the world that do not have access to the network,[89] who sooner or later will get such access.

Addressing scientifically the intensive and extensive increase in the social consequences of the worldwide use of the network of networks — in particular, the Web along with cloud computing, the apps and the mobile Internet — requires prediction as anticipation of the possible future and prescription of the course of action to follow in order to solve specific problems. These need to deal with the problems — structural and dynamic — related to human possibilities of citizens.[90] The applied sciences of design are needed in the three groups of sciences related to Internet as a whole and that use is particularly important in two fields of the second group of disciplines: education and communication, insofar as they shape our social environment.

De facto, the social dimension is considered in the designs of the "sciences of the Internet" (mainly Web science, Network science,[91] and the specific science of the Internet) and in the sciences about emerging properties on the Internet, such as data science. The social dimension is also present in other sciences related to the task of expanding this technological sphere beyond the global Internet.[92] Many of these

---

of networks spanning much of the globe and linking approximately 2.7 billion people, representing 39% of the world's population, by the end of 2013" (Winter & Ono, 2015b, 1). The expansion has continued since then and the current users are estimated at 4 billion.

[88] "Alongside cryptocurrencies like bitcoin, NFTs [non-fungible tokens] are the most visible instantiation of 'web3'." The Economist (2022, 53).

[89] On 22 February 2016, the founder of Facebook pointed out that around 4.000 million inhabitants of our planet still do not have access to the Internet. Mark Zuckerberg made this statement during his presentation at the *Mobile World Congress*, held in Barcelona, available at: http://www.informationweek.com/mobile/zuckerberg-hits-mwc-to-talk-drones-ai-vr/d/d-id/1324403, accessed on 8.8.2016. The situation a few years later is different and the users are around that figure. See footnote number 2.

[90] One of these dynamic problems is the constant interaction between "internal" and "external" factors while using the network. This issue can be seen in the effects of the participation of the citizens in communicating phenomena over the Web. There is already a measurement of such participation over the Web, which has still its methodological problems, cf. Page and Uncles (2014).

[91] The presence of formal sciences is clear in network science, cf. Newman et al. (2006). In addition, Artificial Intelligence, which plays a role in the sciences of the Internet, has its roots in logico-mathematical contributions due to authors like Alan Turing, cf. Hodges (2014).

[92] David Clark has pointed out that there are now other global networks with two features: (a) they use the same kind of technology that the Internet, and (b) they are no directly connected with the network that we know as Internet *sensu stricto*. Among them are content delivery networks (CDNs). These new networks are part of the application development of the artificial system studied here. It happens that these networks are used by cloud computing providers to give better services, which

sciences of the three groups — as is the case of the sciences that look for novelty through the use of the network of networks — frequently develop their task from "inside" to "outside." This happens with economics, information science, education and communication, insofar as these sciences make designs to do things differently from the past, and this increases complexity. It seems likely that this search for novel aims will continue in the future, often with the support of the new contributions in Artificial Intelligence.

Accordingly, there will be new developments in financial economics, novel ways of information recovery, different aspects of on-line education in the future and original forms of audiovisual communication will continue to appear. This scenario requires the prediction of the possible future in every field (in the short, middle or long run) and the prescription to solve the specific problems that may show up. Thereafter, the application of scientific results will enter diverse contexts (micro, meso or macro), which opens the door to new networks related to the main network.

Moreover, it is possible to have in the future a "sectoralization" of the Internet, i.e., a network for each relevant domain: finance, education, industry, communications, etc. In any case, it is a complex system that is open to the future from within, insofar as, in addition to access to the network of networks as such, the users — the citizens — have within their reach peer-to-peer interactions, which are applications that do not depend on centralized services or cloud computing (cf. Clark, 2018, 307). Thus, it seems that we are currently moving towards a richer complex system with multiple networks instead of one network that gives support and is interconnected with several important layers.

6.5 The Role of Prediction and Prescription: Applied Science to Deal with Complexity

Within the setting of complexity, both structural and dynamic, the role of prediction and prescription will be central for the sciences of the Internet. Looking at the future, they will need applied science to deal with complexity in the internal and external perspectives. It seems very likely that the dynamics of the Internet as complex system will bring about many changes (in connectivity, scalability, security, accessibility, privacy, etc.), which might be at the three main layers, without excluding the possibility of new layers (at least in the middle or long run). Thus, besides the "horizontal" or ampliative novelty, there might be "vertical" or transversal novelty. The changes may involve new knowledge and modifications at the ontological levels (micro, meso and macro), which may have repercussion for the aims of the designs of the sciences of the Internet, the processes that will follow them, as well as the expected results and their consequences.

---

protect the cloud computing from attacks made by hackers and give more stability to consumers regarding performance. Cf. Clark (2018, 307).

Hence, dealing with the internal and external dynamics of the three major layers requires prediction of the possible future — in the short, middle or long run — of the Internet. This involves taking into consideration the historicity of this scientific aspect of the Internet, beginning with the internal perspective (aims, processes and results of the designs),[93] which should be followed by the external perspective (social, economic, political, etc.). Furthermore, the technological facet of the Internet as a whole should be considered (cf. Hanseth & Lyytinen, 2010; Uhlig, 2010). In addition, the social dimension, which includes the historical and cultural aspects of the network of networks, should be addressed. This means several kinds of challenges for the future.[94]

Prediction gives knowledge that is needed for scientific design (Gonzalez, 2007a), which is oriented towards specific aims. Thus, some processes are followed in order to reach some results. In the case of the scientific activity of the Internet, prediction can be focused in three directions. (i) In the case of the *Internet in the broad sense*, prediction is needed for the network, the Web and cloud computing, apps and mobile Internet. They require the anticipation of the viability of the enlargement or empowerment of human possibilities, within a defined temporal framework (short, middle or long run). (ii) When the focus is on the *disciplines* that use the Internet to enlarge their field or to reach new thematic territories, prediction can anticipate what can be expected in these cases (economic, informative, educational or communicative). (iii) Based on the *emergence* of new properties in the network itself, prediction can anticipate the possible future, as happens with the new phenomena generated by the Internet itself (which receive the attention of data science, for example) [cf. Cao, 2017a, b].

After the prediction made in any of these three options of scientific activity comes prescription. Some patterns are needed for solving the specific problems at stake. This is the realm of applied science, which is successful once the practical problems are solved[95] (such as connectivity or security). These might be issues related to the network, the Web or cloud computing, problems related to the scientific disciplines that use it (e.g., economic, informative, educational or communicative), or new aspects originated by the emergent properties of the Internet in the broad sense (like data science). In this regard, each science related to the main layers needs to anticipate and, thereafter, guide actions to solve the specific problems in question.

There may be quite different degrees in the reliability of the anticipation of the future, as happens with the concepts of "foresight," "prediction," "forecasting," and "planning." They are based on the number of variables available and the level of control in their knowledge. Thus, they can express different reliability of phenomena

---

[93] See, in this regard, Hooker (2011, 215).

[94] "The scale, topology, and power of decentralized information systems such as the Web also pose a unique set of social and public-policy challenges. Although computer and information science have generally concentrated on the representation and analysis of the information, attention also needs to be given to the social and legal relationships behind this information" (Berners-Lee et al., 2006, 770).

[95] This can be seen in economics, cf. Gonzalez (1998).

in the short, middle or long run (Gonzalez, 2015a, 47–72; especially, 65–72). Meanwhile, prescription is made based on some values, which are oriented towards what is preferable,[96] instead of what is merely preferred. In addition, prescription can be related to planning, which is different from prediction and includes the calculation and distribution of time.

Looking at the "internal" perspective, the interdependence between prediction and prescription in the sciences of Internet is noticeable. For prediction in this ambit, Artificial Intelligence plays a role in a large number of cases, which has repercussions not only for the network, but also for the Web and cloud computing (cf. The Economist, 2020, 15). The interdependence is also important within the "external" perspective (the social dimension) of the three main layers. The problems of complexity in the Internet make this interdependence explicit for the future, insofar as the structural complexity and the dynamic complexity are related to social factors (cultural, political, economic, legal, ecological, etc.) around the use of this system, which is imbued with historicity.

Undoubtedly, prediction is central in the epistemology and methodology of *applied sciences*, such as the sciences of the Internet, as a guide for thinking about the patterns of how to solve specific problems within certain contexts (national or international), which can lead to a planning (i.e., calculus and distribution of actions, usually according to a well-organized schedule). These patterns commonly have a social dimension in this network of networks. *De facto*, the activity of the Internet as such is intertwined with other human activities that use this network as a technological platform for acting.

Also, prediction is a tool for the *application of sciences*, i.e., for the implementation of the patterns or action guides singled out according to the various contexts. This is the case in the three groups of sciences in this sphere[97]: (a) in the sciences of the Internet (Web science, Network science, the specific science for the Internet, etc.), (b) in scientific disciplines that use this network (economics, information science, education or communication), and (c) in the studies that deal with the emergent properties of this network (data science).

6.5.1 Task of Prediction and Complexity: From the Scientific Activity to the Social Dimension

For the future of the network of networks, prediction plays a key role to tackle future aspects of the epistemic and ontological modes of complexity pointed out earlier,

---

[96] This is very clear in the case of applied economics, cf. Sen (1986).

[97] This is different from administrative action or policy-making in the strict sense carried out on the basis of the knowledge provided by the experts. The decision in this case would be administrative or strictly political, even though it is based on the application of knowledge given by scientists (as has been seen in the various countries on the occasion of Covid-19).

which we can find in structural and dynamic factors of this complex system. Prediction is carried out as it is usually done in the sciences of design, which are applied sciences. Thus, prediction provides knowledge for the design of aims and processes, to prefigure the possible results (cf. Gonzalez, 2007b, 2012). The task of prediction is diversified in the ontological, epistemological and heuristic versions, according to the goals of the research made: (I) the anticipation of a future phenomenon or event, (II) to infer the existence of something not yet known or (III) to consider a margin of possibilities of events or occurrences ordinarily in the short, middle or long run.

From a philosophico-methodological viewpoint, prediction cannot be reduced to epistemological or ontological features. It has also semantic, logic, methodological, axiological and ethical characteristics, because language, structure, methods, values in general and ethical values matter to scientific prediction (cf. Gonzalez, 2015a, 13–20; see also Gonzalez, 2010). In the case of the Internet — in the broad sense — prediction is especially related to anticipation of the possible future in order to solve the problems of complexity.[98] In this regard, it should consider the three focuses emphasized here, which are scientific, technological, and social:

(1) In the *scientific branch*, prediction anticipates the possible future (imminent, short, middle, long and very long run) of structural and dynamic traits based on the available knowledge (with variables known with a different degree of certainty), which is the previous step to presenting prescription (it offers the practical patterns for solving specific problems, commonly in the short, middle or long run). The future of the network of networks depends to a large extent on scientific creativity based on prediction for the technological infrastructure (cf. Gonzalez, 2013a), the Web and cloud computing, apps and mobile Internet.

(2) Concerning the technological outcome, prediction is the base for the scientific knowledge that uses the technology of the Internet for its development. Thus, the *technological facet* of the network is fed with scientific predictions (know that), which accompany the specific technological knowledge about instruments and artifacts (know how). In addition, the third kind of knowledge related to technology is evaluative (know whether) [cf. Gonzalez, 2015b]. This knowledge about what is worthy reflects on the ends sought by the technology at stake, which, here, is the network of networks.

(3) Prediction in the *social dimension* of the Internet depends on the "internal" developments of the network (scientific and technological), which are the base for the anticipation of the possible uses by agents (individual or social), business firms, corporations, states, international organizations, etc. In addition, the social dynamics itself enable us to predict future demands (such as economic, informative, educational or communicative) on behalf of the social entities. Science data can also benefit from prediction in the sphere of the social dimension of this network of networks.

---

[98] A methodological study of prediction to face complex systems, in general, is in Nicolis and Nicolis (2012).

Following the *scientific branch*, prediction is key for applied science in the network of networks as well as for application of the scientific contents available through the Internet. Both depend on the number of variables known related to the future and the degree of control of the variables in question. The reliability of the knowledge of the future is key, and will be used as a guide for prescription in order to solve the problems of structural complexity and dynamic complexity of this network of networks.

Regarding the *social dimension* of the Internet in the broad sense, there are a large number of issues related to the users of the network, organizations (Meta with Facebook, Alphabet with Google, Amazon, Microsoft, Twitter, etc.), and the public regulation of the network (including neutrality regarding the governance) [cf. Kang, 2017]. Thus, following the social dynamics of the Internet, there are two main aspects to be considered, as pointed out by Christopher S. Yoo:

(i) The changes in the *technological and economic environments*, such as the increases in the number and diversity of the Internet users, the changes in the nature of Internet usage, the diversification of transmission technologies and end-user devices, and the upsurge in the complexity of business relationships. (ii) The variations in terms of *policy implications*, such as the changes in the level of standardization, the decline of informal governance, the acceptance of new functions into the core of the network, the growing complexity of Internet pricing, the increase in the presence of intermediation, the maturation of business firms and corporations, etc. (Yoo, 2012).

For the analysis of the future of the Internet, the intellectual trajectory proposed here is from internal to external: it goes from scientific creativity to the social dimension through the mediation of the technological innovation. But it is also possible to think of an alternative proposal about the future of this network of networks, along the line from external to internal. This involves social demands feeding the scientific designs of the technology platform, the Web or the cloud computing and the apps.

Nowadays the focus in the research is often on the technological platform seen from its social expressions, such as (a) the information shared by the social agents, (b) the relevance of the culture, and (c) the relations of power (insofar as having information involves having power) [Winter & Ono, 2015c]. With these three aspects, the philosophical approach pays attention to some features of the impact of technology of the Internet for the users that have access to the network.

However, on the one hand, there is a scientific branch, which is diversified in a number of disciplines and usually works in interdisciplinary terms,[99] which are open to transdisciplinary options (Cao, 2017a, 59). This scientific activity is crucial for the future development of the Internet and has repercussions in its temporal aims (commonly, in the short, middle or long run). On the other hand, it happens that the actors in the future of the network of networks are not reduced to citizens,

---

[99] The interdisciplinarity appears in each science of the Internet as well as in the interrelations between them, cf. Tiropanis et al. (2015).

business firms and the nations, because there are a number of other agents, which are transnational, such as the "multi-agent systems," which are related to information and policy-making (Floridi, 2011, 299–300 and 306.).

Altogether, they increase the structural complexity and the dynamic complexity of the Internet, which is no longer a mere platform for information and communication, because of several facts: (1) the Internet generates new forms of sciences of the artificial, whose goals are increasingly more sophisticated, as has happened with Web science[100]; (2) it contributes to the advancement of the knowledge by means of new expressions of technology in order to get new objectives, where Artificial Intelligence commonly has a role; and (3) it creates artificial societies, within a new environment (a virtual world) with an increasing repercussion in the daily lives of the countries all around the world (as it happens with important "social networks," which are actually artificial setting of interpersonal relationship).

6.5.2 Task of Prescription in the Face of Complexity

Besides the philosophico-methodological role of prediction — as an aim, a test, a guide of science and orientation for its application (Gonzalez, 2015a, 10–22 and 321–326)[101] — there is the task of prescription to deal with complexity of the network of networks. In this regard, the sciences of the Internet require prescription as applied sciences, in general, and as sciences of design, in particular (cf. Gonzalez, 2007a, 2017a). Offering the adequate prescription — the patterns that actually can solve problems — becomes more difficult when dealing with complex systems, especially when the issue in question has dynamic complexity, as is the case in the network of networks.

Prescription requires an estimation of what is good or convenient instead of what is to be dismissed or avoided. Thus, the role of *values* concerning ought to do is crucial for prescription in order to advise patterns for action. These values can be initially internal (cognitive, methodological, etc.) or external (social, cultural, etc.). These values to prescribe the recommended course of action cannot be limited to applied science, because they are also focused on the application of science in the diversity of the contexts of use.[102]

---

[100] See, for example, the historical reconstruction of the Web science and the prognosis made in Hall et al. (2016).

[101] Regarding the methodological role of prediction in the network of networks, the problem of the relationship between prediction and the predicted is again raised. In this new scenario, the debate arises in these terms: (a) certain platforms have the capacity to make predictions based on data on the use of the network by users, and (b) these predictions can become a reality — go from correct to true — due precisely to the mechanisms available to these companies to change user habits.

[102] This case of the sciences of the Internet can have similarities with the case of economics. In the sciences of design, such as economics, complex dynamics often receives attention, mainly in the sphere of macroeconomics (e.g., market mechanisms, business cycles, economic growth, economic

Two different directions can follow prescriptions in this complex system: internal and external. In the first case, the attention goes primarily to the sciences of the Internet, conceived as scientific disciplines that enlarge human possibilities, which have repercussion of the technological facet of the Internet. When it is the second case, then prescription looks at the social dimension of the Internet as a whole, which is relevant for the scientific side as well as for the technological facet.

There are two general kinds of values in the scientific side related to prescriptions: (i) the *internal values* of "the Internet as scientific activity," which reflect what is worthy of merit in the scientific activity of developing the network of networks, and (ii) the *external values* of "the Internet as a scientific activity among others," which corresponds to what is worthy of merit in the scientific activity of a human activity intertwined with many other human activities.

Internal values of "the Internet as scientific activity" are efficacy and efficiency, the search for certainty in the patterns of action, appropriateness between means and end (and the diminishing of risk), competitiveness in the face of viable alternatives, the acceptable correlation between cost — human, social, and economic — and benefit, etc. Meanwhile, external values of "the Internet as a scientific activity among others" have a large range of possibilities, according to the facet of the Internet emphasized: social, cultural, educational, economic, communicative, political, industrial, etc.

Among these external values are those related to relevant aspects: (a) the psychosocial dimension of agents (trust, reliability, privacy,[103] etc.), (b) the socio-cultural traits of individuals, groups or organizations (satisficing needs of communication, achieving some expectations, social wellbeing, etc.), (c) the ethical values in the social perspective (dignity, integrity, solidarity, etc.), (d) the protection of environment, (e) the increase in freedom in the public domain and the reduction of inequalities, etc.

When the focus of values is in the second case mentioned — the sphere of the *social dimension* of the Internet — then we have also internal and external values, which can affect the three main layers: the network, the Web and cloud computing, practical applications (apps) and the mobile Internet. "Internal" here refers to organizations, public or private, that develop the background technological infrastructure, the provision of services to users, the development of the Web, the advancement of the cloud or the creation of new apps through start-ups. "External" deals here with the social, cultural, economic, political environment, etc. that is related to the complex system of the network of networks.

Thus, within the sphere of the *social dimension* of the Internet, the internal values for prescription are based on the previous existence of predictive knowledge. They

---

development, etc.), where there are usually more factors involved than in the realm of microeconomics. On the role of prescription in economics, see Gonzalez (2015a, 326–338).

[103] When the Web science was launched, Berners-Lee and collaborators emphasized that "we want to be sure that it [the Web] supports the basic social values of trustworthiness, privacy, and respect for social boundaries" (Berners-Lee et al., 2006, 769). See also World Wide Web Foundation (2019).

include accessibility of the goal, the coherence of the means, the suitability regarding the agents, the relation between cost — not only financial but also energy and time — and benefit, etc. Meanwhile, the external values for prescription in this network are then the social and ethical evaluations, such as the repercussion for individual agents (privacy, respect for the person, etc.), social groups (work, absence of social discrimination, capacity of integration in the labor work, etc.), cultural expressions (of majorities and minorities, traditional or new ones, etc.) that lead to personal or social development, the neutrality of the network, etc.[104]

6.5.3   Evaluation and Meta-evaluation of Studies of the Future

Along with epistemological and ontological aspects, there are methodological traits of prediction and prescription of applied sciences to deal with complexity of the network of networks. Methodologically, the role of prediction and prescription can be analyzed within the agenda of future studies. This is the setting of design sciences, insofar as they seek to enhance human possibilities (cf. Simon, 1996, 4–5 and 111–113), because they look to the future. Moreover, this is what happens with the sciences of the Internet, which are primarily sciences of the artificial — with an interdisciplinary component — and are developed as applied sciences. On this basis, these disciplines can be evaluated within the framework of future research that is carried out from applied science and leads to the application of science. It is then the results of the application of science that lead to modifying, continuing or discarding what applied science has proposed to solve specific problems.

Certainly, the evaluation of future science can be enhanced from the development of theories based on the idea of design. This is what some authors have proposed: "A design theory describes the constructs and principles for form and function to solve a class of problems, which in our case is the evaluation of future studies" (Piirainen et al., 2012, 465). However, the *design* can be in both poles of the process. That is why a methodological distinction should be made between (i) design as a *starting point* (of something scientific, in this case) and (ii) design as a *material realization* of something previously conceived:

From the intellectual point of view, "design" is initially a *construct* or set of constructs, which gives rise to a model applicable in a practical domain, because of certain aims (the solution of specific problems). This is followed by processes (procedures or methods), to arrive at some (in principle, expected) results. Meanwhile, when the design is understood as a *product* and is something tangible, then it is a prototype. Therefore, design can be both what appears at the beginning and what is finally obtained after following several steps from the methodological point of

---

[104]There are also the values of the social responsibility of the business firms and public corporations in the use of the virtual world of infosphere. Cf. Floridi (2014, 217–220).

view. But the constructs of the beginning and the artifact of the end have a different ontological status.

A second distinction is needed between what is methodological (evaluation) and what, strictly speaking, is meta-methodological (meta-evaluation). Because one thing is the evaluation of the *results* themselves of the studies of the future and another is the design of a framework theory that serves to create a *taxonomy* for evaluating the results. What is methodological is when the results themselves of the predictions made and the prescriptions carried out are evaluated. It is then that what seemed to be initially correct — when it was enunciated — is seen whether or not to have actual content to solve the specific problems posed. Meanwhile, the meta-methodology is to design a theory on how to *evaluate* the evaluations of the predictions made and the prescriptions carried out, mainly in order to accept or not a methodological conception.

Hence, there are two different options: (a) we can evaluate the results obtained in the sciences of the Internet, such as the reliability of the predictions made in order to solve the specific problems of the technological infrastructure, the Web or cloud computing, apps and the mobile Internet, or (b) we can design a theory in the framework of the studies of the future on how to evaluate the evaluations made on reliability of the predictions, those made regarding the specific problem-solving at hand. The relations between the former (methodological) and the latter (meta-methodological) may have to address similar problems to those of Imre Lakatos in the relations between his methodology of scientific research programs (MSRP) [Lakatos [1970], 1978a] and his methodology of historiographical research programs (MHRP) [Lakatos [1971], 1978b], which is a second order methodology of scientific research programs (Lakatos [1971], 1978b, 131–132). The former — MSRP — was mainly focused on prediction and its content of progress (theoretical, practical and heuristic), whereas the latter — MHRP — was oriented to the evaluation of methodologies based on the history of science.[105]

When the focus is *methodological*, several criteria can be considered in applied research to evaluate the results obtained: (1) the level of achievement of the aims of the scientific design that was developed to solve a particular problem posed. In the case of the network of networks, this practical component is measured in comparative terms, according to the degree of adequacy between the results achieved and the goals initially set. (2) Quality in the processes used — procedures and methods — which is measured by the capacity to obtain reliable data regarding the problem posed, so that they provide information that can be structured and generate genuine knowledge. (3) Role of values, in general, and of ethical values, in particular, when thinking about the ends, means, results and consequences of the proposed applied research.

---

[105] In MHRP Lakatos sought to assess the validity of methodological orientations on the basis of the findings of the history of science, which is certainly the most questioned part of his second philosophico-methodological period. Cf. Gonzalez (2001, 2014).

The latter is particularly relevant for prediction in the context of applied science, since in applied research prediction does not serve as a scientific test but as a *guide to action*, so it is required to prescribe the guidelines for action that really allow problems to be solved. Therefore, the assumptions underlying the designs are important when planning predictions and not only the available knowledge (number of known variables and degree of control of known variables). In this sense, the *Contract for the Web* is very clear when it comes to setting out the underlying assumptions that are to guide future developments of the Web (cf. World Wide Web Foundation, 2019). From the point of view of values, the first thing is that which concerns the ends, which correspond to evaluative rationality (cf. Rescher, 1988). Then, the values with respect to the means to be used and, thereafter, to consider not only the expected results but also the consequences that can be inferred from them.

Hence, the evaluation criteria must also look at the prescriptions, which always incorporate — either implicitly or explicitly — values for practical action. (I) This involves that predictions cannot be mere "artifacts" that are conceived as potentially useful projections of the future, but rather as a prelude to actions with content that affect people, society and, in principle, nature as well. (II) How predictions are made cannot be reduced to epistemological or methodological issues, since the knowledge used and the processes used require a broader context of interpretation, where the principle of having a public service purpose also counts (Gonzalez, 2020e). Predictions serve to learn from the possible future, which in the network of networks is crucial (and not only in the Internet of Things). (III) Predictions must incorporate values to raise scenarios of a possible future that is sustainable for society, so that the common good is a criterion for advancing technological infrastructure, the Web and the cloud and practical applications (apps).

What seems clear is that quality, validity and credibility of future studies do not depend only on epistemological, methodological and ethical criteria, contrary to what some authors think (cf. Piirainen et al., 2012, 464). There are also semantic, logical, ontological and axiological factors to be considered in studies regarding the future when the complexity of the system under study is involved (cf. Gonzalez, 2020d). In addition, what we predict with regard to the possible future cannot be reduced to the extrapolation of existing structures, since scientific creativity and technological innovation are involved, which introduces the factor of historicity, which is not reduced to evolution insofar as we need also to contemplate revolution. Milestones such as the design of the Web is not an evolution based on an extrapolation but rather a vertical or transversal novelty that not even Tim Berners-Lee's own supervisor at CERN could see its scope and future projection.

6.6   Coda: Dealing Scientifically with the Future
of the Internet

To sum up, the future of the Internet as a whole — through the three layers of the
network, the Web and cloud computing, apps and the mobile Internet — is a key
aspect of our society within a new historical period (hyperhistory) [Floridi, 2014,
1–24 and 185]. It can be thought of in temporal terms, such as the short, middle and
long run. It is also a future that can be considered in ontological terms, such as the
micro, meso, and macro levels. These temporal and ontological aspects concern
primarily a mainly human factor of the Internet as complex system, which has been
considered here: the scientific activity to enlarge the capabilities and characteristics
of the network by means of scientific creativity, which is intertwined with technological innovation[106] and is related to the social dimension.

Initially, this creativity for this complex system is in the realm of the sciences of
the artificial. It diversifies into three main kinds of scientific undertakings: (i) the
disciplines characterized as "sciences of the Internet," (ii) the already available
disciplines that now use the network as a support for new developments (with
longitudinal and transversal novelty), and (iii) the sciences that deal with the
emergent properties due to the use of the network (such as data science). Thereafter,
there are social consequences,[107] because of the use of the technological platform for
an increasing number of personal, professional and social aims.

Besides the scientific realm of the Internet and the social dimension of it, which
will shape our social future to a large extent (education, communications, finances,
etc.), there is a third possible approach: the research led by technological innovation
of the platform of this network. This technology is in a dynamic interaction with the
sciences of the Internet and the social branch of it (individuals, groups, organizations, etc.).[108] Thus, the future of the Internet is at least threefold. Here the focus of
attention has been mainly on the future based on scientific approach. This activity
involves structural and dynamic problems (epistemological and ontological, with
internal and external aspects), but it is crucial for its centralization or decentralization
and the development of the Internet as network of networks in terms of novelty
(horizontal and vertical).

---

[106]To a large extent, this relation between scientific creativity and technological innovation can be
characterized as reciprocal relationality.

[107]"Social media analysts look to understand, mathematically and socially, the trends being seen on
the Web as reflected through information shared on social networking, web sites and mobile
applications" (Hall et al., 2016, 2).

[108]An additional aspect is "socio-informatics" or a perspective on the design that is based on the
practice and the use of artifacts of information technology. This approach assumes that the
computer-supported cooperative work and human-computer interaction in the future will be socially
embedded. Thus, if most of the computer applications will be socially embedded, they will be kinds
of infrastructures for the development of these very social practices that they are designed to
support. Cf. Randall et al. (2018).

Dealing scientifically with such a future of the Internet requires methodologically being able to predict — ontologically, epistemologically or heuristically — the possible outcomes of the emerging trends in this network, as the founders of the Web science have done so far.[109] Thereafter, there is the need for patterns of action of a synthetic kind,[110] as is usual in the prescriptions made in the field the sciences of the artificial, because they are primarily sciences of design. This involves the use of applied science and learning from the application of science in order to know which methodologies are really solving problems in the practical domain (micro, meso or macro).[111]

Acknowledgements This paper has been written within the framework of the research project FFI2016-79728-P supported by the Spanish Ministry of Economics, Industry and Competitiveness (AEI) and the research project PID2020-119170RB-I00, supported by Spanish Ministry of Science and Innovation (AEI). A previous version of this paper was presented at the Conference on *Grappling with the Futures*, organized by Harvard University and Boston University. It was also presented at the Congress entitled *For a Bottom-Up Epistemology*, organized by the University of Bologna. I am especially grateful to Donald Gillies and Thanassis Tiropanis for their comments on the more developed version of this paper.

References

Ackland, R. (2013). *Web social science: Concepts, data and tools for social scientists in the digital age*. SAGE.

Allmer, J. (2019). Towards an Internet of science. *Journal of Integrative Bioinformatics, 16*(3), 1–6. https://doi.org/10.1515/jib-2019-0024

Askitas, N., & Zimmermann, K. F. (2015). The Internet as a data source for advancement in the social sciences. *International Journal of Manpower, 36*(1), 2–12.

Barabási, A. L. (2013). Network science. *Philosophical Transactions of the Royal Society A, 371*(1987), 1–3. https://doi.org/10.1098/rsta.2012.0375

Berners-Lee, T. (1999). *Weaving the Web*. Texere Publishing.

Berners-Lee, T., & O'Hara, K. (2013). The read-write linked data Web. *Philosophical Transactions of the Royal Society A, 371*(1987), 1–5. https://doi.org/10.1098/rsta.2012.0386

Berners-Lee, T., Hall, W., Hendler, J., Shadbot, N., & Weitzner, D. J. (2006). Creating a science of the Web. *Science, 313*(5788), 769–771.

Bishop, R. C. (2007). *The philosophy of social sciences*. Continuum.

Bowler, P. J. ([1983] 2009). *Evolution: The history of an idea*. University of California Press; 25th Anniversary edition with a new preface. University of California Press.

---

[109] See, for example, the remarks about the future in Berners-Lee et al. (2006, 769–771, especially, 770), and Hall et al. (2016, 1–4, especially, 3–4).

[110] These patterns of action move at the micro, meso and macro levels, where the meso dimension is not usually highlighted and is especially important for practical activity.

[111] An eloquent example is in the area of the social sciences: "as polls are increasingly undermined as ways of forecasting the results of democratic elections, analysis of social media conversations is proving to be more accurate methodology for such forecasts, despite the fact that social media users do not present a representative sample of the total electorate" (Hall et al., 2016, 4).

Brynjolfsson, E., & McAfee, A. (2011). *Race against machine: How the digital revolution is accelerating innovation, driving productivity, and irreversibly transforming employment and the economy*. Digital Frontier Press.

Cao, L. (2017a). Data science: Challenges and directions. *Communications of ACM, 60*(8), 59–68.

Cao, L. (2017b). Data science: A comprehensive overview. *ACM Computing Surveys, 50*(3), art. 43, 1–42.

Clark, D. D. (2018). *Designing an Internet*. The MIT Press.

Dean, J., Corrado, G. S., Monga, R., Chen, K., Devin, M., Le, Q. V., Mao, M. Z., Ranzato, M. A., Senior, A., Tucker, P., Yang, K., & Ng, A. Y. (2012). Large distributed deep networks. In *Neural information processing systems, NIPS2012*. Available in: https://research.google.com/archive/large_deep_networks_nips2012.html. Accessed 11 Mar 2016

Dutton, W. H. (Ed.). (2014). *The Oxford handbook of Internet studies*. Oxford University Press. 2013 (reprinted in 2014).

Feldmann, A. (2007). Internet clean-state design: What and why? *ACM SIGCOMM Computer Communication Review, 37*(3), 59–64.

Floridi, L. (2009). Web 2.0 vs. the semantic web: A philosophical assessment. *Episteme, 6*(1), 25–37.

Floridi, L. (2011). *Philosophy of information*. Oxford University Press.

Floridi, L. (2014). *The fourth revolution – How the infosphere is reshaping human reality*. Oxford University Press.

GDPR. (2016, May 4). General data protection regulation. *Official Journal of the European Union, 59*, 1–88. Available in https://gdpr-info.eu. Accessed 19 Nov 2019.

Gonzalez, W. J. (1998). Prediction and prescription in economics: A philosophical and methodological approach. *Theoria: An International Journal for Theory, History and Foundations of Science, 13*(32), 321–345.

Gonzalez, W. J. (2001). Lakatos's approach on prediction and novel facts. *Theoria: An International Journal for Theory, History and Foundations of Science, 16*(42), 499–518.

Gonzalez, W. J. (2007a). Análisis de las Ciencias de Diseño desde la racionalidad limitada, la predicción y la prescripción. In W. J. Gonzalez (Ed.), *Las Ciencias de Diseño: Racionalidad limitada, predicción y prescripción* (pp. 3–38). Netbiblo.

Gonzalez, W. J. (2007b). La contribución de la predicción al diseño en las Ciencias de lo Artificial. In W. J. Gonzalez (Ed.), *Las Ciencias de Diseño: Racionalidad limitada, predicción y prescripción* (pp. 183–202). Netbiblo.

Gonzalez, W. J. (2008a). Rationality and prediction in the sciences of the artificial: Economics as a design science. In M. C. Galavotti, R. Scazzieri, & P. Suppes (Eds.), *Reasoning, rationality, and probability* (pp. 165–186). CSLI Publications.

Gonzalez, W. J. (2008b). Evolutionism from a contemporary viewpoint: The philosophical-methodological approach. In W. J. Gonzalez (Ed.), *Evolutionism: Present approaches* (pp. 3–59). Netbiblo.

Gonzalez, W. J. (2010). *La predicción científica: Concepciones filosófico-metodológicas desde H. Reichenbach a N. Rescher*. Montesinos.

Gonzalez, W. J. (2011a). The problem of conceptual revolutions at the present stage. In W. J. Gonzalez (Ed.), *Conceptual revolutions: From cognitive science to medicine* (pp. 7–38). Netbiblo.

Gonzalez, W. J. (2011b). Conceptual changes and scientific diversity: The role of historicity. In W. J. Gonzalez (Ed.), *Conceptual revolutions: From cognitive science to medicine* (pp. 39–62). Netbiblo.

Gonzalez, W. J. (2012). La vertiente dinámica de las Ciencias de la Complejidad. Repercusión de la historicidad para la predicción científica en las Ciencias de Diseño. In W. J. Gonzalez (Ed.), *Las Ciencias de la Complejidad: Vertiente dinámica de las Ciencias de Diseño y sobriedad de factores* (pp. 73–106). Netbiblo.

Gonzalez, W. J. (2013a). The roles of scientific creativity and technological innovation in the context of complexity of science. In W. J. Gonzalez (Ed.), *Creativity, innovation, and complexity in science* (pp. 11–40). Netbiblo.

Gonzalez, W. J. (2013b). The sciences of design as sciences of complexity: The dynamic trait. In H. Andersen, D. Dieks, W. J. Gonzalez, T. Uebel, & G. Wheeler (Eds.), *New challenges to philosophy of science* (pp. 299–311). Springer.

Gonzalez, W. J. (2014). The evolution of Lakatos's repercussion on the methodology of economics. *HOPOS: The Journal of the International Society for the History of Philosophy of Science, 4*(1), 1–25.

Gonzalez, W. J. (2015a). *Philosophico-methodological analysis of prediction and its role in economics.* Springer.

Gonzalez, W. J. (2015b). On the role of values in the configuration of technology: From axiology to ethics. In W. J. Gonzalez (Ed.), *New perspectives on technology, values, and ethics: Theoretical and practical* (Boston studies in the philosophy and history of science) (pp. 3–27). Springer.

Gonzalez, W. J. (2017a). From intelligence to rationality of minds and machines in contemporary society: The sciences of design and the role of information. *Minds and Machines, 27*(3), 397–424. https://doi.org/10.1007/s11023-017-9439-0. Available at https://link.springer.com/article/10.1007/s11023-017-9439-0. Accessed 6 Oct 2017

Gonzalez, W. J. (2017b). Artificial Intelligence in a new context: 'Internal' and 'external' factors. *Minds and Machines, 27*(3), 393–396. https://doi.org/10.1007/s11023-017-9444-3. Accessed 6 Oct 2017.

Gonzalez, W. J. (2018a). Internet en su vertiente científica: Predicción y prescripción ante la complejidad. *Artefactos: Revista de Estudios sobre Ciencia y Tecnología, 7*(2), 2nd period, 75-97. https://doi.org/10.14201/art2018717597

Gonzalez, W. J. (2018b). Complejidad dinámica en Internet como plataforma de información y comunicación: Análisis filosófico desde la perspectiva de Ciencias de Diseño y el papel de la predicción. *Informação e Sociedade: Estudos, 28*(1), 155–168.

Gonzalez, W. J. (2019). Internet y Economía: Análisis de una relación multivariada en el contexto de la complejidad. *Energeia: Revista internacional de Filosofía y Epistemología de las Ciencias Económicas, 6*(6), 11–36. Available at: https://abfcfc9a-c7ef-4730-b66e-0a415ef434c0.filesusr.com/ugd/e46a96_b400af5a739e4310a31b7e952244745d.pdf. Accessed 1 Apr 2020

Gonzalez, W. J. (2020a). La dimensión social de Internet: Análisis filosófico-metodológico desde la complejidad. *Artefactos: Revista de Estudios de la Ciencia y la Tecnología, 9*(1), 2nd period, 101–129. https://doi.org/10.14201/art2020101129. Available at: https://revistas.usal.es/index.php/artefactos/article/view/art2020101129. Accessed 27 Apr 2020

Gonzalez, W. J. (2020b). Electronic economy, Internet and business legitimacy. In J. D. Rendtorff (Ed.), *Handbook of business legitimacy: Responsibility, ethics and society.* (pp. 1–19, and printed version, pp. 1327–1345). Springer. https://doi.org/10.1007/978-3-319-68845-9_84-1

Gonzalez, W. J. (2020c). Pragmatic realism and scientific prediction: The role of complexity. In W. J. Gonzalez (Ed.), *New approaches to scientific realism* (pp. 251–287). De Gruyter. https://doi.org/10.1515/9783110664737-012

Gonzalez, W. J. (2020d). Levels of reality, complexity, and approaches to scientific method. In W. J. Gonzalez (Ed.), *Methodological prospects for scientific research: From pragmatism to pluralism* (Synthese library) (pp. 21–51). Springer.

Gonzalez, W. J. (2020e). The Internet at the service of society: Business ethics, rationality, and responsibility. *Éndoxa, 46,* 383–412.

Gonzalez, W. J. (2021). Tipos de diseño, innovaciones democráticas y relaciones internacionales. In A. Estany & M. Gensollen (Eds.), *Diseño institucional e innovaciones democráticas* (pp. 37–52). Universidad Autónoma de Barcelona-Universidad Autónoma de Aguascalientes.

Gonzalez, W. J. (forthcoming-a). The Internet as a complex system articulated in layers: Present status and possible future. In W. J. Gonzalez (Ed.), *The Internet and philosophy of science* (Routledge studies in the philosophy of science). Routledge.

Gonzalez, W. J. (forthcoming-b). Biology and the Internet: Fake news and Covid-19. In W. J. Gonzalez (Ed.), *The Internet and philosophy of science* (Routledge studies in the philosophy of science). Routledge.

Gonzalez, W. J., & Arrojo, M. J. (2019). Complexity in the sciences of the Internet and its relation to communication sciences. *Empedocles: European Journal for the Philosophy of Communication, 10*(1), 15–33. https://doi.org/10.1386/ejpc.10.1.15_1. Available at https://www. ingentaconnect.com/contentone/intellect/ejpc/2019/00000010/00000001/art00003. Accessed 6 July 2019

Graham, G. (1999). *The Internet: A philosophical inquiry.* Routledge.

Greenstein, S. (2015). *How the Internet became commercial. Innovation, privatization, and the birth of a new network.* Princeton University Press.

Hall, W., Hendler, J., & Staab, S. (2016, December), *A manifesto for Web science @10* (pp. 1–4). Available at http://www.webscience.org/manifesto. Accessed 16 May 2018.

Hanseth, O., & Lyytinen, K. (2010). Design theory for dynamic complexity in information infrastructures: The case of building internet. *Journal of Information Technology, 25*(1), 1–19.

Hendler, J. (forthcoming). The future of the Web. In W. J. Gonzalez (Ed.), *The Internet and Philosophy of Science.* Routledge.

Hendler, J., & Berners-Lee, T. (2010). From the semantic Web to social machines: A research challenge for AI on the World Wide Web. *Artificial Intelligence, 174*(2), 156–161.

Hendler, J., & Hall, W. (2016). Science of the world wide web. *Science, 354*(6313), 703–704.

Hodges, A. (2014). *Alan Turing: The enigma.* Vintage Books/Random House.

Hooker, C. (2011). Conceptualising reduction, emergence and self-organisation in complex dynamical systems. In C. Hooker (Ed.), *Philosophy of complex systems* (pp. 195–222). Elsevier.

Isaacson, W. (2014). *The innovators.* Simon and Schuster.

Kang, C. (2017). *F.C.C. Repeals net neutrality rules.* Available in: https://www.nytimes.com/201 7/12/14/technology/net-neutrality-repeal-vote.html. Accessed 15 Dec 2017. It was published in the paper edition of the NYT: KANG, C., F.C.C. reverses rules requiring net neutrality. *New York Times,* 15.12.2017, page A1.

Küng, L., Picard, R. G., & Towse, R. (Eds.). (2008). *The Internet and the mass media.* SAGE.

Lakatos, I. ([1970] 1978a). Falsification and the methodology of scientific research programmes. In I. Lakatos, & A. Musgrave (Eds.), *Criticism and the growth of knowledge* (pp. 91–196). Cambridge University Press, 1970. Reprinted in Lakatos, I., *The methodology of scientific research programmes: Philosophical papers, vol. 1,* edited by J. Worrall and G. Currie (pp. 8–101). Cambridge University Press, 1978.

Lakatos, I. ([1971] 1978b). History of Science and its Rational Reconstructions. In R. C. Buck, & R. S. Cohen (Eds.), *In memory of R. Carnap, P.S.A. 1970* (pp. 91–135). Reidel, 1971. Reprinted in Lakatos, I., *The methodology of scientific research programmes: Philosophical papers, vol. 1,* edited by J. Worrall and G. Currie, (pp. 102–138). Cambridge: Cambridge University Press, 1978.

Leiner, B. M., Cerf, V. G., Clark, D. D., Kahn, R. E., Kleinrock, L., Lynch, D. C., Postel, J., Roberts, L. G., & Wolff, S. (1997). The past and future history of the Internet. The science of the future technology. *Communications of the ACM, 40*(2), 102–108.

Leiner, B. M., Cerf, V. G., Clark, D. D., Kahn, R. E., Kleinrock, L., Lynch, D. C., Postel, J., Roberts, L. G., & Wolff, S. (2009). A brief history of the Internet. *ACM SIGCOMM Computer Communication Review, 39*(5), 22–31.

Liu, F., Shi, Y., & Li, P. (2017). Analysis of the relation between Artificial Intelligence and the Internet from the perspective of brain science. *Procedia Computer Science, 122,* 377–383.

Makridakis, S. (2017). The forthcoming Artificial Intelligence (AI) revolution: Its impact on society and firms. *Futures, 90,* 46–60.

McShea, D. W. ([1991] 1998). Complexity and evolution: What everybody knows. *Biology and Philosophy, 6*(1991), 303–324. Reprinted in D. Hull, & M. Ruse (Eds.), *The philosophy of biology,* Oxford University Press, 1998, pp. 625–649.

Meeker, M. (2019). *Internet Trends 2019*, Report published on 11 June 2019, 334 pages. Available at: https://www.bondcap.com/pdf/Internet_Trends_2019.pdf. Accessed 22 July 2019.

Meyer, E. T., Schroeder, R., & Cowls, J. (2016). The net as knowledge machine: How the Internet became embedded in research. *New Media and Society, 18*(7), 1159–1189.

Newman, M., Barabási, A.-L., & Watts, D. J. (2006). *The structure and dynamics of networks*. Princeton University Press.

Nicolis, G., & Nicolis, C. (2012). *Foundations of complex systems: Emergence, information and prediction*. World Scientific.

Niiniluoto, I. (1993). The aim and structure of applied research. *Erkenntnis, 38*(1), 1–21.

Niiniluoto, I. (1995). Approximation in applied science. *Poznan Studies in the Philosophy of the Sciences and the Humanities, 42*, 127–139.

Ornes, S. (2016). The Internet of Things and the explosion of interconnectivity. *Proceedings of the National Academy of Sciences of the United States of America, 116*(4), 11.059–11.060.

Page, K. L., & Uncles, M. D. (2014). The complexity of surveying Web participation. *Journal of Business Research, 67*, 2356–2367.

Park, J. H. (2019). Advances in future Internet and the industrial Internet of things. *Symmetry, 11*(2), 1–4. https://doi.org/10.3390/sym11020244

Piirainen, K. A., Gonzalez, R. A., & Bragge, J. (2012). A systemic evaluation framework for future research. *Futures, 44*(5), 464–474.

Priestley, M., Sluckin, T. J., & Tiropanis, T. (2020, April 11). Innovation on the Web: The end of the S-curve? *Internet Histories. Digital Technology, Culture and Society*, 1–24. https://doi.org/10.1080/24701475.2020.1747261

Randall, D., Rohde, M., Schmidt, K., & Wulf, V. (2018). Socio-informatics—Practice makes perfect? In V. Wulf, V. Pipek, D. Randall, M. Rohde, K. Schmidt, & G. Stevens (Eds.), *Socio-informatics. A practice-based perspective on the design and use of IT artifacts* (pp. 1–20). Oxford University Press.

Rescher, N. (1988). *Rationality: A philosophical inquiry into the nature and the rationale of reason*. Clarendon Press.

Rescher, N. (1990). *A useful inheritance. Evolutionary aspects of the theory of knowledge*. Rowman and Littlefield.

Rescher, N. (1995). *Process metaphysics*. State University of New York Press.

Rescher, N. (1998). *Complexity: A philosophical overview*. Transaction Publishers.

Rescher, N. (1999). *Razón y valores en la Era científico-tecnológica*. Paidós.

Schönwälder, J., Fouquet, M., Rodosek, G. D., & Hochstatter, C. (2009). Future Internet = Content + Services + Management. *IEEE Communications, 47*, 27–33.

Schredelseker, K., & Hauser, F. (Eds.). (2008). *Complexity and artificial markets*. Springer.

Schultze, S. J., & Whitt, R. S. (2016). Internet as a complex layered system. In J. M. Bauer & M. Latzer (Eds.), *Handbook on the economics of the Internet* (pp. 55–71). Edward Elgar.

Sen, A. (1986). Prediction and economic theory. In J. Mason, P. Mathias, & J. H. Westcott (Eds.), *Predictability in science and society* (pp. 3–23). The Royal Society and The British Academy.

Shadbolt, N., Hall, W., Hendler, J. A., & Dutton, W. H. (2013). Web science: A new frontier. *Philosophical Transactions of the Royal Society A, 371*(1987), 1–6. https://doi.org/10.1098/rsta.2012.0512

Simon, H. A. (1995). Artificial Intelligence: An empirical science. *Artificial Intelligence, 77*(1), 95–127.

Simon, H. A. (1996). *The sciences of the artificial* (3rd ed.). The MIT Press. (1st ed. in 1969, and 2nd ed. in 1981).

Simon, H. A. ([1986] 1997a). Chapter 14: The impact of electronic communications on organizations. In R. Wolff (Ed.), *Organizing industrial development* (pp. 251–256). Walter de Gruyter, 1986. Reprinted in Simon, H. A., *Models of bounded rationality*. Vol. 3: *Empirically grounded economic reason* (pp. 145–162). The MIT Press, 1997.

Simon, H. A. ([1990] 1997b). Prediction and prescription in systems modeling. *Operations Research, 38*(1990), 7–14. Reprinted in Simon, H. A., *Models of bounded rationality*. Vol. 3: *Empirically grounded economic reason* (pp. 115–128). The MIT Press, 1997.

Simon, H. A. (2001). Complex systems: The interplay of organizations and markets in contemporary society. *Computational and Mathematical Organizational Theory, 7*, 79–85.

Simpson, G. G. (1949). *The meaning of evolution: A study of the history of its significance for man.* Yale University Press.

Siow, E., Tiropanis, T., & Hall, W. (2018). Analytics for the Internet of things: A survey. *ACM Computing Surveys, 51*(4), 1–35. https://doi.org/10.1145/3204947

Strevens, M. (2003). *Bigger than chaos: Understanding complexity through probability*. Harvard University Press.

The Economist. (2018). More knock-on than network. How the Internet lost its decentralised innocence. Special Report: Fixing the Internet in *The Economist*, v. 427, n. 9098, June 30th, 2018, pp. 5–6.

The Economist. (2019). The Internet's next act. You ain't seen nothing yet. Section *Leaders*, June 8th 2019, pp. 14–15.

The Economist. (2020). Google grows up. Section *Briefing Alphabet*, August 1st 2020, pp. 14–17.

The Economist. (2022). The future of cyberspace. Rewebbing the net. Section *Business*, January 29th, pp. 53–54.

Tiropanis, T., Hall, W., Crowcroft, J., Contractor, N., & Tassiulas, L. (2015). Network science, Web science, and Internet science. *Communications of ACM, 58*(8), 76–82.

Uhlig, S. (2010). On the complexity of Internet traffic dynamics on its topology. *Telecommunication Systems, 43*(3), 167–180.

Varghese, B., & Buyya, R. (2018). Next generation cloud computing: New trends and research directions. *Future Generation Computer Systems, 79*(3), 849–861.

Vasuki, K., Rajeswari, K., & Prabakaran, M. (2018). A survey of current research and future directions using cloud-based big data analytics. *International Research Journal of Engineering and Technology, 5*(5), 3841–3844.

Winter, J., & Ono, R. (Eds.). (2015a). *The future Internet: Alternative visions*. Springer.

Winter, J., & Ono, R. (2015b). Introduction to the future of Internet: Alternative visions. In J. Winter & R. Ono (Eds.), *The future Internet: Alternative visions* (pp. 1–16). Springer.

Winter, J., & Ono, R. (2015c). Conclusion: Three stages of the future Internet. In J. Winter & R. Ono (Eds.), *The future Internet: Alternative visions* (pp. 217–224). Springer.

World Wide Web Foundation. (2019, November). *Contract for the Web. A global plan of action to make our online world safe and empowering for everyone*. Available at: https://contractfortheweb.org. Accessed 25 Nov 2019.

Wright, A. (2011). Web science meets Network science. *Communications of ACM, 54*(5), 23.

Yap, K. L., Chong, Y. W., & Liu, W. (2020). Enhance handover mechanism using mobile prediction in wireless networks. *PLoS One, 15*(1), 1–31. https://doi.org/10.1371/journal.phone.0227982

Yin, H., Jiang, Y., Lin, C., Luo, Y., & Liu, Y. (2014). Big data: Transforming the design philosophy of future Internet. *IEEE Network, 28*(4), 14–19.

Yoo, C. S. (2012). *The dynamic Internet: How technology, users, and businesses are changing the network*. AEI Press.

Zhang, J., & Zhang, W. (2017). Future of law conference: The Internet of Things, smart contracts and intelligent machines. *Frontiers of Law in China, 12*(4), 673–674.

Part III
New Analyses of Probability and the Use of Mathematics in Practice

Chapter 7
From Logical to Probabilistic Empiricism: Arguments for Pluralism

Maria Carla Galavotti

Abstract Starting from the 1960s, the conception of the philosophy of science forged by logical empiricism was gradually superseded by a new way of doing philosophy of science characterized by a progressive opening to the pragmatic components of science. On the one hand, the so-called post-positivist movement urged the need to consider the historical and sociological aspects underlying the formation of scientific knowledge, while on the other, the semantical view of theories prompted a shift of interest from theories to models and called attention to all aspects of experimentation. A decisive step in this direction was taken by Patrick Suppes, who labeled his own perspective "probabilistic empiricism" to emphasize the crucial role of probability and statistical methods in science. The turn of the 1960s brought about a pluralistic attitude regarding both scientific methodology and the spectrum of disciplines of interest for philosophers of science. Pluralism is currently a widespread position that different authors develop in different ways. A peculiarity of Suppes' probabilistic empiricism is the strict link he establishes between pluralism and the statistical methodology for experimentation and forming hypotheses. After an outline of the major features of the logical and probabilistic versions of empiricism, followed by some remarks on pluralism, this article draws attention to the desirability of adopting a bottom-up approach that assigns crucial importance to the context, in tune with Suppes' constructivist and pragmatical attitude. With an eye to the foundations of statistics, it is argued that pluralism and the adoption of a bottom-up approach are necessary choices in the light of two examples: cluster analysis, and the use of probability and statistics in courtrooms, which is discussed in the light of a controversial legal case.

Keywords Logical empiricism · Probabilistic empiricism · Pluralism · Statistics in courtrooms

M. C. Galavotti (✉)
Department of Philosophy and Communication, University of Bologna, Bologna, Italy
e-mail: mariacarla.galavotti@unibo.it

© The Author(s), under exclusive license to Springer Nature Switzerland AG 2022   147
W. J. Gonzalez (ed.), *Current Trends in Philosophy of Science*, Synthese Library
462, https://doi.org/10.1007/978-3-031-01315-7_7

7.1   Logical Empiricism

Let us start by summarizing the key issues characterizing the view of philosophy of science developed by the movement known as logical empiricism – also called logical positivism, or neo-positivism. Developed before World War I by the so-called first Vienna Circle, logical empiricism flourished in the 1920s and 1930s in various European countries before the political situation of the late 1930s forced its representatives to move outside Europe. It can be said that after World War II logical empiricism spread worldwide, the perspective developed in the 1920s and 1930s becoming so widely accepted by philosophers of science in the late 1940s and 1950s as to deserve being called the "received view". To be sure, it was a rich and multifaceted perspective, which cannot be depicted in a few words; nevertheless, it is useful to summarize the central theses that shaped it. The following list is obviously not meant to be exhaustive, but merely functional to a comparison between logical empiricism and a more recent way of doing philosophy of science.[1]

1. Logical empiricists cultivated the project of developing a theory of meaning capable of determining whether a proposition is true or false. The criterion of *verifiability*, later weakened to *confirmability*, played the twofold role of allowing the meaning of a proposition to be ascertained and providing a criterion of demarcation between scientific and non-scientific (pseudo-scientific, metaphysical) statements.

2. Central to the received view is the ideal of the unity of science to be realized both methodologically and linguistically. This project goes hand in hand with the conviction that physics represents the paradigm of 'good' science, namely the most sophisticated from a methodological point of view and which, being expressed in a language based on perceptual experience, can serve as an inter-subjective means of communication. The program known as *physicalism*, promoted especially by Otto Neurath in the 1930s, was rooted in that conviction.

3. According to the logical empiricist approach, the main task of philosophy of science is the *rational reconstruction* of scientific knowledge, including scientific theories, regarded as the major accomplishment of science, and the inferential processes used by scientists. Performing a rational reconstruction entails exhibiting the logical structure of such entities and processes, rearranging them in a consistent scheme, filling conceptual gaps or specifying connections that are not explicitly stated in the reports made by scientists. Strictly functional to rational reconstruction is the distinction between a *context of discovery* and a *context of justification*, as explicitly stated by Hans Reichenbach in *Experience and Prediction* (1938). For Reichenbach, in order to clarify such a distinction

> we might say that it corresponds to the form in which thinking processes are communicated to other persons instead of the forms in which they are subjectively performed. The way, for

---

[1] For an extensive account of logical empiricism, see Stadler ([2001] 2015).

instance, in which a mathematician publishes a new demonstration, or a physicist his logical reasoning in the foundation of a new theory, would almost correspond to our concept of rational reconstruction; and the well-known difference between the thinker's way of finding this theorem and his way of presenting it before a public may illustrate the difference in question. I shall introduce the terms *context of discovery* and *context of justification* to mark the distinction. Then we have to say that epistemology is only occupied in constructing the context of justification (Reichenbach, [1938] 1966, 6–7).

The idea is to keep the social, historical, and psychological aspects of theory formation and the experimental methodology responsible for data production separate from the precision and rigour characterizing the final formulation of theories. While all the elements leading to the statement of a theory belong to the context of discovery, rational reconstruction, namely the process aiming "to have thinking replaced by justifiable operations" (Reichenbach, [1938] 1966, 7) is the object of the context of justification. Confined to the context of justification, philosophy of science aims at clarifying the *logical structure* of scientific knowledge. So conceived, the kind of analysis performed by philosophers of science is only concerned with the syntactical aspects of scientific theories, supplemented by a theory of meaningfulness of theoretical statements. The final, abstract formulation of theories ought to be analysed apart from the process leading to it, including the methodology for the collection and organization of empirical findings.

4. Part of the received view is a conception of scientific theories as *complex spatial networks*, as depicted by Hempel's metaphor of nets whose knots represent theoretical terms, connected by definitions and/or hypotheses included in the theory. The net floats on the plane of observation to which it is anchored by means of rules of interpretation, allowing the scientist both to ascend from observable data to the theoretical network, and to descend from the latter to the former (see Hempel, [1952] 1969). Embodied in this view is a clear-cut distinction between theoretical and observational language.

It is worth mentioning that this view of theories, together with the distinction between theoretical and observational terms, found the resistance of Karl Popper, who instead proposed an epistemology "by conjectures and refutations" based on the tenet that a theory can never be verified, only falsified. According to Popper, theories are "*genuine conjectures* – highly informative guesses about the world which although not verifiable [...] can be submitted to severe critical tests" (Popper, ([1963] 1968, 115) aimed at finding a negative case, which would suffice to falsify them (see Popper, 1959).

7.2   Beyond Logical Empiricism

Partly under Popper's influence, several authors, including Norwood Russell Hanson, Thomas Kuhn, Imre Lakatos, Paul Feyerabend, and many others, triggered a decisive turn in philosophy of science. In a nutshell, this turn brought about an opening to the historical, psychological, and sociological elements surrounding the formation of scientific knowledge. Attention was also given to scientific growth, which ceased to be seen as a cumulative process to become a story of conflicts between old and new theories to solve the anomalies produced by some body of experimental data within some dominant theoretical framework. A vast literature took shape, focused on various problems concerning scientific theorizing, first and foremost the shift from old to new theories. The separation between the theoretical and observational components of science became obsolete.

Yet, the structure of observational data and the methodology for their collection and systematization remained by and large outside the sphere of interest of philosophers of science. The same holds for the statistical methodology involved in the interplay between data and hypotheses, including updating scientific hypotheses in the light of new evidence, the control of hypotheses, and the like. To account for these issues, philosophy of science had to open to further components in the context of discovery, namely measurement, probability, and statistics.

One of the first to stress the need to pay serious attention to these issues was Ernest Nagel. In *The Structure of Science* (1961), Nagel developed an innovative approach to philosophy of science resulting from a combination of empiricist and pragmatist components.[2] A pupil of Nagel at Columbia from 1947 to 1950, Patrick Suppes imparted a decisive turn on philosophy by promoting a novel approach, which he named *probabilistic empiricism*. In a paper published in 1962 titled "Models of data", followed in 1967 by "What is a scientific theory?", Suppes pioneered the *semantic view* of theories – also supported by authors like Bas van Fraassen, Ronald Giere, Frederick Suppe, (to mention but a few) – which has gradually become predominant. Its key idea is that a theory is linked to empirical phenomena by means of models that can be shown to preserve a certain structure under certain operations. As stressed by Suppes, the relationship between the final, highly abstract formulation of a theory and the experimental level is far from straightforward: "a whole hierarchy of models stands between the model of the basic theory and the complete experimental experience" (Suppes, 1962, 260). A hierarchy of models corresponds to a hierarchy of problems that are typically encountered at different levels of analysis, in the course of the complex procedure of comparing theories with experiments. Such a comparison requires the adoption of statistical techniques including measurement, experimental design, estimation of parameters, tests of goodness of fit, identification of exogenous and endogenous variables, and much more. A detailed analysis of the procedure points not only to a plurality of models, but also to the dependence between models, which is such that in a hierarchy of models characterized by an increasing level of abstraction, one does not move from top to bottom, but *from bottom to top*. This means that given a model of data, which exhibits the statistical structure of a phenomenon under investigation, one looks for a theoretical model that fits it. There is a patent gap between the

---

[2] A similar approach was foreshadowed in Cohen and Nagel ([1934] 1936). For Nagel's contribution to the philosophy of science see Neuber and Tuboly (2022).

received view and the position heralded by Suppes, who maintains that to analyze scientific theories it is not enough to clarify their syntactical structure and give semantical criteria of meaningfulness to theoretical terms, neglecting the methodology for experimentation and theory formation. According to Suppes, this kind of analysis is erroneously missing from the traditional approach taken by philosophers of science "who write about the representation of scientific theories as logical calculi" and "go on to say that a theory is given empirical meaning by providing interpretations of coordinating definitions for some of the primitive or defined terms of the calculus" (Suppes, 1967, 63–64). Equally mistaken is "to try to give a classification of theoretical and empirical terms internal to a theory, at least for developed theories with any diversity of range and application" (Suppes, 1988, 23).[3]

By contrast, Suppes calls attention to *empirical structures*, making them an object of investigation no less important than logical structures. This urges the need to take into account the methodology of experimentation: especially measurement, together with the statistical techniques "for constructing and abstracting empirical structures in all domains of science" (*Ibid.*, 33). Suppes called his own approach *probabilistic empiricism* in the conviction that "it is probabilistic rather than merely logical concepts that provide a rich enough framework to justify both our ordinary ways of thinking about the world and our scientific methods of investigation" (Suppes, 1984, 2). Probabilistic and statistical conceptual tools become essential in order to account for the methodology of experimentation and the nature of explanatory and predictive models.

For Suppes, embracing probabilistic empiricism implies focusing on models rather than laws, putting strong emphasis on context, taking a pluralistic attitude towards methodology, and broadening the spectrum of disciplines of concern to philosophy of science. Putting probability and statistics in the foreground forces the adoption of a pluralistic stance, aware that different disciplines raise different problems. For instance: highly theoretical disciplines such as micro-physics face the problem of interpreting data collected through sophisticated pieces of experimental apparatus; in fields like biology and neuropsychology there is the need to integrate the data obtained by means of different experimental techniques; in medicine and law it is essential to distinguish between general and particular (probabilistic) causation, and particular problems affect all uses of statistical techniques. If the analysis of such aspects of scientific investigation is today inbuilt in philosophy of science, it is worth noting that Suppes' groundbreaking work played a non-secondary role in promoting this kind of attitude. Although Suppes' pluralism dates back to the early 1960s, it is discussed in some detail in a paper delivered at the 1978 PSA meeting titled "The plurality of science" (1981). The author's pluralism marks a decisive step towards a philosophy of science freed from the ideal of the unity of science, aware that "the language of the different branches of science are diverging rather than converging as they become increasingly technical" (Suppes, [1981] 1993a, 44). Unity of science cannot be achieved even on the methodological

---

[3] More on this topic in Suppes (1993b, 2002).

level, for "it is especially the experimental methods of different branches of science that have radically different form. [...] Even within the narrow domain of statistical methods, different disciplines have different statistical approaches to their particular subject matter" (Suppes, [1981] 1993a, 47–48). It is worth noting that Suppes was led to take a pluralistic stance by looking at scientific theories in connection with experimentation; in other words, his pluralism is part of the lesson to be learned from a careful analysis of the role played by statistical methods in connection with experimentation, theory formation and appraisal.

Suppes' contribution to the philosophy of science is framed by his pragmatist perspective, in the wake of Charles Sanders Peirce, William James, John Dewey, and Ernest Nagel, this latter strongly influenced by the pragmatist outlook. In that spirit, Suppes maintains that

> scientific activity is perpetual problem solving. [...] Like our own lives and endeavors, scientific theories are local and are designed to meet a given set of problems. As new problems arise new theories are needed, and in almost all cases the theories used for the old set of problems have not been tested to the full extent feasible nor been confirmed as broadly or as deeply as possible, but the time is ripe for something new, and we move on to something else (Suppes, [1981] 1993a, 54).

Within the framework of Suppes' probabilistic empiricism, context acquires a decisive importance in the sense that many crucial notions are accounted for in a constructive, context-dependent fashion. This holds for notions like evidence, causality, probability, and rationality, all of which cannot be cooped up in a single characterization given once and for all. Pluralism becomes the cornerstone of a "new metaphysics" which ought to supersede the "chimeras" of the received view of philosophy of science: completeness of knowledge, absolute certainty, determinism, and the unity of science (see Suppes, 1984).

7.3   Kinds of Pluralism

Pluralism is presently a widely shared attitude among philosophers of science. Just to mention a few examples: Nancy Cartwright's "dappled world" of disciplines, ranging from physics to economics, ruled by different bodies of laws with varying scopes of application (see Cartwright, 1999); Hasok Chang's "active normative epistemic pluralism", namely "the doctrine advocating the cultivation of multiple systems of practice in any given field of science", where by a "system of practice" the author means "a coherent and interacting set of epistemic activities performed with a view to achieve certain aims" (Chang, 2012, 260); Sandra Mitchell's "integrative pluralism" which deems the plurality of descriptive and explanatory models in contemporary biology a result of the complexity of biological systems, and speaks in favor of integrating them since they are partial accounts of the phenomena under investigation (see Mitchell, 2003). Other versions of pluralism are embraced in connection with explanation (De Regt, 2017), the interpretation of probability (Gillies, 2000), induction (Norton, 2021), and causality (Gillies, 2019; Campaner & Galavotti, 2007, 2012; Galavotti, 2008).

Various aspects of pluralism in connection with several disciplines ranging from physics and biology to mathematics, the social sciences, economics, and others, are considered by the papers collected in the volume *Scientific pluralism* edited by Stephen Kellert, Helen Longino and Kenneth Waters (2006a). In their introduction, the editors (who acknowledge the importance of Suppes' pioneering work) distinguish between "plurality in the sciences" and "pluralism about the sciences", the former being the view that "the multiplicity of approaches that presently characterizes many areas of scientific investigation does not necessarily constitute a deficiency" (Kellert et al., 2006b, x) whereas pluralism

> is a view about this state of affairs: that plurality in science possibly represents an ineliminable character of scientific inquiry and knowledge (about at least some phenomena), that it represents a deficiency in knowledge only from a certain point of view, and that analysis of metascientific concepts (like theory, explanation, evidence) should reflect the possibility that the explanatory and investigative aims of science can be best achieved by sciences that are pluralistic, even in the long run (Kellert et al., 2006b, ix–x).

The book puts forward arguments in favor of pluralism from different perspectives such as perceptual psychology (Ronald Giere), economics (Esther-Mirjam Sent), biology (Kenneth Waters), the literature on the evolution of sex (Carla Fehr), behavioral studies (Helen Longino), quantum physics (Michael Dickson), the foundations of mathematics (Geoffrey Hellmann and John Bell), and more. Towards the end of the introduction, the editors write that

> like physicists trying to answer the most fundamental questions about the physical world, philosophers should acknowledge that there might not be answers to many of the most fundamental questions about science. Might the debate between Bayesians and their foes be futile, not simply because of lack of compelling evidence, but also because neither approach can (in principle) offer a comprehensive account of the basis of scientific inference? (Kellert et al., 2006b, xxvi–xxvii)

7.4   Pluralism in the Foundations of Statistics[4]

Indeed, in recent years a pluralistic attitude to the foundations of statistics has been embraced by a few authors working in the field, albeit not the majority. An advocate of pluralism, Christian Hennig compares two different methods for the statistical quantification of evidence, namely tests of significance, usually associated with the frequency interpretation of probability, and the Bayesian method, usually associated with the subjective interpretation, coming to the conclusion that "the frequentist assumptions about the world outside seem to stand on a more or less equal footing with the Bayesian ones about rational reasoning" (Hennig, 2009, 50). Hennig refers

---

[4]The author thanks Christian Hennig for calling her attention to cluster analysis and its relevance for the topic of this paper.

to the assumption underlying frequentism to the effect that one can produce indefinitely long sequences of repeatable experiments, sufficiently identical and independent, upon which frequencies can be calculated. Furthermore, the probabilities obtained on the basis of such frequencies are taken by frequentists as approximations of the true, unknown probabilities characterizing phenomena. By contrast, the Bayesian approach rests on "the crucial assumption that the individual can always be forced to bet either in favour of or against an outcome, according to her specified betting rates" (Hennig, 2009, 49).[5] For his part, Hennig takes a constructivist viewpoint that does not contend the superiority of one statistical method or interpretation of probability over another. The choice of a particular method is left to the context in which a given problem is addressed, in the awareness that "different approaches have different merits and fulfil different aims" (Hennig, 2009, 51). In other words, statistical methods are "vindicated" in relation to the purpose they are meant to accomplish, and in consideration of the nature of the available data.

Hennig focuses on statistical modeling and identifies the main purpose of statistics with the representation and appraisal of the strength of evidence. He warns against the widespread tendency to "quantify evidence in a unified way regardless of the subject matter" (Hennig, 2009, 44), which, in his view, goes hand in hand with the conviction that the mechanical application of statistical methods to data uncritically taken as "given" can produce "objective" results. By contrast, he recommends a bottom-up analysis that starts from the context in which data are collected and moves on to the formation of models for their representation, and to the methods devised for the quantitative appraisal of evidence. Each step of this process requires a number of assumptions that must be spelled out and justified in view of the aim of inquiry.

A similar attitude informs joint work with Andrew Gelman suggesting that one should discard the traditional opposition between "subjective and objective in statistics" in favor of the view that "multiple perspectives and context dependence are actually basic conditions of scientific inquiry, which should be explicitly acknowledged and taken into account by researchers. We think that this is much more constructive than the simple objective–subjective duality" (Gelman & Hennig, 2017, 993; see also Hennig (2010). In this spirit, Hennig and Gelman propose to go beyond the view that Bayesian analysis is only concerned with subjective beliefs and orthodox statistics has only to do with random samples, the idea being that both methodologies can have a wider application: what is important is "to clarify the foundation for using the mathematical models for a larger class of problems" (Gelman & Hennig, 2017, 993).

A good example of a statistical method that requires assuming a pluralistic, bottom-up attitude is given by cluster analysis. 'Cluster analysis' covers an array of techniques for arranging data into groups that can be used for various purposes

---

[5]While it is questionable that subjective probability and the betting scheme are inextricably entrenched as described by Hennig, there is no doubt that the Bayesian model of rationality faces a few objections, extensively discussed in the literature.

such as the organization of large masses of data for the sake of comparison, as well as prediction and understanding of a phenomenon. Here is a list, borrowed from Henning, of tasks for which clustering is widely used in several areas ranging from biology and medicine to economics, archeology, climate sciences, and many more:

- Exploratory data analysis looking for 'interesting patterns' [. . .] potentially creating new research questions and hypotheses,
- information reduction and structuring of sets of entities from any subject area for simplification, effective communication, or effective access/action such as complexity reduction for further data analysis, or classification systems,
- investigating the correspondence of a clustering in specific data with other groupings or characteristics, either hypothesized or derived from other data (Hennig, 2015, 55).

Statisticians agree that there is no univocal way of defining a cluster, nor any single way of grouping data into clusters: "depending on the application, it may differ a lot what is meant by 'cluster', and cluster definition and methodology have to be adapted to the specific aim of clustering in the application of interest" (Hennig, 2015, 55). There are various methods for extracting clusters from data, and the choice of one technique depends on the nature of data, as well as on the purpose for which clustering is made. In conclusion, cluster analysis is "strongly dependent on contexts, aims and decisions of the researcher" (Hennig, 2015, 61).[6]

7.5   On Probability and Statistics in the Courtrooms

This section addresses some issues concerning the use of probability and statistics in court. The adoption of statistical measures in cases of litigation and/or criminal offence raises problems that make it necessary to adopt an approach that is both transparent, in the sense of highlighting the nature of the data used and the assumptions made, and pluralistic, namely able to combine quantitative and qualitative components.

Aided on the one hand by the plethora of identification techniques available – fingerprints, DNA evidence, marks on bullets, etc. – and on the other, by the increasing number of databases collected over recent decades by the police, commercial firms, epidemiological and medical institutions, probability and statistics have become increasingly important for both litigation and criminal trial. The topic is a matter of hot debate among those involved in forensic science, law, and statistics, and is attracting growing attention also on the part of philosophers of science. Statistical methods in court are typically used for the representation of evidence. As observed by Colin Aitken:

scientific evidence requires considerable care in its interpretation. There are problems concerned with the random variation naturally associated with scientific observations.

---

[6]On cluster analysis see Everitt et al. (2011).

There are problems concerned with the definition of a suitable reference population against which concepts of rarity or commonality may be assessed. There are problems concerned with the choice of a measure of the value of evidence (Aitken, 1995, 4).[7]

In addition to the problems connected with the nature of evidence pointed out by Aitken, the adoption of statistical measures in court clashes with the difficulty of conveying their meaning to those involved.

Elsewhere I have discussed two cases of "miscarriage of justice" prompted by bad use of statistics, namely those of Sally Clark and Lucia de Berk (see Galavotti, 2012); here the focus of attention is the homicide case "*R v T*". In 2010, a ruling of the Court of Appeal for England and Wales invalidated a conviction by rejecting the testimony of an expert who, to assess the probative value of footwear mark evidence, had made use of the likelihood ratio (LR). Biedermann, Taroni and Champod summarize the crucial aspects of *R v T* as follows:

the examining forensic scientist analyzed and compared marks recovered from a murder scene that occurred in 2008. The marks were recovered from the tiles of the bathroom floor and the top surface of the toilet below the bathroom window. The shoes of the principal suspect were also provided for comparison purposes. The footwear mark examiner obtained (inked) test prints under controlled conditions from the submitted shoes. The two sets of elements – material from the crime scene and reference material from the suspect – were found to correspond in terms of general sole pattern, size and level of wear. However, due to the quality of the marks, no fine acquired features were visible. The general pattern of the sole corresponded to a Nike model Multicourt III that, according to the manufacturer's information, was produced and sold between 2000 and 2002. Based on particular aspects of the sole design, that is how the structural elements of the design were spatially positioned to each other, and by reference to soles of various sizes produced in the model series of interest, the examiner was in a position to indicate that marks had been left by a pair of shoes of size EU 47.5 (US 13). It happened that this also was the size of the suspect's shoes. More generally, the marks showed very limited amount of wear and so did the shoes of the suspect (Biedermann et al., 2012, 261).

Relying in part on LR calculations, the expert reached the conclusion that the mark at the crime scene was compatible with the shoes of the suspect. The rejection of the expert's conclusion by the Court of Appeal re-opened the ongoing debate between supporters and opponents of the use of statistics in court. The *R v T* case is discussed at length in several publications, including a special issue of the journal *Law, Probability and Risk* published in 2012, entirely devoted to it. There, a number of authors take their cue from it to address some issues regarding the use of the Bayesian and LR approaches (which to some extent were unduly conflated in the debate). The most debated issues concern the choice of the reference class, the availability and representativeness of data, the choice of alternative hypotheses to be compared by means of the LR, the most convenient way of communicating the LR value (with numerical or qualitative values), and the possibility of misunderstanding statistical measures in general, particularly the value of the LR. The desirability, or rather the need to combine the quantitative values derived from statistical methodology with qualitative considerations is also a matter of debate.

---

[7] See, among others, Dawid (2005) and Taroni et al. (2006).

In general terms, the LR allows comparison of the weight of a given body of evidence with alternative hypotheses: in the case at hand the hypothesis that the mark on the crime scene was left by the defendant and the hypothesis that the mark was left by someone else's shoe. To calculate the LR one must specify the alternative hypotheses to be compared and fix their probabilities. These are delicate operations that depend on choices and assumptions to be justified case by case. Regarding the alternative hypotheses to be compared:

> while the prosecution hypothesis might generally be reasonably clear, on some occasions there will be no obvious 'defense hypothesis' available to the investigators. When those accused offer no explanation or defense, it is incumbent upon the investigators/prosecution to develop an alternative hypothesis that can accommodate innocent possibilities and/or explain the limitations of any hypotheses relied upon (Ligertwood & Edmond, 2012, 366).

For instance, in a homicide case like $R \ v \ T$ a plausible innocence hypothesis would be that the marks left at the crime scene came from a burglar, provided that information supporting this hypothesis was available (see Biedermann et al., 2012, 263).

It should be noted that the expert in the $R \ v \ T$ case arrived at the probability values by means of the LR considering four elements: the sole pattern, the shoe size, the degree of wear and the fact that the defendant's shoe showed none of the damage marks detected in the shoe print found at the crime scene. As observed by William Thompson, one can see from the expert's calculations

> exactly what conclusions he reached about the relevant conditional probabilities and how much weight he assigned to each factor in his overall evaluation. It is clear, e.g., that he judged that *if* the shoe print was made by a shoe other than the defendant's there was a 20% chance it would have the same sole pattern, a 10% chance it would have the same size, and a 50% chance it would have a compatible degree of wear. It is also clear that he assigned relatively little weight to the failure to find in the defendant's shoe the damage marks seen in the shoe print (Thompson, 2012, p. 352).

Without going into the details of the calculations made by the expert, suffice it to say that they led to a LR slightly less than 100. According to a widely used qualitative scale devised for interpreting LR values in court, a likelihood ratio in the range 1–33 is considered "weak", "fair" in the range 33–100, "good" in the range 100–330, "strong" in the range 330–1000, and "very strong" if greater than 1000 (see Robertson &Vignaux, 1995, 12; Evett, 1991). In $R \ v \ T$ the expert's report (tacitly) matched this grading by "saying simply that his analysis found moderate scientific support for the conclusion that the shoe print was made by the defendant's Nike trainer" (Thompson, 2012, 351). Now, while some authors hold that qualitative values like those represented by this scale should be preferred to numerical ones for use in court, others challenge this view on account of the lack of precision of such scales, which they regard as a source of diverging interpretations. For instance, William Bodziak claims that "In general, the verbal equivalent scale is arbitrary and has no clear basis from which it is derived" (Bodziak, 2012, 285).

Another viewpoint maintains that frequency values should be preferred to the LR as a means of conveying the strength of forensic evidence to juries and judges, especially when – as occurred in $R \ v \ T$ – an innocence hypothesis is not suggested by the defense and one has to resort to matching probabilities, namely the probability that a given piece of evidence found at the crime scene is to be ascribed to an individual taken at random from the reference population. Among the supporters of this viewpoint, Andrew Ligertwood and Gary Edmond, write: "we suggest that rather than expressing forensic science evidence as a simple likelihood ratio indicating support for a particular evidential source that it be expressed in a manner – such as a frequency – that accommodates the possibility of innocent explanations" (Ligertwood & Edmond, 2012, 365–366). Their idea is that a frequency value is easier to understand on the part of those involved in a trial, and more generally by lay persons, and "are better able to accommodate explanations for the evidence that are consistent with doubt or innocence" (Ligertwood & Edmond, 2012, 366).

The advisability of presenting statistical values like the LR to jurors and judges nurtures hot debate. While few would question the fruitfulness of the use of statistics by forensic scientists, some hold that it is preferable not to make juries a part of the calculations made by experts, particularly of numerical values. The reason for maintaining that numbers should not be brought to court lies with the risk that they be misunderstood. In the case under consideration, "the justices who wrote R v T [...] confused the likelihood ratio with a posterior probability. They took the expert's statements about the relative probability of the evidence under the relevant propositions to be statements about the probability that the propositions are true. As a result, they mistakenly interpreted the expert's likelihood ratio as an opinion about the probability of the source" (Thompson, 2012, 349–350). As observed by Thompson, such a mistake is a version of the infamous "prosecutor's fallacy", which has plagued many trials in various countries.

In general terms, the prosecutor's fallacy amounts to confusing a probability value with the probability of the defendant's being guilty. A typical case obtains when a match probability p is interpreted as the probability that the defendant is not guilty. The conclusion is then drawn that the probability of guilt is $(1 - p)$. For instance, take a match probability p $(M \mid -G) = 1/10,000,000$, where '$M$' stands for a trace found on the murder scene, and '$-G$' stands for the assertion that the defendant is not responsible for it (namely the trace was left by an individual chosen randomly from the reference population). The fallacy originates by confusing that probability with p $(-G \mid M)$, namely the probability that the defendant is not guilty, given the piece of evidence found at the murder scene, and concluding that the probability of the defendant being guilty is 1–1/10,000,000. By so doing, the confusion between the probability that a certain trace was left by an unknown individual randomly chosen from the population and not by the defendant, and the probability that the defendant is not guilty, given the piece of evidence found at the murder scene, results in a very high probability of the defendant being guilty.[8]

Another thorny issue regards the size and representativeness of the database used to determine base rates. Indeed, the choice of the appropriate reference class is a problem affecting all applications of statistical methods. Ideally, a suitable reference

---

[8] A few examples can be found in Dawid (2008). See also Galavotti (2012).

class for the calculation of base rates should take into consideration all relevant variables and rely on carefully collected data. Fulfilment of such a requirement obviously depends on the availability of sufficient and reliable databases and raises a problem that can only be addressed in a context-sensitive fashion. In *R v T*, the court "was justifiably concerned about the absence of a sufficient database" (Thompson, 2012, 353) supporting the expert's report. Indeed, databases stored by the police (usually used in criminal cases) are often insufficient, or in some respects unreliable, as are those collected by commercial firms or various institutions.

Prior to the size and accuracy of the database, the choice of variables guiding the selection of the relevant database is also problematic. In their long and detailed discussion of *R v T*, Biedermann, Taroni and Champod call attention to a wealth of considerations that bear relevance to this problem. For instance, they observe that one might ask whether there is reason to observe the shoe pattern in question "more often among people who commit the kind of crime of interest (in the relevant locality) or among innocent suspects" (Biedermann et al., 2012, 269). In addition, one might think there is reason to believe that sport shoes would be preferred by true offenders because they are more suitable for leaving the crime scene swiftly, even if sport shoes may obviously be chosen simply for their look, or because they are trendy; "one could consider athletic shoes so commonly worn by people in the age range of burglars that the suggested difference [between offenders and innocent people] is negligible" (Biedermann et al., 2012, 269).

Strong supporters of the use of statistics in court, Biedermann, Taroni and Champod maintain that "the fact that arguments may point in different directions is not a problem in principle. [...] this does not represent any hindrance for the formal framework of probability. On the contrary, the very important point is that probability is a flexible template that can accommodate assignments motivated by a broad scope of arguments" (Biedermann et al., 2012, 269). That said, towards the end of their commentary of *R v T*, they speak in favor of a pluralistic attitude according to which the LR formulae "should not be approached in a purely 'mechanical' perspective, that is [...] on a numerical level, to the exclusion of other modes of analysis" (Biedermann et al., 2012, 273). Furthermore, "even if numerical assignments can actually be made, general and qualitative considerations may still be valuable and beneficial for consideration because they can help to place evidence evaluation appropriately in context. That context pertains to the understanding of a scientist's evaluation among recipients of information. It is verbal and argumentative in nature. Numbers may be a part of that, but not a necessary one" (Biedermann et al., 2012, 274).

A similarly pluralistic position is embraced by Peter Tillers in "Trial by Mathematics – Reconsidered" (2011), meant as a contribution to the ongoing debate for and against the use of statistics in courtrooms dating back to the early 1970s. In an influential paper published in 1971 titled "Trial by Mathematics", Laurence Tribe strongly objected to the use of probability and statistics in court, holding that it undermines the moral nature of the two pillars of criminal trial, namely the presumption of innocence and the BARD standard. Tribe's attack prompted a broad debate between supporters and opponents of statistical methods, with special emphasis on the Bayesian method. Whereas Tribe is especially concerned that the hypothesis of guilt should not be expressed in terms of a probability value calculated by means of Bayes' rule, in the debate that followed many authors, including Dennis Lindley, David Kaye, Richard Lempert and others, defended the adoption of the Bayesian method in court as a heuristic device aimed at helping the parties involved in a trial to interpret evidence. As emphasized by Lempert, the Bayesian method can promote understanding of a cluster of issues related to relevance, such as "the meaning of logical relevance" and "the principle that only relevant evidence is admissible" (Lempert, 1977, 1031). This is also the attitude prevailing among the discussants of the *R v T* mentioned in this section.

After a broad summary of the debate, Tillers argues that the opposition to Bayesian – and formal methods in general – is largely due to a number of mis-understandings, like

> to suppose that storytelling is inconsistent with Bayesian or mathematical analysis of evidence with cardinal numbers; [. . .] to suppose that Bayesian analysis is equivalent to objective or statistical analysis; [. . .] to suppose that formal or mathematical analysis is necessarily 'mechanical' or that formal mathematical analysis necessarily amounts to an 'algorithm'; [. . .] to suppose that the debate over trial by mathematics was really only a debate about the uses of mathematics in or about fact finding in the legal process rather than (as Tribe himself recognized) a debate about the broader question of uses and limits of formal argument about evidence, factual inference and factual proof in legal proceedings (Tillers, 2011, 170).

To counteract this sort of thinking, Tillers advocates a flexible, pluralistic attitude, according to which there is no reason why the Bayesian method should not be used in connection with qualitative considerations, because it "is in fact the case that most complex argument about inferences from evidence rests on almost innumerable personal or subjective judgments" (Tillers, 2011, 171). For instance, according to Tillers' perspective, Bayesian analysis would not be inconsistent with the "story model for jurors decision making" proposed by Nancy Pennington and Reid Hastie as a model of "the cognitive strategies that individual jurors use to process trial information in order to make a decision prior to deliberation" (Pennington & Hastie, 1993, 192). In fact, most upholders of the usefulness of statistics do not recommend it as a means for taking decisions in court, but rather as a tool for helping decision makers by conveying information on the weight of evidence. From this standpoint, it seems feasible to combine it with accounts of facts aimed to suggest alternative explanations.

Increasing attention is being paid to Bayesian networks as an effective tool for combining information of different kinds. As emphasized by Franco Taroni, Alex Biedermann, Silvia Bozza, Paolo Garbolino, and Colin Aitken in *Bayesian Networks for Probabilistic Inference and Decision Analysis in Forensic Science*, a Bayesian network – which makes extended use of the LR – can provide a model of a case under study that includes both quantitative and qualitative variables, which might, for instance, represent alternative hypotheses put forward by the prosecution and the defense. Although going into the details of this method would fall beyond the scope of this paper, it is worth mentioning that the authors distinguish three kinds of propositions (describing the hypotheses considered in order to model a given situation), forming a three-level hierarchy: (1) the source level, (2) the activity level, and (3) the crime level. At each level of this hierarchy different kinds of information are taken into account, such as evidential data but also extra scientific elements, which depend on a number of delicate choices and evaluations. While at the source level only information obtained from careful examination of material recovered from the crime scene and control samples is considered, at the activity level "evaluation cannot meaningfully be conducted without a framework of circumstances [....] This will require the expert to examine possible scenarios of the case" (Taroni et al., 2014, 95); finally, the crime level typically involves extra scientific considerations such as "whether or not a crime occurred or whether or not an eyewitness is reliable" (Taroni et al., 2014, 97). According to its supporters, by embedding probabilities in graphical representations Bayesian networks allow a better understanding of the situation under scrutiny by clarifying not only the relationships among different evidential bodies, but also the underlying assumptions made by the experts.[9]

The preceding discussion of *R v T* was meant to show the broad spectrum of problems arising with the application of statistical methods in legal processes. After a few decades of bitter debate for and against the use of statistics in courtrooms, its supporters have come to a flexible and pluralistic position according to which statistical methodology should not be applied mechanically, but rather combined with qualitative considerations, and placed in context.

7.6   Concluding Remarks

A most prominent aspect of the shift from the way of doing philosophy of science typical of logical empiricism and the way it is done today amounts to adopting a wide-ranging pluralism. A pioneer of the pluralistic approach, Suppes – who labeled his own perspective probabilistic empiricism – strongly emphasized the need for the philosophy of science to focus on the role played by probabilistic and statistical tools in the methodology of experimentation, model formation, explanation, and prediction. As observed in Sect. 7.2, Suppes was led to assume a pluralistic stance by looking at scientific knowledge in connection with probabilistic and statistical methodologies. Together with pluralism, a careful analysis of the role played by statistical methods in a variety of contexts suggests assuming a bottom-up perspective which starts from data and moves on to model construction, hypotheses formation, testing, predictive inference, etc. A fundamental feature of the bottom-up approach – as described by Suppes and endorsed in this paper – is the recommendation that all assumptions made at every step of such a process be sorted out and

---

[9]For the application of Bayesian networks to footwear marks evidence, see Taroni et al. (2014), 172–176.

justified based on the nature of the data available and in view of the purposes of the research. Cluster analysis, on the one hand, and statistical methodology for the representation and appraisal of evidence, on the other, exemplify the need to adopt a pluralistic and bottom-up attitude.

Acknowledgements The author is grateful to Paolo Garbolino, Donald Gillies, and Christian Hennig for comments and suggestions on an earlier version of this paper.

References

Aitken, C. (1995). *Statistics and the evaluation of evidence for forensic scientists.* Wiley.

Biedermann, A., Taroni, F., & Champod, C. (2012). How to assign a likelihood ratio in a footwear mark case: An analysis and discussion in the light of *R v T. Law, Probability and Risk, 11*(4), 259–277.

Bodziak, W. (2012). Traditional conclusions in footwear examinations versus the use of the Bayesian approach and likelihood ratio: A review of a recent UK appellate court decision. *Law, Probability, and Risk, 11*(4), 279–287.

Campaner, R., & Galavotti, M. C. (2007). Plurality in causality. In P. Machamer & G. Wolters (Eds.), *Thinking about causes. From Greek philosophy to modern physics* (pp. 178–199). University of Pittsburgh Press.

Campaner, R., & Galavotti, M. C. (2012). Evidence and the assessment of causal relations in the health sciences. *International Studies in the Philosophy of Science, 26*(1), 27–45.

Cartwright, N. (1999). *The dappled world. A study of the boundaries of science.* Cambridge University Press.

Chang, H. (2012). *Is water H2O? Evidence, Realism and Pluralism.* Springer.

Cohen, M. R., & Nagel, E. ([1934] 1936). *An introduction to logic and scientific method.* Harcourt, Brace and World, 1936[2].

Dawid, P. A. (2005). Probability and statistics in court, appendix online to the second edition of T. Anderson, D. Schum and W. Twining, *Analysis of evidence.* Cambridge University Press. http://tinyurl.com/7q3bd

Dawid, P. A. (2008). Statistics and the law. In A. Bell, J. Swenson-Wright, & K. Tybjerg (Eds.), *Evidence* (pp. 119–148). Cambridge University Press.

de Regt, H. (2017). *Understanding scientific understanding.* Oxford University Press.

Everitt, B. S., Landau, S., Leese, M., & Stahl, D. (2011). *Cluster analysis* (5th ed.). Wiley.

Evett, I. W. (1991). Interpretation: A personal odyssey. In C. G. G. Aitken & D. A. Stoney (Eds.), *The use of statistics in forensic science* (pp. 9–22). Ellis Horwood.

Galavotti, M. C. (2008). Causal pluralism and context. In M. C. Galavotti, R. Scazzieri, & P. Suppes (Eds.), *Reasoning, rationality and probability* (pp. 233–252). CSLI Publications.

Galavotti, M. C. (2012). Probability, statistics, and law. In D. Dieks, W. J. Gonzalez, S. Hartmann, M. Stoeltzner, & M. Weber (Eds.), *Probability, laws, and structures* (pp. 401–412). Springer.

Gelman, A., & Hennig, C. (2017). Beyond objective and subjective in statistics. *Journal of the Royal Statistical Society A, 180*(4), 967–1033.

Gillies, D. (2000). *Philosophical theories of probability.* Routledge.

Gillies, D. (2019). *Causality, probability, and medicine.* Routledge.

Hempel, C. G. ([1952] 1969). *Fundamentals of concept formation in empirical science* (2nd ed.). University of Chicago Press.

Hennig, C. (2009). A constructivist view of the statistical quantification of evidence. *Constructivist Foundations, 5*(1), 39–54.

Hennig, C. (2010). Mathematical models and reality: A constructivist perspective. *Foundations of Science, 15*(1), 29–48.

Hennig, C. (2015). What are the true clusters? *Pattern Recognition Letters, 64*, 53–62.

Kellert, S., Longino, H., & Waters, K. (Eds.). (2006a). *Scientific pluralism. Minnesota studies in the philosophy of science* (Vol. XIX). University of Minnesota Press.

Kellert, S. H., Longino, H. E., & Waters, C. K. (2006b). The pluralist stance. In S. H. Kellert, H. E. Longino, & C. K. Waters (Eds.), *Scientific pluralism, XIX Minnesota studies in the philosophy of science* (pp. vii–xxix). University of Minnesota Press.

Lempert, R. (1977). Modeling Relevance. *Michigan Law Review, 75*(5/6), 1021–1057.

Ligertwood, A., & Edmond, G. (2012). Discussion paper: A just measure of probability. *Law, Probability and Risk, 11*(4), 365–369.

Mitchell, S. (2003). *Biological complexity and integrative pluralism*. Cambridge University Press.

Nagel, E. (1961). *The structure of science*. Harcourt, Brace & World.

Neuber, M., & Tuboly, A. T. (Eds.). (2022). *Ernest Nagel: Philosophy of science and the fight for clarity*. Springer.

Norton, J. D. (2021). *The material theory of induction*. University of Calgary Press. http://hdl.handle.net/1880/114133

Pennington, N., & Hastie, R. (1993). The story model for juror decision making. In R. Hastie (Ed.), *Inside the juror* (pp. 193–221). Cambridge University Press.

Popper, K. R. (1959). *The logic of scientific discovery*. Hutchinson. English enlarged edition of *Logik der Forschung*. Springer, 1934.

Popper, K. R. ([1963] 1968). *Conjectures and Refutations*. Harper Torchbooks, 1968[3].

Reichenbach, H. ([1938] 1966). *Experience and prediction*. Chicago University Press, 1966[6].

Robertson, B., & Vignaux, G. A. (1995). *Interpreting evidence*. Wiley.

Stadler, F. ([2001] 2015). *The Vienna Circle. Studies in the origins, development, and influence of logical empiricism*. Springer. 1st ed., Springer, 2001.

Suppes, P. (1962). Models of data. In E. Nagel, P. Suppes, & A. Tarski (Eds.), *Logic, methodology and philosophy of science* (pp. 252–261). Stanford University Press.

Suppes, P. (1967). What is a scientific theory? In S. Morgenbesser (Ed.), *Philosophy of science today* (pp. 55–67). Basic Books.

Suppes, P. (1984). *Probabilistic metaphysics*. Blackwell.

Suppes, P. (1988). Empirical structures. In E. Scheibe (Ed.), *The role of experience in science* (pp. 23–33). De Gruyter.

Suppes, P. ([1981] 1993a). The plurality of science. In P. Asquith, & I. Hacking (Eds.), *PSA 1978*. East Lansing: Philosophy of science association (Vol. 2, pp. 3–16). Reprinted in Suppes, P. (1993). *Models and methods in the philosophy of science: Selected* essays (pp. 41–54). Kluwer.

Suppes, P. (1993b). *Models and methods in the philosophy of science: Selected essays*. Kluwer.

Suppes, P. (2002). *Representation and invariance of scientific structures*. CSLI Publications.

Taroni, F., Aitken, C., Garbolino, P., & Biedermann, A. (2006). *Bayesian networks and probabilistic inference in forensic science*. Wiley.

Taroni, F., Biedermann, A., Bozza, S., Garbolino, P., & Aitken, C. (2014). *Bayesian networks for probabilistic inference and decision analysis in forensic science*. Wiley.

Thompson, W. (2012). Discussion paper: Hard cases make bad law – Reactions to R v T. *Law, Probability and Risk, 11*(4), 347–359.

Tillers, P. (2011). Trial by mathematics – Reconsidered. *Law, Probability and Risk, 10*(3), 167–173.

Tribe, L. (1971). Trial by mathematics: Precision and ritual in the legal process. *Harvard Law Review, 84*(6), 1329–1393.

Chapter 8
Instrumental Realism – A New Start for the Philosophy of Mathematics and the Philosophy of Science

Ladislav Kvasz

Abstract Instrumental realism is a type of scientific realism which emphasizes the importance of scientific instruments for the acquisition of scientific knowledge. According to instrumental realism, we have epistemic access to reality but this access is often indirect, mediated by means of instruments. The development of science is accompanied by the introduction of new instruments which open up access to phenomena to which we had formerly lacked epistemic access. The knowledge gained by means of the new instruments usually results in the refinement of our theories. Nevertheless, sometimes the new experimental results contradict our theories and force us to engage in their correction. The results obtained by means of particular instruments are integrated, or synthesized, into a unified instrumental, linguistic and theoretical practice. We can distinguish four kinds of such synthesis – instrumental, relational, compositional and deductive. All four kinds of synthesis require idealization, which in instrumental realism is understood to be a linguistic reduction.

Keywords Instrumental realism · Empiricism · Wittgenstein · Pictorial form

The term *instrumental realism* was introduced in 1991 by Don Ihde to describe a philosophical position towards which a group of five American philosophers of science allegedly converged: Robert Ackermann (1985), Hubert Dreyfus (1972), Ian Hacking (1983), Don Ihde (1991) and Patrick Heelan (1983). They were united by an interest in the instrumental dimension of science and held a realistic position in the philosophy of science. The alleged convergence seems to have been wishful thinking on Ihde's part rather than an actual fact. The above authors are so different in their focus and tradition – Heelan was developing a phenomenological approach to science, Dreyfus worked in the philosophy of artificial intelligence, Ihde in the philosophy of technology, with Hacking and Ackermann in the philosophy of science – that despite undeniable common features, in the end the differences prevailed and a joint project did not emerge. Nobody other than Ihde is known to

L. Kvasz (✉)
Charles University of Prague, Prague, and Czech Academy of Sciences, Prague, Czechia

© The Author(s), under exclusive license to Springer Nature Switzerland AG 2022 165
W. J. Gonzalez (ed.), *Current Trends in Philosophy of Science*, Synthese Library 462, https://doi.org/10.1007/978-3-031-01315-7_8

subscribe to his instrumental realism so this term, having not been adopted, can be considered free again. One of the reasons for the failure of Ihde's attempt was the lack of radicalism in formulating the principles of the new position and the insufficient clarification of the role of instruments in the production of scientific knowledge. Ihde was aware of the differences in the approaches of the above authors: thus, instead of trying to sharpen the principles of instrumental realism under the threat of possible disagreement from these four colleagues, he began to soften them. Speaking for a group of five distinct individuals, he blurred and obscured the essence of instrumental realism in tailoring it to fit each one of them. In the end, what he created failed to attract anyone. It is a pity, because the idea that instruments play a key role in science opens up a perspective from which many problems in the philosophy of science can be solved.

The aim of this paper is to focus on elucidating the epistemological role of instruments. I will seek to consider the concept of instrumental practice as broadly as possible, so that in addition to scientific instruments such as telescopes and barometers, I will include geometric constructions using compasses and rulers, and tools of symbolic representation such as the decimal positional system in arithmetic or matrix symbolism in linear algebra (see Kvasz, 2019). The use of a compass and ruler in geometry shall be held as epistemologically analogous to the use of a telescope in astronomy. The systematic juxtaposition of the instrumental practice of mathematics with the instrumental practice of science will enable the revelation of a number of assumptions implicit in the use of instruments which, due to focusing only on the instrumental practice of science, have remained hidden from Ihde. I believe that interest in instrumental realism can be revived by giving it a broader scope, including both mathematics and the empirical sciences, and granting it thus a greater degree of coherence.

Instrumental realism is a realistic position according to which we acquire knowledge about a transcendent world which is largely independent of us whilst we remain part of the world about which we acquire such knowledge. Our presence as a cognitive subject in such a world has far-reaching consequences. These consequences range from contingent features, such as the relationship between ten as the basis of the positional system in arithmetic and the fact that we have ten fingers, or between the size of units such as a meter and the approximate length of our step, to much deeper aspects, for instance the relationship between the syntactic and semantic structure of the language of scientific theories and the syntactic and semantic structure of natural language. One such deeper aspect stems from the fact that scientific instruments are designed as aids for the improvement (i.e., the expansion and refinement) of sensory perception, and as such they bear traces of the sensitivity and scope of perception of our sensory organs. Gradually, however, the instrumental practice of science has considerably expanded, and it is therefore useful to distinguish four levels at which scientific instruments can influence the process of knowledge acquisition. These levels can be tentatively referred to as: (1) epistemic contact; (2) the stabilization of epistemic contact; (3) the extension of epistemic contact; and (4) the idealization of epistemic contact.

8.1   Epistemic Contact with Reality

Classical empiricism was born in the seventeenth century, when it was natural to identify epistemic contact with sensory experience. However, today, when scientific knowledge relies on a wide range of diverse scientific instruments, insisting on the interpretation of sensory experience as the only source of adequate knowledge is limiting because it hinders a true understanding of scientific practice. Therefore, I propose that we distinguish *epistemic contact* from sensory experience. In the context of everyday life, of course, sensory experience remains the main source of epistemic contact. However, in scientific practice our senses are systematically supplemented and enhanced by dozens of scientific instruments such as the telescope, microscope, barometer, thermometer, camera, voltmeter, spectroscope, X-ray, radar, sonograph, tomograph, cloud chamber and many more. In science, instrumental practice creates epistemic contact with a substantially wider range of phenomena than our sensory perception. One of the main reasons I propose the concept of *epistemic contact* is to allow for the full breadth, accuracy and subtlety of instrumental practice in science.

The second reason is that instrumental realism seeks to transfer Frege's arguments against psychologism from logic to epistemology.[1] According to Frege, although logical thinking is an activity of the human mind, the subject matter of logic is not the laws of human thinking but the *objective logical relationships between propositions*. Analogous with this, the subject matter of epistemology should not be the actual knowledge acquisition as the performance of an epistemic subject (nor how this subject as a rational being should acquire knowledge) but *objective epistemological relations between theories and reality*. The basic element that establishes such a relationship is epistemic contact. In our view, epistemic contact was that component of sensory experience for the sake of which classical empiricists considered sensory experience to be the basis of knowledge. In sensory experience we have direct and immediate epistemic contact with reality. Instrumental realism accepts this fact and adheres to the principle that epistemic contact is the basis and starting point of all knowledge. Therefore, instrumental realism is a kind of empiricism.

However, unlike classical empiricism, instrumental realism does not limit epistemic contact to sensory experience. It does not interpret epistemic contact as *a subjective event in the consciousness of a knowing subject*, as the empiricists understand it. Instrumental realism tries instead to interpret epistemic contact as *an objective event in the physical world*, of which epistemic instruments constitute an important element. The purpose of the transition from interpreting epistemic contact as sensory experience to interpreting it as a physical event (or more precisely, to a combination of a physical event and a mental experience) is to put on a par, as far as

---

[1]Frege was one of the founders of formal logic. I believe that an analogous formalization is taking place now in epistemology. The rejection of psychologism is thus an important component of formal epistemology.

possible, instrumental and sensory contact.[2] Of course, in the end someone has to read the position of the pointer of the instrument. However, what is important from the viewpoint of instrumental realism is that epistemologically significant aspects of the epistemic contact become incorporated into the physical event, as a result of which the reading of the pointer itself becomes trivial, often to the extent that machines are able to do it. It cannot be completely trivial – there will always be positions that are ambiguous; background noise, thermal fluctuations and material fatigue can never be completely suppressed. However, under normal circumstances, the act of measurement consists of two parts: an objective event in the outside world and a subjective reading of its outcome.

From this point of view, sensory experience is a special case in which the role of a measuring device is fulfilled by our bodily organs and the reading is done using nerve receptors. Thus, instrumental realism is not a rejection of empiricism. On the contrary, its goal is to expand and generalize empiricism. The first step on this path is to replace sensory experience with epistemic contact as the basis of knowledge. Sensory experience is the most common kind of epistemic contact, but it is only one kind. According to instrumental realism, instrumental contact is another kind of epistemic contact that is, in many respects, equivalent to sensory experience.

Thus, *the first role of scientific instruments is to create an epistemic contact with reality*. As an example, we can take the telescope that enabled Galileo to discover the moons of Jupiter and to determine those periods of their orbit that he wanted to use in navigation; to discover mountains on the Moon and to measure their altitude; to discover spots on the Sun and to determine from their motion the speed of rotation of the Sun. These phenomena are below the threshold of visual resolution in normal observation, so the telescope is needed to establish epistemic contact with them. We can say that the telescope establishes a fully-fledged epistemic contact with these phenomena, because it allowed in the case of each of them the recognition of a certain aspect of the phenomenon (periods of orbit, altitude, speed of rotation), which had hitherto been unknown (Ihde, 2011). Thus, the telescope not only revealed the particular phenomenon, but it rendered it open to further investigation.

The fact that in instrumental realism epistemic contact is considered to be an event in the physical world and instrumental contact is considered, as an authentic form of epistemic contact, has direct consequence for the understanding of the objectivity of knowledge. On the one hand, an event in the physical world can be observed by several people, so that instrumental contact is automatically

---

[2] In the case of sensory experience, sensory organs such as the eye, ear or tactile receptors play an important role. These organs can be understood as "protein instruments", and so instrumental realism can be considered a generalization of classical empiricism, in which, in addition to the protein instruments with which evolution has equipped us, we also use instruments composed of metal, glass and other materials, instruments that were created in the process of technical development. It is important to realize that the act of epistemic contact is always composed of a *bodily component* (formed by an event in the physical world) and a *cognitive component* (formed by an event in the mind of the knowing subject). In science, more and more aspects of epistemic contact are gradually being shifted from the cognitive component to the bodily component.

intersubjective. Moreover, instrumental contact is not as prone to illusions as sensory knowledge. If, during instrumental contact, problems occur such as those caused by the chromatic error in Galileo's telescope lenses, it is possible to *change the principle* upon which the instrument is built. Newton using a parabolic mirror instead of one of the lenses reduced considerably the chromatic error. Similarly we may *technically improve* some part of the instrument, as for instance making the lens not from a single piece of glass but laminating several kinds of glass to compensate for the chromatic aberrations. Or we may overcome the problem of chromatic error by means of *technological progress*, such as Carl Zeiss by discovering that molten glass could be stirred as it solidified, and such homogenizing significantly reduces the chromatic error. However, instead of improving a given instrument, it is possible to add another instrument to eliminate the errors. Such an additional instrument does not create a new epistemic contact, but only improves the functioning of the instrument that already establishes such a contact. I have in mind color filters that eliminate chromatic aberrations. This brings us to the stabilization of the epistemic contact.

8.2 Stabilization of the Epistemic Contact

In instrumental realism, we understand epistemic contact as an instrumentally constituted event in the physical world. However, this contact is often unstable and so there is a need to stabilize it. There are many ways to stabilize epistemic contact with reality. As an example, we can take how Galileo's use of musical rhythm to stabilize the length of time intervals in his inclined plane experiments. In 1604 he developed a method for measuring the velocity of the accelerating motion of a ball on an inclined plane (with an inclination of less than $2°$) by recording the positions of the moving body at multiples of equal time intervals of about half a second. Galileo measured these short time intervals using a musical instrument, because when listening to music, we are sensitive to slight deviations from the rhythm and thus we can compare relatively accurately short time periods that are a multiple of the tact (see Drake, 1975). The use of a musical instrument in measuring the velocity of accelerating motion is thus an example of the stabilization of epistemic contact – listening to a musical instrument in an experiment allows the experimenter to more accurately determine the length of time intervals and thus stabilizes epistemic contact with the velocity of the motion. The ***stabilization of epistemic contact is thus the second role of instruments in the process of knowledge acquisition***.

Scientific instruments make it possible to *adjust the conditions of observation* so that the aspects of interest stand out better, such as Galileo's use of an inclined plane to study motion, thereby reducing the gravitational acceleration to such an extent that the increase in velocity became accessible to measurement; *eliminate interfering influences* that make it difficult to recognize the phenomenon under investigation (for example, the air pump allows one to rarefy the air and thus reduce the resistance of the medium); and *isolate the investigated phenomenon* from nearby phenomena with which it is often confused. In order to better understand the role of instruments in the stabilization of epistemic contact, we present three examples of such stabilization – the Renaissance perspective, Copernican astronomy and Hamiltonian mechanics.

8.2.1   The Construction of a Perspectivist Picture

In Gothic painting the painters sought to depict buildings, trees and landscape from the point of view from which they were most typical, so that the viewer could easily recognize them. Therefore, the individual elements in the painting were often represented from different perspectives. The Renaissance, on the other hand, brought an effort to display the whole picture, including all its details, from a single, fixed, motionless point of view. The painter's eye no longer wandered around in the country but had the discipline to observe the world from a single, fixed point. Albrecht Dürer illustrated a trick by means of which it is possible to fix the point of view in this way (see Fig. 8.1).

If we push a nail into the wall, attach a rope to it, tie a thin tube to the end of the rope, stretch the rope and look through the tube like the painter on Dürer's engraving, all the rays of light passing through the tube will meet at the point where the nail is pushed into the wall. Thus, the point of view from which the painting in (Fig. 8.1) is painted is not identical with the painter's eye but is the point at which the nail is attached to the wall. It is the point from which the perspective of the painting is constructed.

In the case of Gothic painters, as well as in the case of the Renaissance painter captured on Dürer's engraving, the epistemic contact has the form of immediate visual perception – the painter paints what he sees. However, in the case of the Renaissance painter this immediate epistemic contact is stabilized, the Renaissance

Fig. 8.1 Albrecht Dürer (1471–1528), a picture from the book *Underweysung der messung mit dem zirckel vn richt scheyt* (1525)

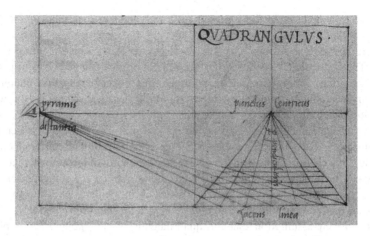

Fig. 8.2 Leon Battista Alberti (1404–1472), from the book *De Pictura* (1435)

painter is constructing his painting so that all its parts are related to a single, fixed point so that they are depicted as if they were seen by an observer from that point. As a result, a painting using such a stabilization of its epistemic contact carries much more relevant information than a spontaneous representation created without such stabilization. In a Renaissance painting we can determine the distance, relative size and mutual orientation of the depicted elements.[3] From a Gothic painting depicting the same elements, we can estimate the distances, sizes and orientation with much less certainty.

In order to achieve a precise representation of the spatial coordination of various elements of architecture, Renaissance painters drew on their canvas tiles running into the depth of space. They placed afterwards the individual figures and objects in this *"pavimento"* so as to capture their mutual spatial relationships. The rule for the apparent reduction of tiles as they run into the depths of space was discovered by Leon Battista Alberti and called *constructioni legitima* (see Fig. 8.2). We can perceive the apparent reduction in the sizes of the tiles. However, the exact rule of

---

[3] We are faced with the dilemma of whether these additional aspects, which appear in the painting due to the stabilization of epistemic contact, should be attributed to the contact itself, and thus to claim that we see them immediately (and insist that was the case for centuries before the discovery of perspective, only no one noticed), or we should consider them as phenomena that are constituted only by the stabilization of epistemic contact (and thus did not exist before the Renaissance, so there was nothing to notice). In the case of perspective, we tend to prefer the first interpretation and to claim that, for example, the parallels running into the depths of space converged even before the Renaissance, only no one noticed. However, with more sophisticated ways of stabilizing epistemic contact, this is no longer so clear-cut. Was the elliptical shape of the orbits of the planets of the solar system a fact for pre-Copernican astronomy as well, but no one noticed it before Kepler, or did the elliptical shape of the orbits arise only within the heliocentric system? Instrumental realism adheres to scientific realism, and thus considers the elliptical shape of orbits to be an objective feature of reality, to which, however, we have gained an epistemic access only by means of Copernican astronomy.

this reduction appears in the representation only due to the stabilization of the epistemic contact.

There are several reasons why I have discussed Renaissance painting in such detail. The first is that the concept of space did not originate in geometry or physics, but in painting. From Renaissance painting, this concept was first adopted by (projective) geometry and then by (Newtonian) physics. The second reason for my excursion into Renaissance painting is that it nicely illustrates a number of formal aspects that the stabilization of epistemic contact introduces into a representation. Similar aspects can be found in other examples of stabilization of epistemic contact. These are: *the point of view*, which in Dürer takes the form of a point where a rope is attached to the wall, whereas in Alberti it has the form of an eye that is presented on the left edge of the sketch; *the horizon*, which in the Alberti takes the form of a horizontal line passing through the *punctus centricus*; and *the relation of identity*, i.e. the implicit assumption that the quadrilaterals of the *"pavimento"* represent identical tiles. The third reason is that in Renaissance paintings, when compared with their Gothic equivalents, the epistemic gain clearly stands out, i.e. the amount of detailed information that the stabilization of epistemic contact brings is obvious. These features – the emergence of new concepts (space), formal aspects (point of view, horizon, relation of identity) and epistemic gain – are a characteristic feature of the stabilization of epistemic contact.[4]

8.2.2 Construction of the Heliocentric Model of the Solar System

The planets (especially Venus) have been observed by humans since ancient times. Copernicus came up with the idea that the motion of the planets we observe in the sky is a composite motion, reflecting both the movement of the observer and the movement of the planet. Therefore, he proposed *stabilizing the epistemic contact* (which since ancient times has been observation with the naked eye) by relating all positions of the moving planet to the motionless Sun. It was not as simple as it was for Dürer, for whom it was enough to push a nail into the wall and relate the representation of individual elements of the painting to the position of this nail. Copernicus could not drive a nail into the Sun and use it to construct an image of the planetary system. However, in essence he did something very similar. Epistemic contact with the planetary positions was stabilized by relating all the individual observations obtained at different moments in time to the same, immobile point.

---

[4] The concept of *space* can be added to the *point of view*, the *horizon* and the *relation of identity* as a fourth aspect, accompanying the stabilization of epistemic contact. In the *History of Geometry and the Development of the Form of its Language* (Kvasz, 1998) I relate these aspects to Wittgenstein's notion of the pictorial form from the *Tractatus Logico-Philosophicus* (Wittgenstein, 1921). If this interpretation is correct, Wittgenstein's notion of the *pictorial form* is actually a description of the *stabilization of epistemic contact*.

As in the case of perspective, in astronomy the stabilization of the epistemic contact creates a representation that potentially contains much more detailed information than the original geocentric model. What was discovered as the fruit of Copernicus' stabilization was the elliptical shape of the orbits of the planets, the constant sectorial velocity of their motion, and the relationship $T^2 \sim a^3$ between the square of the orbital period T and the cube of half the orbital diameter a. Kepler's three laws of planetary motion are empirical laws; they were derived from observations with the naked eye, Kepler having used Tycho de Brahe's observational data. However, the influence of the observer's movement had to be deducted from that data. The elliptical shape of the trajectories as well as the invariance of the sectorial velocity appear only due to Kepler's relating all partial observations to the same fixed point, i.e., only after the stabilization of the epistemic contact. Without such stabilization, the data are too fragmented to see anything like Kepler's three laws.

Thus, the stabilization of epistemic contact, consisting in increasing the degree of correlation between the various acts of epistemic contact, makes it possible, as in the case of perspective, to reveal a large number of new details and connections in the empirical data. The conceptual gain of this stabilization is also undeniable, as evidenced by the concept of sectorial velocity.[5] The question of the legitimacy of declaring Kepler's three laws to be empirical facts (or phenomena, as they were called by Newton), even if they cannot be discovered by direct observation, is something we need not herein decide. Instrumental realism understands experience quite broadly and considers the knowledge that arises as a result of the stabilization of epistemic contact to be authentic empirical knowledge.

8.2.3 Hamiltonian Mechanics

In order not to give the impression that the stabilization of epistemic contact occurred only in the sixteenth century, when both Dürer (1471–1528) and Copernicus (1473–1543) lived, we present the stabilization of epistemic contact, upon which Hamiltonian mechanics is based. The epistemic contact on the basis of which Hamilton created a representation of a mechanical system is the same as in Newtonian or Lagrangian mechanics, i.e. the measurement of the positions, velocities and accelerations of individual bodies. However, when we look at how Newtonian, Lagrangian, and Hamiltonian mechanics describes the Earth-Moon system, we find interesting differences. (For more details, see Arnold, [1974] 1978). Newtonian mechanics describes this system as the motion of two bodies in three-dimensional Euclidean space. In contrast, in Lagrangian mechanics the Earth-Moon system is

---

[5] The concept of *sectorial velocity* is the first step on the path to the concept of angular momentum. This concept is therefore related to the concept of *space*, a concept that was born of the stabilization of epistemic contact in painting. In analytic mechanics it can be shown that the law of conservation of angular momentum is related to the invariance of space with respect to rotations.

represented as the motion of a single body in a six-dimensional configuration space, and in Hamiltonian mechanics as the motion of a single body in a twelve-dimensional phase space. We see that these mechanical systems differ from each other with respect of the stabilization of epistemic contact, i.e. in how they incorporate the pieces of information obtained in epistemic contact into the overall representation of the mechanical system.[6] We know from Copernicus' example that a higher degree of the stabilization of epistemic contact enables the representation to contain significantly more details and it also brings the possibility of introducing new concepts. Just as Copernicus' transition to the heliocentric system made it possible to introduce the notion of sectorial velocity and discover the law of its constancy – a law which does not appear in Ptolemy's system – so Hamiltonian mechanics enables the introduction of the notion of phase volume and to discover the law of its conservation in mechanical motion (the formulation of which is not even possible within the framework of Newtonian or Lagrangian mechanics) and also to discover deterministic chaos (see Diacu & Holmes, 1999).

8.2.4   Stabilization of Epistemic Contact and the Pictorial Form

The stabilization of epistemic contact, which we have illustrated above, is certainly not limited to Renaissance painting, Copernican astronomy and Hamiltonian mechanics. On the contrary, it can be said that the stabilization of epistemic contact is a characteristic feature of all scientific knowledge, one which distinguishes the epistemic contact upon which scientific knowledge is founded from ordinary sensory experience. When we look at the three examples above, in each of them there is a certain fixed point against which a corresponding representation of reality is constructed. We propose calling this point a *point of view*. In Dürer and Alberti, a real point of view is used to stabilize the epistemic contact; in the case of Copernicus, one can speak of a point of view in a figurative sense.[7] Further, the origin of coordinates in phase space can be considered a point of view in a generalized sense, since it is a point from which the twelve-dimensional representation of the world is constructed.

When, in analogy with Dürer, for whom the point of view is the point from which the object and its representation on the painting look exactly the same, we understand the point of view as the *point from which we construct the correspondence*

---

[6]In the case of Hamiltonian mechanics, the term 'stabilization of epistemic contact' obtains an almost literal meaning, because by moving from a six-dimensional configuration space to a twelve-dimensional phase space, velocity is turned into a coordinate, i.e. in the representation of the mechanical system it is static, i.e. stable.

[7]The Copernican system is formulated from the point of view of an observer motionless with respect to the Sun, even if such a point of view cannot really be taken, because a motionless observer would fall towards the Sun.

between the object and its image – i.e. between the world and its representation – it is not difficult to see the connection between the stabilization of the epistemic contact as we have described it and the pictorial form as it was introduced by Wittgenstein in the *Tractatus Logico-Philosophicus* (hereinafter "TLP", Wittgenstein, 1921). If we accept this identification, that is, if we interpret Wittgenstein's notion of pictorial form as an explication of the stabilization of epistemic contact, Wittgenstein provides a relatively good guide for identifying the *formal aspects* of this stabilization. TLP 2.2: *A picture has logico–pictorial form in common with what it depicts.* An important feature of this form is that it cannot be expressed in language. TLP 2.172: *A picture cannot, however, depict its pictorial form: it displays it.*[8]

Wittgenstein described two aspects of this form. One is the boundary of language. TLP 5.6: *The limits of my language mean the limits of my world.* This boundary is inexpressible in the language. In a Renaissance painting, this boundary is present in the form of the horizon. Nevertheless, if we entered into the landscape shown by the painting, and we went to the place where the horizon, according to the painting, is located, we would not find anything particular. The horizon is an element of the picture to which nothing really corresponds. The horizon is an illustration of Wittgenstein's distinction between what the image depicts and what it only displays. Although the horizon is clearly visible in the painting – we can see it clearly – it is not there because it depicts some element of reality. The horizon does not belong to the theme but to the form of the painting. According to Wittgenstein, the second aspect of the pictorial form is the subject of language. TLP 5.632: *The subject does not belong to the world: rather, it is a limit to the world.* This means that it is an element of the pictorial form. TLP 5.641: *The philosophical self is not the human being, not the human body, or the human soul, with which psychology deals, but rather the metaphysical subject, the limit of the world – not a part of it.* Like the horizon, the subject too is an aspect present in every representation. Each representation is constructed from a certain point of view, but this point of view is not depicted.

From Wittgenstein we can also take a general rule according to which the pictorial form consists of those aspects of a representation that are not explicitly expressed, but only displayed, or implicitly assumed. If we accept this rule, besides the *point of view* and the *horizon*, the pictorial form also includes the *relation of identity*. The pavement tiles are depicted in the painting as being of different sizes, but we must learn to see them as identical. Furthermore, *space* is only displayed in the painting, which is actually flat. Transferring these four aspects of the pictorial form (*perspective, horizon, identity* and *space*) or the stabilization of epistemic contact (if we accept the identification of stabilization with pictorial form) to other disciplines turns out relatively easy.

---

[8]The sentences of Wittgenstein's *Tractatus Logico-Philosophicus* are numbered using a decimal classification system, whereby sentence TLP 2.2 comments on sentence TPL 2, just as sentence TLP 2.172 comments on sentence TLP 2.17. This numbering makes searching the *Tractatus* very easy. Therefore, it is customary to refer to the sentences of the *Tractatus* by their number, and not (as per other books) by page number.

A second advantage of identifying the stabilization of epistemic contact with Wittgenstein's pictorial form is that in this way we gain a deeper understanding of the development of scientific theories. In the *History of Geometry and the Development of the Form of its Language* (Kvasz, 1998) I proposed interpreting the history of geometry as the *development of the pictorial form* of geometric figures. The history of algebra is described in a similar way in (Kvasz, 2006) and the history of classical mechanics in (Kvasz, 2011). Due to the identification of pictorial form with the stabilization of epistemic contact, the history of the respective disciplines is thus described in these texts as the *development of the stabilization of epistemic contact*. Thus, in geometry, epistemic contact with geometric reality is stabilized in an ever-improving, increasingly sophisticated way, thanks to which increasingly complex features of geometric reality become recognizable.

8.2.5   Pictorial Form and Instrumental Practice

Measurement is a very important type of epistemic contact. A measuring device is an artefact created by man. The interaction of the measuring device with the studied phenomenon is subject to physical laws, the knowledge of which allows scientists to establish an accurate correlation between changes in the studied phenomenon with different sections of the scale The identification of the stabilization of epistemic contact with the pictorial form subsequently makes it possible to use our knowledge of the aspects of the pictorial form in our analysis of the stabilization of the epistemic contact. The creation of a measuring device is of fundamental importance in quantifying a phenomenon (such as temperature, pressure, or pitch) – it allows the introduction of *a zero*, *a scale* and *a unit* into the area of variability of the phenomenon (i.e. into color changes, pressure changes or tone changes). The introduction of the *zero value* of a certain quantity (e.g., of zero temperature), the introduction of *a scale* (the range of temperatures for which the given thermometer is designed) and the introduction of *a unit* have an undeniable conventional component. The zero value of the measured quantity is the reference point with respect to which we perform the measurement. It can therefore be said that the zero point represents, in the language of a given physical discipline, what in the language of geometry represents a point of view, the point to which we relate all particular aspects of the representation. The zero value is thus the first aspect of the pictorial form of the language. Similarly, the scale introduces into the variations of the phenomenon subjected to that measurement a linear passing towards infinity. We have termed a similar passing to the limit in the language of geometry *the horizon* and declared it the second aspect of the pictorial form. Another aspect of the pictorial form is the unit, which is probably the most important element introduced into the phenomenon by the measuring device. For example, in the case of temperature, a measuring instrument (i.e. a thermometer) makes it possible to identify the temperature changes from, say, 2 to 3 degrees Celsius and from 124 to 125 degrees Celsius as the same change. Something like this is absolutely not possible on the basis of sensory experience. Thus, the measurement device introduces into the language an *identity relation*, i.e., a relation that identifies two *different* changes in temperature as different locations of the *same* change.

Elements such as point of view, scale or unit are aspects of the pictorial form of the language of a given discipline and as such have, from an epistemological point of view, a clear function. The pictorial form is a structure that connects the *epistemic subject* of the relevant scientific discipline (which takes the form of a point, i.e., the point from which the representation of reality is constructed) with the *universe of the language* of this discipline, the point of view actually being a representation of the epistemic subject in the universe of the language. In addition, the pictorial form introduces an orientation of the universe of the language with respect to the epistemic subject. Thus, in geometry, the horizon introduces the horizontal direction; the direction in the painting that is perpendicular to the horizon specifies the vertical direction; and the direction connecting the point of view with the vanishing point specifies the direction of the depth of space. Similarly, the identity relation introduces into the universe of the language a structure that makes it possible to recognize two representations as representing the same object at different locations.

8.3 Extension of the Epistemic Contact

In addition to the tasks of creating an epistemic contact with reality and stabilizing this contact, scientific instruments have a third important role. This is to *extend* the epistemic contact beyond the boundaries of our sensory experience and to *supplement* it with epistemic contact with phenomena that we have no opportunity to perceive in sensory experience. To fulfill this task, it is important to realize that scientific instruments do not exist alone, but usually they exist in communities which together create an instrumental practice.

8.3.1 Atmospheric Pressure as an Example of Extended Epistemic Contact

At the beginning of the seventeenth century, when pumping water from deep shafts, miners discovered that there was a maximum depth from which water could be pumped. If this limit was exceeded, the water column would tear off and the water could not be pumped up. Torricelli studied this phenomenon experimentally, but instead of water he took mercury, which is torn off much earlier, i.e. can maintain a much shorter column. In his experiments, he created a vacuum in a glass tube above the mercury column. The height to which the mercury column rose in the tube could be accurately measured, and a barometer was gradually formed from such a tube.

The atmospheric pressure with which Torricelli created an epistemic contact in his experiments is a phenomenon that is not accessible through sense perception. We cannot perceive it, and the ancient Greeks, for example, had no idea that something like atmospheric pressure existed at all. While heat is accessible by means of sense perception, and so the thermometer can be interpreted as an instrument that refines the common experience of heat, *the barometer extends our epistemic contact with the world* and opens access to a completely new phenomenon – the phenomenon of atmospheric pressure.

The barometer was created by standardizing the Torricelli experiment. Therefore, if we want to understand what measurement is, we must not forget what an experiment is. An experiment is a purposeful examination of an aspect of a certain phenomenon using an *artificial situation* (such as an inclined plane, in the case of Galileo's investigation of the accelerated motion of falling bodies, or the tubes filled with mercury, in the case of Torricelli's investigation of the tearing off of the water columns experienced when pumping water). A measurement is created out of this artificial situation, by means of *standardization* of the objects, relationships and procedures that constitute it. In the case of a barometer, we fix the dimensions of the tube, standardize the numerical scale which we attach to the tube, specify the quantity of mercury and determine the range of temperatures in which the device can be used. This ensures the reproducibility and the intersubjectivity of measurement. Phenomena such as atmospheric pressure, cathode rays, ultrasound and X-rays are not accessible to sensory experience. They are accessible only by means of an artificial situation; nevertheless, by standardizing this situation, these inaccessible phenomena can be transformed into reproducible phenomena, and thus permanently integrated into the intersubjectively shared world of science. Standardization is often followed by *encapsulation*, i.e. by closure of the artificial situation inside a certain portable object, which thus becomes a measuring device. Thus, the artificial situation becomes 'portable' and the new device can enter the process of observing other phenomena (as in thermodynamics, where we combine determination of pressure with measurement of temperature and volume of gas to represent its state).

As long as physics studies those realms of phenomena to which we have immediate sensory access, it is possible to interpret experiments and measurements as a refinement of the picture of reality that is presented by the senses, i.e. merely as a means to stabilize the epistemic contact. For example, in free fall we are unable to decide on the basis of sensory perception whether the fall is uniform or accelerating, but in general we know this phenomenon intimately from everyday experience. An experiment makes it possible to determine that this is an accelerating motion, so it can be said that it has refined and stabilized the image of free fall that we have created on the basis of sensory experience. In the case of temperature, the interpretation of measurement as a refinement of the sensory image, which we obtain in direct epistemic contact, is a bit problematic. We can measure temperatures of bodies such that if we touched them, our hand would be charred: it is not really possible to imagine a sensory image which the measurement would only refine. In the case of atmospheric pressure, the situation is even worse. Falling pressure initially manifests itself in the sensory realm via a headache and ends in the destruction of our organs after the gases bound in the body fluids exceed their boiling point. We cannot speak of measuring pressure as a refinement of sensory experience. What the pressure of 0.1 atm. means in the sensory realm is definitely beyond the reach of the human imagination. Given such atmospheric pressure, we would be long dead.

Thus, measurement not only stabilizes the perception of the phenomena of sensory experience, but also makes it possible to extend the epistemic contact with physical reality beyond the boundaries of sensory perception. What confronts us beyond the boundaries of our sensory experience is often quite different from what we are used to in ordinary life. The scientific image of the world is constantly adapting to the results obtained in experiments and measurements, and thus it gradually moves away from the image of our world that we have created on the basis of sensory experience.

8.3.2 Sensory Experience Versus Instrumental Practice

Every aspect of our surroundings that is accessible to our senses is usually perceived by a single sense organ. We perceive color by sight, temperature by touch, and pitch by hearing. Under normal circumstances, it is not possible to feel the colors, see the pitch or hear the temperature. Therefore, as a rule, we are unable to verify the data obtained by one sense organ using another organ, and this inability is the basis of various sensory delusions. In contrast, it is often possible to measure a certain physical quantity using several physical principles. Thus, the temperature can be measured by the thermal expansion of liquids such as alcohol or mercury, but also by spectral analysis of thermal radiation, or by any other physical quantity that is related to temperature by a known law of physics. It is important that the various methods' measurements overlap so that they can be calibrated with each other, allowing the results obtained by one method of measurement to be checked using the results obtained by other methods.

If the principles upon which the individual measurement methods are based are independent (i.e. there is no physical law to relate these principles, such as the independence of the thermal expansion of liquids from the spectral distribution of thermal radiation) yet the results obtained by these methods strongly agree, then the result of the measurement can be considered, with a sufficient degree of certainty, an objective feature of reality. In instrumental realism the *objectivity of knowledge* is understood as the independence of this knowledge from the specific method by means of which it was obtained.

In addition to its uniqueness, the perceptions of each sensory organ are limited. To our senses, only a limited range of temperatures, colors or tones is available. We are not able to perceive those wide variations in these phenomena that fall beyond the scope of our senses. We are not able to distinguish very high temperatures, we cannot perceive a whole range of colors and we do not hear the highest pitched tones.

In contrast, measurement opens up, thanks to the plurality of measurement methods, access to a virtually unlimited range of variations in any phenomenon. In the case of temperature, we can measure this quantity from almost absolute zero up to billions of Kelvins; we can measure pressure from the very low pressures studied by vacuum physics to the very high ones used in geophysics; we are able to register electromagnetic radiation across practically its entire range reaching from very short waves of gamma radiation to very long radio waves; and in the case of sound waves we can register them almost at any frequency. So, unlike our sensory organs, which show us temperatures, colors, and sounds only on a limited scale, instrumental practice opens epistemic accesses to these phenomena over virtually their *full range of variability*.

In addition to uniqueness and limitation, sensory experience has a qualitative character. Whether it is a visual perception of a certain color, a tactile perception of a certain temperature or an acoustic perception of a certain tone, this perception has a unique character. It is very difficult to compare one's perceptions of two colors, or two temperatures, or two tones, either within oneself or with another observer. In contrast, the measurement, upon which the interpretation of epistemic contact is based in instrumental realism, is quantitative in nature. Phenomena that are the subject matter of scientific research allow a number of variations: the color may have different degrees of brightness and intensity; the temperature may increase or decrease; the sound may change its pitch and color. A measuring instrument is a device that converts changes of a certain phenomenon into changes in length in a regular, standard and reproducible manner. We can mention a thermometer, a barometer or a spectroscope: all of these devices have a scale attached to them, by means of which they transform changes in temperature, pressure or color into changes in length. This *quantifies the phenomenon* (of color, of temperature, or of sound), i.e., provides a regular, standard and reproducible correlation between *aspects* of the phenomenon and *numbers*.

8.3.3 The Symbolic Extension of Epistemic Contact

Quantification enables the changes in an aspect of the studied phenomenon to be perceived with *accuracy* and *clarity*, which is not at all conceivable in sense perception due to the qualitative nature of the sense perception. Much of this accuracy and clarity can be considered part of the stabilization of epistemic contact, and we have addressed it in the previous section. However, the fundamental sharpening and refinement of the phenomenon by means of the values of the measured quantities enables the embedding of the phenomena in a network of mathematical relations, expressing empirical laws such as Galileo's law of free fall, Snell's law of refraction and Hooke's law of elasticity.

The establishment of such empirical laws as well as the introduction of physical constants make it possible to relate various aspects of the studied phenomenon; such laws and constants are thus essential to the further stabilization of the epistemic contact.[9] Epistemic access to an aspect of a phenomenon is stabilized not only, as per Copernicus, by organizing the pieces of information acquired within a particular situation in which we have epistemic contact with the aspect. Thanks to the network of mathematical relations, it is possible to stabilize the epistemic contact with the studied phenomenon indirectly, by stabilizing the contact with any other phenomenon with which the studied phenomenon is connected by means of an empirical law. Thus, if we have two quantities, one of which we can measure with much higher precision than the other, and these quantities are connected by means of a mathematical law, we can use the high precision of the measurement of one of these quantities to stabilize the epistemic contact with the other.

In addition to its undeniable contribution to the stabilization of epistemic contact with individual phenomena, the network of mathematical relations enables the extension of the epistemic contact in a fundamental way. Perhaps one of the most striking examples of this was the 'weighing of the Earth' by Lord Cavendish. In reality, he measured the gravitational attraction between two bodies in a laboratory, but because this force is related by means of a mathematical formula to the mass of the Earth, his experimental results enabled him, for the first time in history, to establish epistemic contact with the mass of the Earth.

With the case of the stabilization of epistemic contact we have already argued that a perspectivist picture, the Copernican heliocentric system and Hamiltonian mechanics *enrich* the representation of reality with a number of additional details, such as spatial relations between the depicted objects, the elliptical shape of planetary trajectories and the conservation of the phase volume in motion. The network of mathematical relations, which can be introduced thanks to the quantitative character of measurement, brings even more possibilities for such *enrichment*. We can form different powers of the measured quantities and construct their formal combinations, for which we have no counterpart in sensory experience. We can try to include these formal combinations in our epistemic contact with reality.

---

[9]These constants are not entirely determined by nature; conventions are also involved. We propose, therefore, to consider the constants by means of which the empirical laws are formulated as another aspect of the pictorial form. Thus, we propose that one consider the empirical law as a fact about the world and the relevant constant as an aspect of the pictorial form. The constant is displayed by the empirical law. Its numerical value has no factual significance, just like the position of the horizon in a painting does not. The value of a constant depends on the choice of units for the individual quantities. On the one hand, a constant contributes to the stabilization of the epistemic contact with reality; yet on the other hand, it relates several quantities, and therefore appears only in a complex instrumental practice, in which several scientific instruments work side by side. The constants are used to interconnect local instrumental practices associated with these individual devices.

8.3.4   Instrumental Practice Can Offer a Complete List of Quantities

There are numerous holes in our sensory system. The senses offer us a very selective choice of aspects of reality. We do not have a sensory organ for precisely measuring atmospheric pressure, electric charge, or the polarization of light, probably because in the evolution of our species the perception of such aspects of our environment was not important. However, we can very precisely distinguish different shades of color and we are sensitive to a wide range of sounds because the information obtained by these senses was important for our survival. Nevertheless, the evolutionary signif-icance of certain aspects of reality, which was important in the formation of our sense organs, is irrelevant in the construction of the physical image of the world.

Measuring instruments are diverse enough to provide a complete list of quantities that allows a description of the state of a physical system which then forms the basis of the physical description of reality. For example, in the case of thermodynamics we can sensually perceive volume and temperature, but for the third quantity, which together with volume and temperature represents the state of a system, namely pressure, we have no sensory organ to adequately perceive it. Pressure and its changes are, nevertheless, reflected in the behavior of gases and liquids, and so a barometer was created that allowed the measurement of pressure. Similarly, only when scientists mastered the measuring of all the quantities that together determine the state of the electromagnetic field was it possible for them to move from the experimental study of electromagnetic phenomena to the theory of the electromag-netic field. This move brings us to another topic, idealization.

8.4   The Idealization of Epistemic Contact

We have arrived at the final level of our analysis of the use of scientific instruments, which is idealization. The first and second sections of this paper dealt with individual instruments and introduced the concept of *epistemic contact* and its *stabilization*; the third part discussed the plurality of instruments and introduced the concept of an instrumental practice. However, we have not yet asked ourselves the basic question regarding the compatibility of individual elements of *instrumental practice*.[10] I have mentioned that a certain aspect of the studied phenomenon and the scale of its variability are usually simultaneously accessible to several instruments, and that

---

[10] An **instrumental practice** connected to a physical discipline can be defined as a system of measuring devices that allows the measuring of all the quantities necessary for the identification of the state of the physical system studied by the particular discipline. However, this is not the right place for such a definition because thus far we have neither defined the concept of the state of a physical system nor shown from where it comes. It comes from idealization, so we first have to address this notion in more detail.

different aspects of the same phenomenon, which are connected by a mathematical relation, are usually also accessible by the use of different instruments. The question is how to ensure the compatibility and interplay of the various components of the instrumental practice, where different instruments have different resolution, different sensitivity, different accuracy, different scales and different reliability. The same number obtained by different instruments can have a fundamentally different meaning.

In the case of one scientific instrument examined separately from the others, I introduced the notion of the *stabilization of epistemic contact*, and interpreted it using Wittgenstein's notion of pictorial form. In the case of the system of scientific instruments that make up an instrumental practice, I shall proceed in a similar way, introducing the notion of *idealization* and seeking to interpret it using the concept of a language game from Wittgenstein's *Philosophical Investigations*.[11] Idealization can be seen as an irreducible component of instrumental practice. Its purpose is to enable unification, i.e. the synthesis of the components of an instrumental practice into a functioning whole.

There is no room for a detailed description of all aspects of idealization, but in principle it can be said that ***idealization is a linguistic reduction***. In other words, complex and variously conditioned acts of instrumental practice are, during the process of idealization, replaced by their sharp, unambiguous, clear, and simple linguistic representations. In (Kvasz, 2019) I explained this reduction for the case of geometric constructions in terms of ruler and compass, and I am convinced that a similar interpretation is possible for other types of instrumental practice. My interpretation of idealization is based on Wittgenstein's notion of a language game. I consider instrumental practice to be part of a certain language game and that a language game establishes idealization.

A language game has three components. The first is the ***material component***, which in the case of the instrumental practice of geometric constructions has the form of the physical manipulation of a compass and ruler to apply graphite to the surface of a piece of paper. In addition to the material component, there is a ***linguistic component*** that contains a set of terms connected by syntactic and semantic rules. The linguistic component is closely linked to the physical manipulation of individual instruments. In the case of geometry, I propose interpreting geometric figures as part of this linguistic component, i.e. understanding geometric

---

[11] As regards the notion of the *pictorial form*, I violated Wittgenstein's intention in introducing this notion. I narrowed the pictorial form to particular disciplines such as geometry and algebra and introduced a specific pictorial form for each of them, having previously interpreted the history of these disciplines as the development of their pictorial form in (Kvasz, 1998, 2006, 2011), whereas Wittgenstein intended the concept of pictorial form as a monolithic concept, so he would probably reject both narrowing it down to individual disciplines and the idea that the pictorial form changes over time and that the historical development of geometry or algebra is driven by the development of the pictorial form. Nevertheless, I am convinced that what I offered in those three papers is a correct description of their development. It seems that as regards the notion of a *language game* the deviation of my use of this notion from Wittgenstein's intended use is even sharper. I will use the notion of a language game for understanding rather than criticizing idealization.

figures as linguistic expressions with specific syntax and semantics. As in any language, in the linguistic component of a language game tied to an instrumental practice, the rules of syntax and semantics are conventional. Idealization consists in the choice of the rules regulating the syntax and semantics of the linguistic component of a language game so that these rules correspond to the behavior of the ideal objects for the representation of which the language is intended. Finally, the third component of the language game is the *social component* which brings normativity to the language game, i.e. it declares certain moves in the language game to be correct, others to be incorrect. It is at this normative level that idealization is fixed.

In the case of geometry, the idealization consists in the fact that when a geometric image (understood as a physical object, i.e. paper smeared with graphite) is created in accordance with the postulates of Euclidean geometry (understood as the syntactic rules of the language of geometry, i.e. rules allowing a move from one properly formed figure to another, for example, by connecting two of its points) we are obliged (by the normative component of the game) to answer the question of how the two lines intersect, contrary to what we actually see in the figure (i.e. a certain area of specific dimensions and shape) and in accordance with the standards of the language game, that it is to say, that they intersect at a point (i.e. a *semeion* lacking shape, dimensions and parts). This means that in the language game upon which classical Euclidean geometry is based, the instrumental practice of geometric constructions (which is the material practice of physical manipulations with physical objects) is subject to linguistic rules (i.e. rules of correct construction and interpretation) whose syntax and semantics are designed to comply with the standards of the behavior of ideal geometric objects.

This means that *straight lines can be extended to infinity*, that is, beyond the boundaries of countries, of continents, and even of the universe; *from any point it is possible to describe a circle with any radius*, even if the arm of the compass with which we are to do it should be thousands of kilometers long and weigh millions of tons; and *lines and circles intersect at points that have no dimensions, shape or parts*, even if they would be, as a result, completely invisible. In other words, everything in the language game takes place as it would happen in the world of ideal geometric objects. Furthermore, the logical arguments that make up the steps of mathematical proofs are based on the properties of these idealized linguistic representations and not on the properties of the material component of the instrumental practice, in which *straight lines usually end* at the edge of the paper, *circles have relatively small radii* determined by the available compasses, and their *intersections are small areas* covered with crumbs of graphite, having specific dimensions, shape and parts.

I believe that other kinds of instrumental practice have similar properties. For example, Galileo's inclined plane experiments were part of a language game the material component of which used particular slabs, marble balls, and instruments for measuring distance and time. Its linguistic component used different expressions to describe the results obtained in the experiments. The syntactic and semantic rules used in the linguistic component were subordinated to rules according to which quantities have sharp numerical values; certain material manipulations correspond to linguistic operations and relations, so that the whole argument put forward on the third day of Galileo's *Discorsi* can be built by means of this linguistic component. The same goes for those language games created by Newton, Maxwell and Schrödinger. There too the material (i.e. instrumental and artificial), the linguistic (i.e. symbolic and conventional), and the social (i.e. normative) components are combined into a whole that corresponds to the idealized reality.

The purpose of idealization is synthesis, i.e. the interconnection of individual components of scientific practice into a meaningful and functional whole. There are four kinds of synthesis. The first kind of synthesis is *instrumental synthesis*, i.e. the synthesis of individual instruments into a complex instrumental practice. One way in which the individual instruments are incorporated into the common instrumental practice is by turning the results obtained by means of the particular instrument into numbers which are governed by the rules of the linguistic component of the language game and subject to its standards. For example, it is assumed that if the velocity of a body is measured, its derivative (i.e. the acceleration) can be found by means of the standard rules of the calculus, although the measurement certainly cannot give the values of velocity with that precision – 'for each epsilon there is a delta' – as is required by the definition of derivation.

The second kind of synthesis is *relational synthesis*, which makes it possible to combine quantities obtained by individual instruments into relations expressing empirical laws. As an example, we can take Galileo's law of free fall or Snell's law of the refraction of light. We must idealize here because the very existence of such laws presupposes that individual quantities have accurate and clear values which enter into the formula expressing the law. In reality, it is far from obvious that we are able to determine the values of one quantity (such as time) at those points which correspond precisely to the values of the other quantity (such as distance). It is much more likely that the points for which we determine the values of one quantity fall somewhere in between the points for which we are able to determine the values of the other quantity and so at the empirical level no formal relation between the quantities is established.

The third kind of synthesis is *compositional synthesis*, which in the case of physics consists in the fact that to every physical system a mathematical object called a *state* can be assigned. The existence of the state of a physical system means that its behavior can be separated from its history. Knowledge of its state is sufficient for the determination of its future behavior. This is not possible for biological systems. A biological system carries information about its past in the form of a genetic code. For physical systems nothing like this is needed: all the information about the system that is relevant for the prediction of its further behavior is condensed into the notion of its state. Needless to say, the state represents the physical system in an idealized way – it omits all the information irrelevant to the prediction of the future behavior of the system while all information necessary for that prediction is taken to be absolutely precise and sharp.

The fourth kind of synthesis is *deductive synthesis*, which in physics has the form of the equations of motion, which are differential equations that connect the state of a system at a given moment with its state at later moments. This equation makes it possible to calculate the temporal evolution of the state, i.e., to determine the future behavior of the system. Whether we consider Newton's second law, the equation of the vibrating string, Fourier's equation of heat propagation, the Navier-Stokes equations of hydrodynamics, Maxwell's equations of electrodynamics or Schrödinger's equation of quantum mechanics, they are all the equations of motion. In order to fulfill the role of *deductive synthesis*, i.e., to allow us to derive from the state at the present moment the future states, these equations must be idealized – the values of all parameters must be sharp.

8.5   Instrumental Realism and the Philosophy of Science

Distinguishing the four roles that instruments play in science – epistemic contact, the stabilization of epistemic contact, the extension of epistemic contact and the idealization of epistemic contact – allows us to approach several questions in the philosophy of science in a new way.

One such question is the question of **idealization**. Angela Potochnik, in *Idealization and the Aim of Science* (Potochnik, 2017), claims that idealization is rampant in science and goes unchecked. However, this is rather strange, given that in the philosophy of science, idealization is usually defined as "a deliberate simplification of something complex (situation, concept ...) with the intention of achieving at least a partial understanding of the matter. It may contain distortion of the original, or it may simply mean omitting certain components of the complex in order to better focus on the remaining ones" (McMullin, 1985, p. 248). If we take Potochnik seriously, it is a mystery why scientists simplify, distort and omit certain components in a rampant and unchecked way. If they did it occasionally, it could be tolerated, but why almost everywhere?

From the point of view of instrumental realism, it is understandable why scientists use idealization so often. Idealization is a way of synthesizing individual tools into the unified instrumental practice of a certain discipline such as geometry, mechanics or electrodynamics. Taking the example of geometry, which is the simplest, if mathematicians did not idealize lines and circles, the instrumental practice of geometry would fall apart. Euclid's first postulate requires that any two points can be connected by a straight line. However, if we took as points the real physical objects that arise as intersections of two straight lines, a straight line and a circle, and of two circles, it would be a matter of fact that these points would have different sizes and shapes. If the straight lines intersected at a very sharp angle, their intersection would be a considerably stretched and pointed region. If we wanted to draw a circle passing through the point thus obtained, which should go roughly in the perpendicular direction to the two intersecting straight lines, it is not clear upon which criterion we should base our knowledge of the appearance of the point through which the circle would pass. If the line forming the circle were not to be so thick as to to cover the entire point under consideration (which is thin but long), we would have to formulate complex criteria to determine when the circle had passed through the intersection of two such lines.

Idealization is a neat solution to this problem. We declare the intersection of straight lines to be a Euclidean *semeion* and we consider a circle as a line that has no width. Then all kinds of lines have a single kind of intersection and there is a clear criterion for when a circle passes through such a point. What is present on the paper (that is, the material component of the language game) is only an imitation of a real point and a real circle. The language game makes it possible to establish rules and standards for how to use these approximate physical objects to obtain knowledge concerning the ideal geometric reality. It suffices to subject the syntactic and semantic rules of that instrumental practice to the norms valid for ideal geometric objects. That is, whenever someone asks, "What is the intersection of two lines?" we answer in a manner contrary to what we see, and in accordance with the standards of geometry, that it is a point without shape and magnitude. In all constructions and proofs we exclusively use such idealized assumptions.

It is similar in mechanics. When we want to determine the state of a mechanical system (i.e. its representation that enters into the acts of compositional and deductive synthesis of the language of mechanics), we have to use measurements of position, velocity, mass and so on. The state of the system is obtained by combining the results of these measurements, i.e. as a kind of intersection. Each measurement has a certain degree of accuracy, just as each line has a certain thickness. The state arises as an intersection of areas determined by individual instruments in individual acts of measurement. The problem here is similar to that in geometry.[12] We can visualize this by imagining that the individual acts of measurement determine certain areas in the configuration or phase space. At their intersections, relatively complicated objects can arise which are hard to use in the steps of compositional and deductive synthesis, i.e. in the construction of a complex system and the calculation of the temporal evolution of its state. The solution is the same as in geometry – idealization. Thus (in mechanics) we assume that all quantities have absolutely precise values, and therefore the state is also absolutely precise and it determines the future values of all quantities with absolute precision.[13] So, in order to integrate individual instruments into common instrumental practice, we need to idealize the results of epistemic contact, i.e. quantities obtained in the acts of physical measurement. Idealization enters the game of scientific inquiry at a more elementary level than that level upon

---

[12]In geometry, the *compositional synthesis* is manifested in the *postulates*, which are rules for constructions and the *deductive synthesis* is manifested in the *axioms*, which are rules for conducting proofs. In geometry we idealize, for instance, the intersection of two lines for the same reason as in mechanics we idealize the state, namely that in this way they can be part of further steps of compositional and deductive synthesis.

[13]Quantum mechanics intervened in this problem through the uncertainty principle but physicists found an elegant answer when they attributed the sharpness and unambiguity, i.e. idealization, to the state represented by the wave function, and attributed all the uncertainty to the act of measurement.

which McMullin's definition of idealization rests. Idealization is a prerequisite for the relational, compositional and deductive synthesis of the language of physics.

Acknowledgement I express my acknowledgement of the generous support of the *Formal Epistemology – the Future Synthesis* project, within the framework of the *Praemium Academicum* program of the Czech Academy of Sciences. I would like to thank Michael Pockley whose comments helped to clarify several arguments.

References

Ackermann, R. (1985). *Data, instruments, and theory*. Princeton University Press.
Arnold, V. I. ([1974] 1978). *Mathematical methods of classical mechanics*. Springer, translation in 1978.
Chang, H. (2004). *Inventing temperature: Measurement and scientific progress*. Oxford University Press.
Diacu, F., & Holmes, P. (1999). *Celestial encounters: The origins of chaos and stability*. Princeton University Press.
Drake, S. (1975). The role of music in Galileo's experiments. *Scientific American, 232*(June 1975), 98–104.
Dreyfus, H. L. (1972). *What computers can't do*. Harper and Row.
Galilei, G. (1638). *Dialogues concerning two new sciences*. Translated by H. Crew and A. de Salvio, New York 1914.
Hacking, I. (1983). *Representing and intervening*. Cambridge University Press.
Heelan, P. (1983). *Space perception and the philosophy of science*. University of California Press.
Ihde, D. (1991). *Instrumental realism: The interference between philosophy of science and philosophy of technology*. Indiana University Press.
Ihde, D. (2011). Husserl's Galileo needed a telescope! *Philosophy of Technology, 24*, 69–82.
Kvasz, L. (1998). History of geometry and the development of the form of its language. *Synthese, 116*(2), 141–186.
Kvasz, L. (2006). History of Algebra and the development of the form of its language. *Philosophia Mathematica, 14*(3), 287–317.
Kvasz, L. (2011). Classical mechanics between history and philosophy. In A. Máté, M. Rédei, & F. Stadler (Eds.), *The Vienna circle in Hungary* (pp. 129–154). Springer.
Kvasz, L. (2019). How can abstract objects of mathematics be known? *Philosophia Mathematica, 27*, 314–334.
Kvasz, L. (2020). Mathematical language and the changing concept of physical reality. In W. J. Gonzalez (Ed.), *New approaches to scientific realism* (pp. 206–227). De Gruyter.
McMullin, E. (1985). Galilean idealization. *Studies in the history and philosophy of science, 16*(3), 247–273.
Potochnik, A. (2017). *Idealization and the aim of science*. Chicago University Press.
Wittgenstein, L. (1921). *Tractatus Logico-Philosophicus*. English translation by D. F. Pears and B. F. McGuinness. Routledge 2001.

Part IV
Scientific Progress Revisited

Chapter 9
Scientific Progress and the Search for Truth

Philip Kitcher

Abstract Scientific progress is popularly conceived in teleological terms. The goal is to find the complete true story about nature, or, perhaps more sensibly, the complete truth about the fundamental structure of nature. I argue that scientific progress is better understood as *pragmatic* progress, "progress from" rather than "progress to." It consists in overcoming the difficulties and the limits of the current situation. This approach, pioneered by Thomas Kuhn and Larry Laudan, should be elaborated through explicit recognition of the role the sciences play in human life. Although we normally defer to scientific communities, trusting them to single out problems that are worth tackling (*significant questions*), judgments of significance are always subject to evaluation. Ethical values inevitably frame and constrain the progress of the sciences. As many writers have suspected, an approach of this kind emphases the practical deliverances of scientific inquiry. Does it leave a place for "pure" (or "basic") research? I address this question in two ways. First, everyone should agree that the route to practical success often runs through studies with no pragmatic payoff. More importantly, we should recognize how the idea of any "disinterested search for the truth" comes very late in the history of inquiry. I deploy a speculative genealogy to show how it might have emerged, using my conjectural history to illuminate the supposed ideal of "contemplating the truth about nature." Such reflection, I argue, does not focus on the structure of nature as it is, independently of human cognition and action, but on the organization of a world largely shaped by human goals and values.

Keywords Scientific progress · Pragmatism · Science and values · Truth · Basic research · Knowledge for its own sake

P. Kitcher (✉)
Columbia University, New York, NY, USA
e-mail: psk16@columbia.edu

© The Author(s), under exclusive license to Springer Nature Switzerland AG 2022
W. J. Gonzalez (ed.), *Current Trends in Philosophy of Science*, Synthese Library
462, https://doi.org/10.1007/978-3-031-01315-7_9

9.1 Difficulties About Progress

Talk of progress often elicits an ironic smile or a knowing wink. Sophisticated intellectuals see themselves as having grown beyond the naïve hopes and the breathless pronouncements of those who report human advances. Nevertheless, in one domain, ascriptions of progress are harder to dismiss. In discussions of the natural sciences, the curl of the lip is not so easily sustained. After all, the last millennia, the past four centuries, and the most recent decades seem to have produced a remarkable expansion in human knowledge of the natural world. Haven't we learned much more about all sorts of aspects of the cosmos?

An astute questioner might wonder about the transition from the impersonal formulation to the casual first-person plural. *Who* exactly has learned more? That good question will recur later. For the present, however, let's set it on one side, and focus on other difficulties in celebrations of scientific progress.

More than half a century ago, Thomas Kuhn (1962) exposed problems in the idea of a cumulative process in which the sciences amass knowledge. Focusing on major transitions in the history of science, he identified discontinuities and losses. Nevertheless, at the end of his brilliant monograph – in my judgment, one of the works of twentieth century philosophy most likely to endure – he struggled to find a sense in which the sciences progress. Some part of previous achievements proves – somehow – permanent. Kuhn denied the possibility of explicating that idea in the most straightforward way, as the piling up of more and more truths about nature. For, famously, he worried about the concept of truth as correspondence to reality, and, notoriously, he proposed that scientific revolutions change the world (1962, 169–170, §X). Wrestling with problems that, as we shall see, are deep and recalcitrant, he struggled to find a clear way of expressing the common view: natural sciences are striking examples of progress.

I want to begin with considerations complementary to those voiced by Kuhn. The standard picture rests on a number of confusions. By bringing them into the light, we may see how to do better. How to make progress with progress, that is.

9.2 Against Teleology

An obvious way to conceive scientific progress is to view science as having a goal, and progress as consisting in diminishing the distance to that goal. Progress is *teleological*. Moreover, it appears easy to identify the goal. One of Kuhn's most distinguished successors, Bas van Fraassen, tells us how the goal is usually conceived. According to scientific realism, "science aims to give us, in its theories, a literally true story of what the world is like" (van Fraassen, 1980, 8). Realism is thus characterized for the purpose of introducing the approach van Fraassen favors, according to which scientific theories need only "save the phenomena". What concerns me is not what divides realism from constructive empiricism, but a feature shared by both parties to the debate van Fraassen sets up. How do we make sense of "a literally true story" or the corresponding empiricist notion of an account empirically adequate to all the phenomena?

I envisage two possible interpretations. The first (incautious) reading takes the "literally true story" to be "the truth, the whole truth, and nothing but the truth". Science aims to provide every true sentence about the universe (or perhaps about its observable fraction). We make progress by "filling in the gaps", amassing more true sentences. But is there a totality of truths? (Perhaps the supposed "collection" is too large to be a set?) Given that the set of space-time points has the cardinality of the continuum, and given that there are many physical magnitudes taking values at each of those points, and given all the potential relations among those values, among sets of those values, and so on, it's clear that the "whole truth" consists of some large infinity of sentences – too many to fit into any story we could hope to tell ourselves. Indeed, anything humanity could generate, between the dawn of science and our ultimate extinction, could only be a pathetic fragment of the entire truth. How could we get as close as is humanly possible? Apparently in a very different way from that pursued by the sciences we know: identify the measuring devices that are quickest and easiest to use; send researchers to different places; have them record the values of the measured magnitudes as fast as they can; assemble all the results, and hire teams of expert logicians to derive consequences at the maximal rate.

van Fraassen's interpolated phrase, "in its theories", points to a far more promising version of the teleological view. The goal is to find systematic accounts capable of generating *any* of the truths (realism) or *any* of the truths about observables (constructive empiricism). Instead of some vast totality of truths, science seeks a "total theory". Although this resonates with the proposals of those scientists who dream of "final theories", it is only slightly more realistic than the incautious pursuit of "the whole truth". One of the great lessons of the past forty years in the philosophy of science is the disunity of the sciences (Dupré, 1993; Cartwright, 1999). The idea of some unified hierarchy, in which "special sciences" are reduced to "more basic" sciences, with "fundamental physics" at the apex has been decisively refuted. Classical genetics doesn't reduce to molecular biology, and molecular biology doesn't reduce to biochemistry (to cite just two among many examples of non-reductive relations among adjacent scientific fields; Kitcher 1984, Culp and Kitcher 1989). Moreover, the Newtonian dream (expressed in the Preface to *Principia*), of general laws embracing all phenomena, is radically at odds with the subsequent development of the sciences. Welcome as they are, general laws are relatively rare. Many sciences have to settle for large classes of models, adapted to the cases of particular interest, and even for mutually inconsistent models that, when used judiciously, can be deployed to treat various aspects of specific systems.

To envisage the sciences as directed towards something complete, of the kind presupposed by van Fraassen's influential framing of debates about realism, overlooks the limitations inherent in any human enterprise. The world we inhabit is far too complex for us to cope with all its facets. Human inquiry is inevitably selective. The various domains of scientific research grow in disorderly ways, from problems that once attracted interest and that have now generated derivative questions, supplemented with current efforts at addressing the concerns of the moment. In thinking about scientific progress, we should abandon teleology.

9.3 Pragmatic Progress

The view of progress as inevitably goal-directed is fostered by particular examples. We travel to a specified destination, we build houses according to plans, we modify our eating habits and exercise regimes to gain or lose weight. It's easy to forget that many plausible examples of progress (on similarly small scales) are quite different. Computer technology isn't aimed at some specific long-term goal; it progresses by fixing glitches and expanding limits. Aspiring pianists don't strive to approximate some ideal player who combines all the technical and interpretative virtues.

Kuhn already recognized the point in his first tentative suggestions in the final section of *Structure*, and his later writings began to elaborate it (Kuhn, 1962, §XIII; Kuhn, 2000). Science, he suggested, is driven from behind. Its progress is progress *from*, rather than progress *towards* anything. I'll generalize the point. Besides teleological progress, there's another species, *pragmatic* progress, made by over-coming the problems and limitations of the current state. Mundane instances of progress often belong to that species. So too does the most prominent large-scale example, progress in the sciences.

Kuhn's insight was developed further by Larry Laudan, in one of the most detailed and illuminating discussions of scientific progress ever provided (Laudan, 1977). Yet, written at a time when Kuhn was typically seen as challenging both the progress and the rationality of science, Laudan's treatment of the topic (like my own attempt in Kitcher 1993) highlighted some questions at cost to others. Haunted by the threat of relativism, raised by that fearsome bogeyman Kuhn-Feyerabend (or Feyerabend-Kuhn), philosophers interested in scientific change paid too much attention to theoretical science, to incommensurability and "conceptual problems", to the large scale units (like research traditions), whose rational development and progressive achievements seemed to need defense.

Instead of posing different questions. If progress consists in problem-solving, what exactly is a problem? What counts as solving a problem? And who decides? Those questions are, I suggest, fundamental if we are to make sense of scientific progress. To tackle them, it's worth taking a look at how scientists talk about progress in their fields.

9.4 Significant Questions

Imagine a skeptical observer, in dialogue with a friendly scientist. "What's new?", the observer asks, "What have you done for us lately?" The scientist flashes a few reprints (or maybe brings the electronic versions up on a screen). "See," she says proudly, "we've now figured out how this molecule binds to its cellular target." "So," responds the skeptic, "what's the big deal about that?" "It's an important step forward," declares the scientist, "since it dovetails with work done by my colleagues, enabling us to think about ways in which cells might react under conditions of reduced oxygen." Of course, the skeptic then wonders why that project is an important one, eliciting a further reply from the excited investigator. And so it goes, until, eventually, the conversation terminates, either in specifying some practical benefit ("If we can do all that, then we can treat sickle-cell disease") or some large theoretical question ("We'll understand how early-stage embryos differentiate").

My imaginary dialogue traces a path through what I've elsewhere called a *significance graph* (Kitcher, 2001, 78–80). The particular questions that occupy scientists daily, sometimes even through the course of an entire career, derive their felt importance from their potential contribution to addressing larger questions, envisaged as answerable by combining the efforts of a broader research community. Asked why the most embracing, most fundamental questions, deserve pursuit, representatives of that community might point to practical benefits, the ability to intervene in nature in desirable ways or to predict the course of natural events so as to avoid danger or reach valuable ends. Or they might cite the intrinsic interest of the question, waxing eloquent about satisfying human curiosity: "Doesn't every thoughtful person see the point of understanding how organisms develop?"

Even when it's easy to acquiesce in such judgments of significance, we should ask what underlies them. Does the "fundamental" question arise from some feature of our species – we are the kind of beings who want to know things like that – or is it *objectively* important? Many people, perhaps including most scientists, might opt for the latter view. But I do not see how that idea can be sustained. As noted above, even if the idea of a complete totality of truths about the cosmos is coherent, it is too vast for us to specify. Even if nature has some intrinsic structure, would investigators be concerned to fathom all of it? Survey the actual research programs undertaken in the sciences, past and present, and it becomes clear, very quickly, how the supposedly fundamental questions are a motley assortment, sometimes even a strange mixture, infused with views about values, about ideals for human living and for human societies – human inquiry has been dominated by investigations that seek understanding of ourselves and of those aspects of the environment appearing to have the most important impact on our lives.

Why should it be otherwise? Even if you are an ultra-realist, someone who adopts the Platonic metaphor of a world with joints at which the sciences should carve, it's hard to envisage some of nature's many component parts declaring their own importance, crying out to human inquirers "I'm significant! Look over here! Investigate me!" Pragmatic progress should absorb lessons from pragmatism. Prominent among them is the thought of inquiry as a human venture, undertaken for purposes that vary over time as aims and interests evolve.

Scientific research is inevitably selective. Some "fundamental questions" are picked because, at a particular historical stage, they promise possibilities of meeting practical needs or arouse the curiosity of people in a position to investigate them. For much of what we know of human history (a handful of millennia), those equipped to make such choices are called, anachronistically, "scientists". During the past seven decades or so members of scientific communities have had to cede some of their absolute authority to decide what counts as "fundamental" and "significant". Needing support to pursue their endeavors, they have had to convince outsiders, sometimes governments with more or less right to claim to represent a broader swath of humanity, sometimes private sugar daddies who recruit investigators to pursue a variety of favored ends. Nobody should suppose that those who pay the piper always call the tune. In the many instances where scientists, with their own curiosities and their own ambitions, hope to address questions diverging from those motivating the paymasters to dispense the necessary funds, canny researchers often find ways to continue the lines of inquiry dearest to their hearts. Many prominent geneticists were exemplary in delivering the kinds of information explicitly demanded by the government departments funding the Human Genome Project; yet, besides the long work weeks in the "front room lab", there was often a "kitchen lab" in the back, where experiments into the group's long-term pet projects – the "theoretically fundamental" significant questions – continued apace.

Whether the research agenda is fixed by scientists, by benign governments, by corporate sponsors, or by a complex tussle among individuals and groups with different motivations, it's worth asking whether the resulting distribution of research effort can go astray. Should any of these modes of determination have final authority? Or, to put the question differently, under what conditions does the decision have legitimacy? Some examples from science past and science recent should give us pause.

Did the study of heredity make progress between 1900 and 1925? The quick answer is "Of course". In 1900, three investigators independently rediscovered the "rules" Mendel had published in his neglected paper. Soon after that, the connection between "Mendelian factors" and chromosomes was recognized, crossing-over and recombination were discovered, and Thomas Morgan and his co-workers developed techniques for gene-mapping. By the early 1920s, geneticists had assembled an impressive corpus of results about the bases of heredity for a variety of traits in some particular organisms, the fruit-fly *Drosophila melanogaster* prominent among them.

But that answer is *too* quick. By 1925, the work of the genetics community had become entangled with that of the Eugenics Record Office – Cold Spring Harbor Laboratories still occupies some of the Record Office's land and buildings. The connection gave a whimsical twist to some of the inquiries pursued. Scientists sought genes for thalassophilia, a condition manifested primarily in adolescent boys, causing them to run away to sea. Fears of the Decline of the West, brought about by an ever increasing "load of mutations", shadowed research in far darker ways. Confident in the possibility of identifying genes for criminal tendencies, social deviance, intelligence and "feeble-mindedness", the work of geneticists fostered programs of confinement and involuntary sterilization. It also inspired tests, wrongly inflicted on many people, blighting the educational prospects of large numbers of them – and also denying asylum to desperate refugees, fleeing danger and persecution in Europe, sent back on the basis of their "eugenic profiles" to die in the genocide carried out in their native lands (Kevles, 1985; Gould, 1981).

Because the record of the genetics community is so mixed, containing answers to some questions whose significance is readily endorsed and pursuit of others that are ill-conceived and sometimes socially harmful, the verdict of assured progress is too quick. I'm inclined to see this as a case in which the best assessment retreats from any overall judgment, opting for identifying progress in some respects, offset by regress in others. (Skeptics about progress often take this "balance of incommensurables" to arise everywhere. The generalization is wrong, but they point to a genuine feature of some episodes. To accept the coherence and usefulness of progress talk doesn't entail determinate judgments of progress for any and every comparison of states.) To a lesser extent, similar difficulties affect recent work in biomedicine. Until quite recently, ventures in using molecular biology to devise treatments and cures for diseases were strongly tilted towards conditions affecting people in the affluent world – citizens of the countries in which most of the research was conducted – while the infectious diseases causing lifelong disability and death to enormous numbers of people, particularly women and children, in the global South were grossly neglected. The imbalance was not restricted to the laboratories of private companies, where the scientists employed sought new cosmetic creams rather than drugs, some of them easily manufactured, that would prevent blindness and fatal forms of dysentery. Until recently, the National Institutes of Health directed only a tiny fraction of its budget towards research into tropical diseases. The amount is almost a rounding error, and the funds primarily went to study diseases in countries to which American soldiers might be dispatched. (More recent budgetary information is hard to elicit, since infectious disease research is lumped with research into allergies; moreover, some inquiries into infections are justified through their contributions to "biodefense".)

Whether or not the features I've indicated warrant retracting overall judgments of progress in these cases, they surely cast doubt on claims of final authority to evaluate significance. Neither a scientific research community nor some hybrid entity, composed of investigators and their sponsors, can be granted the privilege of identifying which questions are "fundamental" and which are "significant". Problems aren't identifiable with what any of these groups of potential decision-makers selects for directing inquiry.

9.5 Problems About Problems

Back, then, to *my* supposedly "fundamental" questions: What is a problem? Who decides? And what is it to solve a problem? Let's start with an explicit relativization. A situation is problematic for an individual or for a group just in case, in that situation, the individual or group cannot discern any means to achieve some chosen end. The problematic situation *blocks* the end. A question is significant for the individual or group when answering the question would issue in actions attaining

the end. The answer solves the problem. It *liberates* the end. Fundamentally significant questions are those enabling systematic achievement of a range of legitimate
ends, where no broader systematic success is currently identified.

Sometimes, the correct response to a problematic situation is to abandon an end.
The 1920s genetics community should have abandoned its search for genes for
thalassophilia, and repudiated premature speculations about hereditary bases for
crime, social deviance, feeble-mindedness and racial inferiority; researchers ought
to have protested when asked to produce cosmetic creams rather seeking life-saving
treatments. A related point underlies our everyday judgments of those who complain
about supposed troubles: "*It* isn't a problem," we say; "*you* have a problem, and you
ought to grow up." Just as there are whiners, some people are too stoical. The spartan
lad with the dangling entrails has a genuine problem, whatever his views of his
condition.

Judgments about progress turn on deciding when problems are significant – when
the blocked ends are worthy of liberation. The first requirement is that the goal
sought should be legitimate. Yet, even when that test has been passed, questions still
arise about whether scarce resources are appropriately directed to the end. Sometimes inquiry goes awry at the first step. Research is undertaken to support unjust and
cruel programs, to lure young people into habits harmful to their health or to craft
techniques for controlling an oppressed group. On other occasions, the end pursued
is harmless enough, but investing effort in attaining it would detract from far more
important endeavors. Cosmetic creams can bring mild benefits to relatively
privileged people, but, you might well suppose, the sufferings of the poor are
more urgent. The line between the thoroughly illegitimate and the legitimate-but-
unimportant is plainly a blur. Just how trivial must the effects of the cream be, and
how pressing the plight of the impoverished, before outrage is the appropriate moral
response to the market decisions of Big Pharma?

In the contemporary world, the results of inquiry interact with social institutions,
often to produce effects nobody can foresee. From a scientific investigation, originally directed at an apparently legitimate theoretical aspiration, understanding the
structure of matter say, conclusions may emerge that are readily adapted to pursue
destructive goals. Dürrenmatt's secret agent posing as Einstein expresses his sorrow:
"Ich liebe die Menschen und liebe meine Geige, aber auf meine Empfehlung hin
baute man die Atombombe" (Dürrenmatt, 1980). In retrospect, should we declare the
problems pursued in the development of quantum mechanics and particle physics to
be illegitimate, and the apparent progress of this area of twentieth-century physics to
be illusory? I think not. When institutions interact, genuine progress in one domain
(science, say) can be nullified or reversed by regress in another. Advances in
physics – as well as in other sciences – have generated new problems for politics
and technology policy, posing challenges that governments have only partially and
haltingly met. Because those who posed the original question could never have
predicted the likely outcomes, they, and the discipline to which they contributed,
cannot be assigned responsibility for the harmful consequences. To declare their
problems illegitimate would set too high a bar, demanding an ability to foresee
beyond human powers. Other institutional interactions with a similar structure are

less benign. Researchers who set out to study racial differences can usually predict with high confidence how their findings will be appraised. Slight indications of the "racial inferiority" of some groups will be overblown, apparently egalitarian results will be dismissed as "politically correct" whitewash. However scrupulous the investigators try to be, their pursuit of these questions clearly runs a high risk of continuing the dark history of racial oppression.

How then, should legitimacy, be understood? I shall adopt a perspective on inquiry and on moral progress that I've defended in some detail elsewhere. Inquiry is best conceived as a practice serving the common human good. The best we can do, at any historical stage, is to try to direct research towards ventures improving human lives. A research agenda should distribute resources among the most urgent legitimate ends. Judgments of legitimacy and urgency are those that would be achieved in an ideal deliberation, one featuring representatives of the full diversity of human perspectives, tutored through provision of the best available factual information, and conducted under conditions of mutual engagement, in which all the deliberators seek an outcome each can accept. Implementing that sort of conversation as adequately as we can is the key to progressive morality and progressive ethical life (Kitcher, 2021).

As I have already noted, our capacity to predict the future is limited, and any judgments of legitimacy we make are provisional. Accompanying the pursuit of inquiry should be a commitment to controlling the unexpected opportunities for damage they can generate. Unforeseen dangers to the wider human good are to be met by improving the socio-political institutions through which results of inquiry are translated into (potentially harmful) policies and actions. If the challenge to reconstruct those institutions is impossible to meet, the provisional judgment of legitimacy has to be retracted. It was reasonable once to attribute legitimacy to that form of inquiry, but, in retrospect, we see that we were wrong to do so.

The account I've outlined is often viewed (with some justice) as focusing inquiry on practical benefits, and downplaying the value of "pure knowledge". Of course, viewing science as progressing through the provision of resources for humanity allows for ventures in "basic science" and for the satisfaction of "simple curiosity". For the history of the sciences shows again and again how addressing the large questions at the apex of significance graphs, the issues scientists hail as "fundamental", enables later proliferation of techniques for tackling practical problems. To cite just one obvious example, the route to contemporary medical medicine ran through a series of investigations into seemingly minor features of insignificant organisms – like eye color in *Drosophila* – out of which, decades later, the tools of current genomics emerged. Had the research community attempted to jump directly to applying Mendelian ideas to human diseases, it would almost certainly have failed disastrously. Wise deliberation will understand the importance of following a variety of routes, some of them surely circuitous (Kitcher, 2011, 118–21).

This "partial vindication" of basic science will undoubtedly strike many people as unsatisfactory. Aren't truth and understanding important *for their own sake*? Here we should return to a question raised (and postponed) in my opening section. Important *to whom*? Even if our world contains some people who exemplify the condition Aristotle celebrates at the end of the *Nicomachean Ethics*, if there are sages whose lives are fulfilled in the contemplation of great and important truths, there are precious few of them. Even supposing their bliss attains the highest form possible for our species, ought it to dominate inquiry when vast numbers of others are narrowly confined, with their legitimate aspirations unfulfilled? Perhaps, in the future, educational reforms will swell the class of those for whom reflecting on large discoveries about the natural world is a source of great satisfaction – a significant *component* of what makes their lives go well; it seems unlikely that it will ever emerge as the *summum bonum* for more than a tiny minority. The democratic core of my picture of ethical life requires attending to the frustrations encountered by the many. It is perhaps liberating to compare the appreciation of theoretical science (for its own sake) with the appreciation of high avant-garde art. If you abstract from the potential of basic science to yield a cornucopia of practical techniques, is there any greater justification for supporting high theory in science than for underwriting the ventures of those who are at the cutting edge of the most sophisticated forms of art?

9.6   Claiming Truth

Truth has tiptoed into my discussion. Its entrance marks the fact that problem-solving is often linked to truth. People are inclined to say (as, challenging Laudan, I once did – Kitcher 1993, 130n) that problem solutions are true answers to significant questions. Nevertheless, seeing scientific progress as problem-solving ought to resist that narrowing of the topic. Central to my approach to problems was the (Deweyan) idea of a problematic situation (Dewey, 1938). Finding the question (if there is one) is already a response, an accomplishment. Resolving the problem consists in liberating the (legitimate) end. Sometimes that can be done without some linguistically formulated result. You find a way out of the wood, you discover how to build a stable barrier against the wind, you rescue diseased and withered plants.

Scientific inquiry doesn't invariably aim at propositional knowledge. Research often delivers measuring instruments, techniques, new substances and novel organisms. Developing a mouse model for a human disease can take months or years and be hailed by the research community as an important accomplishment. Providing a means to liberate an end is sometimes accompanied by a description of how something is to be done – a *recipe* shared with other investigators. Yet, like the everyday recipes of the kitchen, the verbal formulation is typically incomplete. Learning how to act in the appropriate way may require significant tinkering, or even an apprenticeship with the author of the new technique.

The practice of a scientific community amasses techniques, used to liberate previously blocked ends and, typically, potentially useful for unblocking similar future ends. Among these techniques are some that deploy symbolic resources – descriptions, equations, diagrams, graphs, and the like. The body of techniques central to such practices includes far more than purely practical skills. There are also hybrids, in which a symbolic representation supplements and guides activity.

Problem-solving sometimes consists in depositing a novel statement or representation in the corpus of techniques, together with some testimonial to its truth or accuracy. Scientific fields advance by providing resources that can be relied on for particular kinds of future projects. The limiting case – unqualified truth – occurs when inquiry liberates an end (or some collection of ends) by introducing a statement that does not interfere with any of the representations or techniques previously adopted, and when all the other potential rival statements meeting the constraints of the problem have been shown to be inadequate. Endorsement of statements as "true enough" recognizes that the statements can be deployed in particular contexts or for specific purposes (Elgin, 2017).

Liberating ends can have costs. Adding a statement recognized as inconsistent with statements previously adopted and put to use in successful problem-solutions requires restricting the circumstances in which some or all of the rivals can be relied on in future problem-solving. The episodes Kuhn identified as scientific revolutions are special instances of this general type. Some end, long blocked, would be liberated if the community used a representation incompatible with parts of accepted lore. Full revolutionaries recognize the need to scrap many apparently successful problem-solutions if the representation is adopted as reliable across the board – and they accept the task of showing how their new practice can be expanded to replace the discarded achievements. A more conservative strategy is to accompany the newly-introduced material with explicit qualifications, limiting its application to specified contexts. Retreat to instrumentalism is one mode of pursuing this strategy.

Scientific inquiry thus takes over – and refines – the simplest pragmatist approach to truth. According to that conception, 'true' is used provisionally to mark statements – and, more generally, 'accurate' is used provisionally to mark other kinds of representations – that have previously helped to liberate some end and can now be relied on for use in achieving other ends, perhaps quite unrestrictedly, perhaps with specified limitations. Pragmatists are happy to concede that the title tentatively given may have to be withdrawn. The pessimistic induction on the history of science holds no terrors for them. For they envisage two possibilities. Perhaps the representation now adopted and put to use will survive in the indefinite future; or perhaps it will have to be replaced by something that proves more consistently reliable. Finding those superior replacements is likely to be promoted through efforts to put the representation to work. Hence arriving at representations that prove "expedient in the long run and on the whole" – that thus count as true according to William James's version of the pragmatic exposition of truth (James, 1907, Lecture VI) – is facilitated by the strategy of systematically adopting and using representations that liberate previously blocked ends.

This apparent vindication of the pragmatist approach needs to be refined to cope with complications emerging from earlier discussions. As James himself emphasized, there are occasions on which the potential costs of false belief – or the potential costs of withholding belief – favor deviating from the attribution strategy (James, 1896). When relying on a representation might lead to great suffering, if the representation were inaccurate, it is right to use it cautiously, or not at all; when immediate action is needed, and no other guidance is at hand, standards for adoption are relaxed. Furthermore, when conflicting representations have enabled researchers to liberate different ends, subsequent practice needs to differentiate the circumstances under which each is properly to be deployed. Incompatible models are often valuable, but the communities using them need to learn the limits of their safe application.

I now want to suggest that the inconveniences of pluralism, arising from the need to monitor the deployment of clashing representations, underlie the search for truth. I'll offer a speculative genealogy to explore how the ancient idea of truth as correspondence to reality might have emerged, and how it came to figure in the widespread view of science as aiming at the "true story" of nature.

9.7 Seeking the Truth: A Myth

Science as we know it develops out of forms of inquiry carried on not only during the periods for which we have written records but also in the very distant human – and pre-human – past. For at least two and a half million years, our ancestors have made tools and used them for a variety of tasks. At some stage in hominin evolution, they discovered how to shape pebbles into all-purpose tools for digging, chopping, and cutting – the "Swiss army knife" of pre-history, also known as the hand axe. In the remote past, a hominin predecessor or some group, discovered the technique of flint-knapping. That technique became *entrenched*, learned by descendants, used in practical projects, passed down across the generations, and transmitted to bands who had not yet arrived at it independently.

Human pre-history testifies to all kinds of investigations, some of them doubtless costly to the investigators. Our predecessors worked out which kinds of naturally occurring plants could safely be eaten, how to control the use of fire, how to transport the water they needed, how to construct various forms of shelter, how to trap large and dangerous animals, how to extract pigments for painting their bodies and the walls of caves, and a host of other things. Inquiry began long before human beings acquired language, accumulating techniques and entrenching them. The beginnings of ancient science (as it is commonly understood) in Mesopotamia and in Egypt presuppose this large corpus of skills. Science grows out of a long prior history of practical inquiry. Any goal of seeking the truth is a latecomer.

Where might that idea have come from? Let's do some myth-making.

The scene: a tract of savannah, some 60,000 years ago. A small band of hunter-gatherers has temporarily settled in the area. Band members engage in daily foraging, and, from time to time, repair their shelters and their carrying equipment. It's useful for them to know where water can be found, where dangerous predators might be encountered, and where particular kinds of materials – useful for building, mending, and other everyday projects can be collected. The band members have learned how to speak to one another, and they understand the advantage of pooling information. Thanks to the division of epistemic labor, they benefit from one another's explorations of the environment.

One member of the band (I'll call her "Jones" – with apologies to Wilfrid Sellars 1963, Chapter 5) is especially adept at exploration. Equipped with a strong visual sense and a good memory, she is able to travel further from the place at which the group reunites. One day, Jones returns from an extensive expedition, bringing with her materials needed for urgent projects. She explains to her fellow band members how she has discovered a relatively distant place in which these resources abound. It's not entirely suitable for a campsite, since water is scarce in the area and snakes are also frequent. Others are eager to collect more of the materials than Jones has brought back, but she is too tired to lead them. She points in the direction she took, but the route has to be extended beyond the visible horizon. How can she guide them?

Taking a stick, she draws in a bare patch of dirt. From the furthest visible landmark, she extends the route, using simple pictures to mark places where it is necessary to turn – a cluster of rocks here, a dead tree there, a brackish pool and so forth. Jones has created the first sketch map. Those wishing to go collecting look carefully at her drawing. They question her about details. One of them even copies the map on nearby ground. Finally, they decide that they are ready, and leave. A few hours later they return, triumphantly bearing the resources they needed.

The practice catches on. The band has already begun the division of labor, recognizing how some are better at fashioning arrows, others at detecting underground water sources, yet others skilled at mending the nets and bags they use for carrying. A new role is created, that of the scout. Jones is deputed to explore the environment, finding places it would be useful to visit, and then to report back to others. After her excursions, she sometimes records her route by drawing a map.

As Jones ages, she is no longer able to travel as far. Her visual acuity also diminishes. Fortunately, other, younger, band members have some of the talents she once enjoyed. Jones trains successors, who can go out in her stead. Although the apprentices usually help in directing successful group foraging, their performances are imperfect. They don't yet prove as reliable as Jones in her prime. Sometimes, in their drawing they are impatient, and the would-be foragers struggle to reach their destination, or return, frustrated and empty-handed. Jones continues to work with them, and one or two of them are eventually able to provide directions that invariably succeed. Those who do not are assigned to different tasks.

In the course of her career as instructor, Jones arrives at an insight. Reflecting on the pattern of successes and failures, she recognizes the importance of a correspondence between the drawing and the environment. When the relations among the lines and the symbols, on the one hand, and the features of the terrain on the other, are appropriately similar, the representation in the sand guides the travelers to the target; when that correspondence is absent, the mission typically fails. Jones has learned empirically that a correspondence between representation and world enhances practical achievements. She first invented the idea of a hybrid inquiry, an inquiry in which a symbolic representation is deployed in action towards a desired end. Through her reflections on when such representations work she understands what later generations will see as a commonplace. At the very end of her life, she explains the observation, first to the scouts who have succeeded her, and then to the entire band. "There needs to be the right correspondence between the lines on the map and the world," she tells them, "our drawings must be accurate." Jones' legacy is the correspondence theory of truth.

9.8 Practical Truth

If Jones actually existed in the human past, I would be flabbergasted. It would even be surprising if the notion of correspondence truth developed out of ventures in cartography alone. For there are several other potential sources of notions of truth and accuracy. The myth I have told shows one way in which a prelinguistic human community might have felt the need to map part of the environment. Mapping might well have begun from attempts to reach different goals. More importantly, there are other plausible candidates for projects inspiring recourse to visual representation. Our predecessors may have drawn plans for the construction of shelters, sketched the proper shape of a tool, used a figure to show how to dig a safe firepit in an area full of combustible material. In all these instances, they could have been led to Jones's commonplace: in a hybrid project drawing on a representation, success requires correspondence (enough) between the representation and the part of nature it is supposed to represent. Although it is likely that the insight only emerged after the acquisition of full language – and thus perhaps in contexts of adjudicating disagreement – I conjecture that the notion of accuracy arose first for visual representations, and was only later imported into the linguistic domain.

The truths (or, more generally, the accurate representations) sought by our ancestors were originally inspired by practical projects. Small groups of human beings, coping with all kinds of uncertainties and daily pressures, lacked the leisure for pondering how best to represent the natural world. Disinterested truth-seeking appears late in the history of human inquiry, figuring as something to be pursued only when people have discovered that accurately representing the features of particular parts of nature can stimulate new and valuable practical ventures. Mapping the heavens and subsequently exploring the motions of heavenly bodies emerges from earlier recognition of gross regularities, useful for organizing mundane activity. The moon waxes and wanes, and awareness of the cycle can help in planning. An imminent full moon is worth waiting for if you want to move camp by night; even though the moon tonight is a bare sliver, tomorrow it will vanish entirely, and the prospects for the hunt will be even better. Charting the pattern of the sun's rising and setting is a harder task, but, once accomplished, it offers the key to predicting the seasons. Agricultural societies, many of them pre-literate, built monuments and crafted artefacts to help them identify the solstices and equinoxes. Star charts enabled the mariners of the ancient world to navigate. These practical projects, whose significance we continue to recognize, were, of course, accompanied by another, one that most educated people today find dubious. Predicting the course of human affairs was an even more ambitious practical endeavor, one supposedly facilitated by accurate calculation of the positions of important stars at particular historical moments. Astrology and divination fueled the search for exact astronomical truths.

Similar stories can be told for the origins of other sciences. Interests in materials, in the motion of projectiles, in crops, in the properties of wild and domestic animals, in soils and strata, generated the fields we now know as dynamics, chemistry, metallurgy, botany, zoology, and earth science. When the sciences are set in the long career of human inquiry, the practical ventures are the roots out of which contemporary assessments of significance grow. Moreover, the later enterprise of methodology, devoted to understanding how findings give evidence for the truth of scientific hypotheses and theories, is born in the simple pragmatist strategy of accepting statements – and, importantly, other kinds of representations – when they aid in liberating goals. Mature science continues to use that simple strategy, refining it by taking a larger view of the goals to be achieved, and thus adopting a more complex view of "what is expedient in the way of belief". Students of science who focus on the elementary features of inquiry, the link between significance and the practical, will acquiesce in the thought that "it is true because it is useful". Yet the tether to the practical is often long – the inquiries of contemporary cosmologists and astrophysicists are at a far remove from the projects of the early agriculturalists. Scholars concerned with advanced theoretical science, with its apparently disinterested search for truth, will have absorbed Jones's insight. They will say instead: "It is useful because it is true." William James believed we can assert either or both (James, 1907, 98). He was correct, although it should be recognized that the senses of 'because' differ between the two utterances.

9.9 Pure Inquiry

When inquiry is seen as a practice whose starting points and directions are inevitably selective, the role of value judgments in that practice cannot be denied. To be sure, the long history of inquiry has led from attempts to fashion techniques for survival to contexts in which a segment of our species can pose questions about the "fundamental structure of nature", and even to prize highly answers to those questions. Yet what we see as "fundamental structures" arise for worlds of experience that have been reconstructed out of a sequence of efforts to intervene and to control, ventures that should have been governed by a commitment to lessening the confinement of human lives.

Science, as it has been practiced for all of recorded human history, has collected symbolic representations, deploying them in hybrid techniques. To the extent that its investigations have been legitimate, the techniques have had the potential to reduce the diminution of human welfare – and, if that potential has been blocked, we should now attempt to remove the obstacles. Becoming fully aware of the ways in which those techniques have shaped the worlds of experience, among which we live and move, opens up for all of us today the task of appraising those worlds from the perspective of our values – and seeing the latter as themselves evolving, partially in response to the restructuring of the world factual inquiry has generated. Progress in the ethical project is inevitably entangled with progress in the sciences, broadly construed.

My version of pragmatism recognizes no absolute foundations on which further human progress can be founded. The future course of inquiry ought now to be guided by selecting projects conforming to the best value-judgments we can derive from the project of determining what is valuable. Equally, continued investigations of values are dependent on the best-established results of inquiry. When that is appreciated, the idea of the search for truth as some supreme end evaporates. The satisfaction of human curiosity is a value, one among many. Today, our species inhabits a world that continues to pose challenges to human survival. Our need for hybrid techniques has not gone away. Progressive science consists in pursuing inquiries worth undertaking, and that requires preliminary diagnostic work to identify the most pressing human problems and the most promising prospects for overcoming them. Eloquent rhetoric about the disinterested search for truth is worthwhile only insofar as it facilitates diagnosis. Otherwise, it is a distraction, a "sentimental indulgence for a few" (Dewey, 1916, 338).

References

Cartwright, N. (1999). *The dappled world*. Cambridge University Press.

Culp, S., & Kitcher, P. (1989). Theory structure and theory change in contemporary molecular biology. *British Journal for the Philosophy of Science, 40*(4), 459–483.

Dewey, J. (1916). *Democracy and education*, reprinted as Volume 9 of The middle works of John Dewey. Southern Illinois Press, 1985.

Dewey, J. (1938). *Logic: The theory of inquiry*, reprinted as Volume 12 of *The later works of John Dewey*. Southern Illinois Press, 1991.

Dupré, J. (1993). *The disorder of things*. Harvard University Press.

Dürrenmatt, F. (1980). *Die Physiker*. Diogenes Verlag.

Elgin, C. (2017). *True enough*. The MIT Press.

Gould, S. J. (1981). *The mismeasure of man*. Norton.

James, W. (1896). The will to believe, reprinted in *The works of William James: The will to believe*. Harvard University Press, 1979.

James, W. (1907). *Pragmatism*, reprinted as *The works of William James: Pragmatism*. Harvard University Press, 1975.

Kevles, D. J. (1985). *In the name of eugenics*. Knopf.

Kitcher, P. (1984). 1953 and all that. A tale of two sciences. *Philosophical Review, 93*(3), 335–373.

Kitcher, P. (1993). *The advancement of science*. Oxford University Press.

Kitcher, P. (2001). *Science, truth, and democracy*. Oxford University Press.

Kitcher, P. (2011). *Science in a democratic society*. Prometheus Books.

Kitcher, P. (2021). *Moral progress*. Oxford University Press.

Kuhn, T. S. (1962). *The structure of scientific revolutions*. University of Chicago Press.

Kuhn, T. S. (2000). *The road since structure*. University of Chicago Press.

Laudan, L. (1977). *Progress and its problems*. University of California Press.

Sellars, W. (1963). *Science, perception, and reality*. Routledge & Kegan Paul.

van Fraassen, B. (1980). *The scientific image*. Clarendon Press.

Chapter 10
The Logic of Qualitative Progress in Nomic, Design, and Explicative Research

Theo A. F. Kuipers

Abstract The leading idea of this paper is twofold: (i) claims that progress has been made can be justified, and (ii) progress in different areas has a common structure. The paper proposes two general sufficient conditions for making justified claims of qualitative progress. The conditions are first presented in their strict form, assuming a specific target. They are then applied to theory-oriented research aiming at nomic truth approximation as well as to design research aiming at a certain product or process.

Next, the two conditions are generalized, instead of assuming a specific target, now only assuming that some specific constraints have to be satisfied. They are again applied to theory-oriented science, but now only as far as it primarily aims at empirical progress in light of constraints provided by evidence. It can be shown that such empirical progress is functional for nomic truth approximation. The generalized conditions are also applied to concept explication, undertaken in philosophy and science, where the constraints are provided by evident examples and conditions of adequacy, leading to explicative progress. Finally, progress in design research is briefly revisited in the generalized perspective.

The paper closes with some suggestions for further research, notably case-studies and elaborating applications to e.g. technological, social and moral progress.

Keywords Progress · Nomic research · Design research · Concept explication · Truth approximation · Empirical progress · Prototype · Ocean cleanup · Symmetric difference

T. A. F. Kuipers (✉)
University of Groningen, Groningen, The Netherlands
e-mail: t.a.f.kuipers@rug.nl

© The Author(s), under exclusive license to Springer Nature Switzerland AG 2022
W. J. Gonzalez (ed.), *Current Trends in Philosophy of Science*, Synthese Library 462, https://doi.org/10.1007/978-3-031-01315-7_10

10.1   Introduction

What do we or, better, what can we sensibly mean by expressions like "Theory Y is (empirically) more successful, relative to the evidence, than theory X", e.g. "Einstein's theory of gravity is more successful than Newton's theory"? That is, what is progress in empirical (in particular nomological) research, briefly, what is empirical progress?

What do we or, better, what can we sensibly mean by expressions like "Prototype y is, relative to the (un)desired functional features, better than prototype x". For example, The Ocean Cleanup project (Boyan Slat) aims at a device to clean-up the plastic soup in the ocean. Recently the team reported that a revised prototype performs much better. That is, what is progress in technological or design research?

What do we or, better, what can we sensibly mean by expressions like "Explication E2 is a better explication of a concept than explication E1", e.g. "Jeffrey's explication of conditionalization is an improvement of the Bayesian one"? That is, what is progress in explicative research?

Regarding the first two expressions, it turns out to be easier to start with the more ambitious expressions: "Theory Y is closer to the (nomic) truth than Theory X", and "Prototype y is, relative to an all-inclusive desired profile, better than prototype x".

The leading idea of this paper is twofold: (i) claims that progress has been made can be justified, and (ii) progress in different areas has a common structure. The claim that qualitative progress has been made occurs in science and society. However, it is frequently contested that such claims can be justified. In this paper we will formulate general sufficient conditions for making such claims, with applications to nomological and technological sciences and to conceptual research in philosophy and science.

Roughly, there are two conditions that are sufficient to characterize progress:

1. *The (generalized) target condition.* There should be a clear set of conceptual possibilities within which either, the strict variant, a particular subset forms the target of the research, e.g. the set of physical possibilities or the set of desired properties. Or, the generalized variant, at least a number of specific constraints can be formulated that have to be satisfied. E.g. that a theory has to explain some specific empirical laws or that an explication of a concept has to respect certain conditions of adequacy.
2. *The (generalized) DSD-condition.* The candidate for progress has to satisfy a specific criterion relative to the target in comparison with its competitor, the so-called (strict or generalized) decreasing symmetric difference (DSD-)condition.

In Sect. 10.2 these two conditions will be presented in general, assuming that there is a unique target, such that the strict DSD-condition for progress can be applied. In Sect. 10.3 this will be applied to theory-oriented research aiming at nomic truth approximation as well as to design research aiming at a certain product or process.

In Sect. 10.4 the two conditions will again be presented in general, but now assuming that some specific constraints have to be satisfied, the generalized target condition, leading to the generalized DSD-condition for progress. In Sect. 10.5 this will again be applied to theory-oriented science, but now only as far as it primarily aims at empirical progress in light of constraints provided by evidence. It can now be shown that empirical progress is functional for nomic truth approximation. The generalized conditions will also be applied to concept explication, with constraints provided by conditions of adequacy and by evident examples, leading to explicative progress. Finally, we will briefly revisit progress in design research from the generalized perspective.

Section 10.6 closes with some suggestions for further research, notably case-studies and elaborating applications to e.g. technological, social and moral progress.

The paper improves on our earlier work both by providing an encompassing, integrative insight in (the logic of) qualitative progress and by introducing for that purpose the notion of 'generalized symmetric difference'. However, some relativizing remarks are in order. First, we have to stress that we are dealing with sufficient conditions of a qualitative nature. There are lots of contexts in which it is plausible to speak about progress on the basis of quantitative conditions, which may or may not automatically be satisfied if our qualitative conditions are satisfied. Second, in many cases the (generalized) conditions will not apply and lead to dispute. So, our two (generalized) conditions constitute a kind of basic or ideal type, providing a standard to discuss the degree to which this is satisfied in a particular case and leaving room for refined versions. Finally, this is mainly a rather abstract paper. Sections 10.2 and 10.4 are on the meta-meta-level and the applications in Sections 10.3 and 10.5 are on the meta-level. However, as a few indicated (tentative) examples and case-studies on the object level illustrate, it is easy to look for further potential examples and case-studies. To be sure, there is a need for detailed case-studies to evaluate the here presented basic logic of qualitative progress and to develop refined logics.

10.2   Two Conditions for Qualitative Progress

10.2.1   The Target and Decreasing Symmetric Difference (DSD-)Condition

In order to characterize evident cases of progress, or regress, in some developmental process the notion of symmetric difference will play the main role. Assume that there is a target subset T of some universe of items U and that T is (supposed to be) known, *the target condition*. Suppose further that the task is to specify candidate subsets X and Y of U by which one tries to reproduce T. A plausible measure for the (qualitative) distance between X and T is $(X-T) \cup (T-X)$, known as the symmetric difference between X and T and usually abbreviated as $X \Delta T$, see Fig. 10.1.

Now suppose that X and Y are subsequent attempts (in this order) to characterize T, so that Y is a successor candidate of X. The Decreasing Symmetric Difference (DSD-) condition requires for 'Y being closer to the target T than X', that is, *progress*, that the symmetric difference of Y with T is a proper subset of that of X with T: $Y \Delta T \subset X \Delta T$. It is easy to check that this condition can be decomposed into:

$Y - T \subseteq X - T$ and $T - Y \subseteq T$
   $- X$ and at least one of them must be a proper subset relation.

Fig. 10.1 The symmetric difference between X and T: $X\Delta T = (X-T)\cup(T-X)$, i.e. the union of the two @-areas

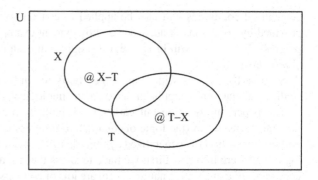

Fig. 10.2 Y is closer to the target T than X: shaded areas empty and at least one #-area non-empty

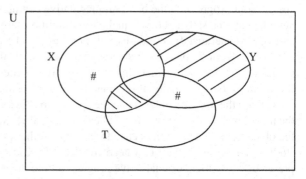

Of course, if the order is reversed, so that X is now considered a successor candidate of Y, the same condition would characterize regression or *regress*, that is, of X relative to Y. In contexts where the target T is changeable, the difference (s) between X and T may also be reason to adapt T to T* in the direction of X. We will call it a *concession* to X, when $X\Delta T^* \subset X\Delta T$.

Figure 10.2 represents the situation that "Y is closer to the target T than X" as follows: shaded areas indicate empty sets and at least one of the two #-areas has to be non-empty.

The DSD-condition is a rather strict, qualitative criterion that will not easily be satisfied. However, we like to propose it as a kind of ideal type of progress. If it is not fully satisfied, it specifies the challenge how to improve the better candidate in such a way that this condition becomes satisfied. Moreover, the DSD-condition is the starting point of concretizations.

10.2.2 Three Concretizations

The strict condition just proposed can be modified in different ways in order to be applicable to more and more interesting cases. Three such concretizations have already been explored in earlier work in the context of truth approximation (the topic of Sect. 10.3.1), viz. quantification, refinement, and stratification. Here they will briefly be indicated in their general form, with references.

To be begin with, there is a plausible way to liberalize the condition to a *quantitative* one when the universe of relevant items U is finite. Instead of imposing the subset relation between the relevant symmetric differences, we may impose the condition that the one symmetric difference is smaller in number than the other, formally: $|Y \Delta T| < |X \Delta T|$. For an elaboration of this liberalization in the context of truth approximation (the topic of Sect. 10.3.1), see (Kuipers, 2017, 2019, Ch. 5).

Another concretization takes into account that not all differences are equally important, e.g. one item in the symmetric difference may be more similar to a target item than another. The so-called *refined* definition of 'closer to the target' is weaker than the 'basic' definition in one respect but stronger in another respect. For an elaboration of this concretization in the context of truth approximation (the topic of Sect. 10.3.1) and empirical progress (the topic of Sect. 10.5.1), see (Kuipers 2019, Ch. 6, and 2020a). A final concretization in the same contexts concerns *stratified theories*: it is assumed that there is a (theory-relative) distinction between an observational and a theoretical vocabulary and on both levels the existence of the nomic truth is assumed. For details, see (Kuipers 2019, Ch. 7, and 2020b).

In the next section we will deal with two straightforward applications of the target- and the DSD-condition: nomic truth approximation and progress in design research. In Sect. 10.4 we will deal with a generalized target and DSD-condition and in Sect. 10.5 two applications, viz., empirical progress and progress in concept explication, and a sketch for refinement of progress in design research.

10.3   Two Applications: Nomic Truth Approximation and Progress in Design Research

The conceptual proposals enable us to explicate two rather different kinds of progress in science, viz. truth approximation (increasing truthlikeness or verisimilitude) and progress in design.

10.3.1   Nomic Truth Approximation

There are at least two kinds of truth that one may be interested in: the truth about what is or was actually the case and the truth about what is nomically (e.g. physically) possible. Here we will focus on nomic truth approximation, also called increasing nomic truthlikeness or increasing nomic verisimilitude or, more specifically, increasing legisimilitude.

The point of departure for characterizing nomological research is a domain and a vocabulary enabling the formulation of a set of conceptual possibilities (also called potential models, i.e. structures of a certain similarity type), indicated by U. Think e.g. of the set of conceptually possible states of an amount of gas regarding volume, pressure and temperature. Moreover, it is assumed here that the representation of the domain in terms of that vocabulary amounts to a unique, (in some way or other) characterizable subset T of U consisting of the nomic possibilities, that is, the nomically possible conceptual possibilities. Depending on the context 'nomically possible' might for example be interpreted as physically, biologically, or economically possible. The complement of T, $cT =_{df} U - T$, represents, of course, the nomic *im*possibilities, that is, conceptual possibilities which cannot be realized, whatever (discipline specific) efforts we might do. T is the, by definition, not yet known target set of research, which may be identified with the (nomic) truth[1] for reasons that will become clear soon. The ultimate aim of theory formation and revision is the actual characterization of T, that is, the demarcation of T from cT.

In the present context it will be useful to focus first, in Popperian spirit, on exclusion theories, i.e. theories that exclude that certain things will happen. Hence, a theory X is a subset X of U with the claim 'all nomic possibilities are included in X', formally: 'T $\subseteq$ X' or, equivalently, 'all possibilities excluded by X are nomic impossibilities', formally: 'cX $\subseteq$ cT', for which reason it is called an exclusion claim. In case the exclusion claim is true (false) the theory itself will also be called true (false).

If T is a subset of Y and Y is a subset of X, formally: T $\subseteq$ Y $\subseteq$ X, Y excludes more nomic impossibilities than X and hence the corresponding claim 'T $\subseteq$ Y' is stronger than 'T $\subseteq$ X', and nevertheless true. In this sense theory T is the strongest true theory, and this is the plausible explication of 'the (nomic) truth' in the present context.

The basic definition of formal progress in nomological research, i.e. nomic truth approximation or increasing legisimilitude, now is just an application of the general definition of progress in terms of the DSD-condition.

Definition: Theory Y is closer to T than theory X iff YΔT $\subset$ XΔT

Recall Fig. 10.2. It now represents the situation that theory Y is closer to the nomic truth than theory X.

Above we focused on exclusion theories, but roughly the same story can be told when we go to 'inclusion' theories. An inclusion theory X claims that all members of X are nomically possible, that is, X $\subseteq$ T, and Y amounts again to progress when it satisfies the DSD-condition relative to X.

Finally, we may deal with two-sided theories <M, P> consisting of an inclusion claim M $\subseteq$ T and an exclusion claim T $\subseteq$ P, with M a subset of P in order to be compatible claims. Progress requires then the satisfaction of at least one of the two corresponding DSD-conditions and the other in a plausible weak sense. In case M = P = X we obtain a 'maximal' theory, with an inclusion and an exclusion claim for the same set, with the consequence that the DSD-conditions are now equivalent.

---

[1]Moreover, it is a deterministic truth (as opposed to a probabilistic truth).

Although the corresponding definitions of 'closer to the truth' among exclusion, inclusion, or maximal theories are plausible, it is not easy to find genuine scientific examples. One might e.g. submit that whatever the true maximal theory is about the nomically possible states of a gas, the theory of Van der Waals is closer the true one than the ideal gas theory. This example requires already the refined version as indicated in Sect. 10.2.2 (Kuipers 2001, 10.4.2; Kuipers 2019, 6.4.2). Similarly, it is possible to argue within the framework of 'the old quantum theory' that, assuming for a moment that Sommerfeld's theory of the atom is the truth, Bohr's theory of the atom is closer to the truth in the refined sense than Rutherford's (Kuipers 2000, Ch. 11). This is called a case of 'potential nomic truth approximation'.

For an elaboration of the nomic application of the DSD-condition and its con-cretizations, see (Kuipers 2019, Chs. 2–7). This elaboration is in terms of the truth- and falsity-content of theories, of which the definitions are the qualitative versions of the quantitative ones proposed in (Cevolani et al., 2011) applied in the nomic context.

Recall that in actual science the nomic truth is of course not known beforehand. On the contrary, it is the great unknown to be searched for. In Sect. 10.5.1 we will deal with empirical progress, which can more or less convincingly be established and which can be used to argue abductively for the claim that nomic truth approximation has also been obtained.

10.3.2   Progress in Design Research[2]

In design research we aim to design, by way of tentative prototypes, a certain product (or process) having certain desired properties or features and lacking other, undesired ones. Design research abounds in modern science, not only in typically applied and technological sciences, such as pharmacy and mechanical engineering, but also within more basic disciplines. Think of the well-established synthetic chemistry and, more recently, synthetic biology. We start with the simplest version of this application of our logic of progress.

10.3.2.1   Basic Application

Let U indicate the space of relevant (logically independent) features, that is, desired and undesired features. Let D indicate the set of desired features, the desired profile, and hence U-D the set of undesired features. D is of course the target. Every prototype x will have an operational profile, $O(x)$, the set of (relevant) features it actually has and hence $U-O(x)$ is the set of features it lacks. Now one may attribute to every prototype

---

[2]This section, except the Ocean Cleanup example, is strongly based on Kuipers (2013), notably Section 3 on design theories. Section 2 of that article deals with design laws and Section 4 with design research programs.

the equality claim $O(x) = D$ and one may compare different prototypes in terms of the 'degree' to which they realize the desired properties and avoid the undesired ones. This leads to the plausible definition of (qualitative) progress in design research. The idea is of course that the new prototype satisfies (set-theoretically) more desired features and lacks more undesired features, that is, it satisfies the DSD-condition.

Definition
Prototype y is, relative to D, an improvement compared to prototype x iff the symmetric
    difference between $O(y)$ and D is a proper subset of that between $O(x)$ and D,
    formally: iff $O(y)\Delta D \subset O(x)\Delta D$.

Another option appears when, for some reason or other, $O(x)$ turns out to be in some respects at least as attractive as D. This opens the way to adapting D in the direction of the $O(x)$. This occurs for example quite often in pharmaceutical research, as pointed out by Rein Vos in his aptly called book *Drugs Looking for Diseases* (Vos 1991). Formally, it amounts to replacing D by D* such that $O(x)\Delta D^* \subset O(x)\Delta D$. This type of transition we called in general already (doing) a *concession*.

It is plausible to see technological progress as similar to progress in design, that is, progress not between prototypes but between products that come on the market. For example, new cars with some new desired properties while keeping all desired properties of their predecessors (and hence, in its simplest representation, the reverse for the undesired properties).

10.3.2.2 Some Refinements

So far for the simplest application. To represent, usually very complex, cases of design research, e.g. drug design, speech technology, food technology, traffic technology, one needs compound representations. The aim of (complex) design research can be adequately represented within a space of possibly relevant properties by property profiles of (prototypes of) the products to be made and by two, mutually orthogonal, distinctions (Kuipers et al., 1992; Kuipers 2001, Ch. 10):

– desired and operational profiles,
– structural and functional profiles.

The first distinction we introduced already, still assuming a restricted space of strictly relevant properties. It will be adapted after clarifying the second.

For the second distinction it is presupposed that the space of possibly relevant properties can be subdivided into a functional and a structural subspace, roughly the things you can(not) do or might (not) wish to do with a certain device versus the ways in which this is or might be realized. For example, in car technology there are on the one hand structural properties like wheel diameter, and engine power, and, on the other, functional properties like maximum speed and range. For the functional subspace we still assume that all its properties are relevant, that is, either desired or undesired.[3] A functional profile collects the total of functional properties of a (potential) product, thereby implying that it lacks the other properties in the subspace of functional properties. Similarly, a structural profile collects the total of structural properties of a (potential) product, thereby implying that it lacks the other properties in the subspace of structural properties.

Returning to the first distinction, that between desired and operational profiles, we implement it in the second distinction. A desired functional profile collects the total of desired functional properties of an intended product, thereby implying that the other functional properties are undesired. A desired or, as we prefer, an intended structural profile collects the total of intended structural properties of an intended product, thereby implying that the other structural properties are unintended. An operational functional/structural profile collects the total of actual functional/structural properties of (a prototype of) a product, thereby implying that the other properties in the relevant subspace of properties are absent.

In sum, and assuming that there is already a prototype of a product, say x, the two distinctions lead to four kinds of profiles:

- an operational functional profile $(OF(x))$ and an operational structural profile $(OS(x))$,
- a desired functional profile (DF) and an intended structural profile (IS).

On this basis, design theories can be formulated, tested, and improved. A design theory presupposes the above distinctions and deals with claims of the following type, leaving a contextual reference implicit:

Imposing intended structural profile IS causes functional profile F.

This type of claim has two versions, an actual and an intentional one, depending on whether it refers to an existing or an intended artefact:

Operational structural profile $OS(x)$ causes (operational) functional profile $OF(x)$,
formally implying, for all x, if $OS(x)$ then $OF(x)$
Intentional:
Imposing intended structural profile IS will cause desired functional profile DF,
formally implying: for all x, if $IS(x)$ then $OF(x) = DF$.

We define a design theory more specifically as a tuple of the form <IS, DF> with the corresponding intentional claim, which is further on simply called the claim of the theory. A design theory is typically science-based when its claim is based on theoretical considerations, here called the underlying theories, which make testing of the design theory, see below, also testing of these underlying theories.

Figure 10.3 characterizes the problem state of design research at a certain moment, assuming that there is already a prototype (x).

Testing the claim of a design theory <IS, DF> is of course done by making a prototype, x, having the structural profile IS and checking whether $OF(x) = DF$ holds. As the picture already suggests, the prototype may or may not be such that $OS(x) = IS$.

---

[3] In Sect. 10.5.3 we will indicate a way to liberalize this assumption as well.

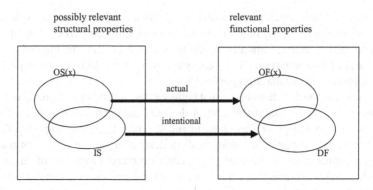

Figure 3: A problem state of design research. Legend:

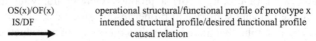

OS(x)/OF(x)   operational structural/functional profile of prototype x
IS/DF   intended structural profile/desired functional profile
                      causal relation

Fig. 10.3 A problem state of design research
Legend:
OS(x)/OF(x)   operational structural/functional profile of prototype x
IS/DF   intended structural profile/desired functional profile
→   causal relation

Let us start with the second case, OS(x) ≠ IS, in which case the theory does not claim that OF(x) = DF. However, assuming the theory is true, it is plausible to try to make progress by making a new prototype having a structural profile that is more similar to IS, that is, a prototype y that satisfies the DSD-condition relative to IS in comparison with x. Prototype y may be expected, assuming the theory is true, to have a functional profile that is more similar to DF, that is, a functional profile that satisfies the DSD-condition relative to DF. The option hinges upon the heuristic default principle that more similar structural profiles cause more similar functional profiles. To be precise, assuming the design theory, if y is a structural improvement of x relative to IS, OS(y)ΔIS ⊂ OS(x)ΔIS, y is likely to be a functional improvement of x relative to DF, OF(y)ΔDF ⊂ OF(x)ΔDF.

To elaborate a specific example requires quite some space and it may even be more appealing to quote a real life example of claimed progress in design. The Ocean Cleanup project, conceived of and started in 2012 by Boyan Slat, is aiming at a device to clean-up the plastic soup in the ocean. In 2018 a first prototype (001/A) was put to test and showed quite some problems. On October 2, 2019, the research team reports that an improved prototype (001/B) performs much better. The team's text, see below, can be rephrased in our terminology, including the structural-functional distinction, and will exemplify a case of progress in design as presented. Moreover, it also indicates that further development is needed to realize additional desired properties. The inserted abbreviations are as follows: sp: structural property, (u)dfp: (un-)desired functional property. Here is the quotation.

NEW DESIGN LEADS TO IMPROVED PERFORMANCE

The aim of System 001/B was to trial modifications, which addressed known complications, primarily aimed at correcting the inconsistent speed difference between the system and the plastic (udfp). Consistency was achieved by slowing down the system with a parachute sea anchor (sp), allowing for faster-moving plastic debris to float into the system (dfp). Once this main challenge was resolved, prominent plastic overtopping was observed (udfp) – becoming the next technical challenge to solve. Due to the modularity of System 001/B, a modification to increase the size of the cork line (sp) was designed and implemented while the system was offshore. With the new cork line, minimal overtopping is now being observed (dfp), thus allowing the system to capture and concentrate the plastic.

THE MISSION CONTINUES

Despite the early success of System 001/B, there is still much work to do. With new learnings and experience derived from the successful deployment of System 001/B, The Ocean Cleanup will now begin to design its next ocean cleanup system, System 002; a full-scale cleanup system that is able to both endure and retain the collected plastic for long periods of time (dfp).

Once fully operational, The Ocean Cleanup will return plastic to land (dfp) for recycling. The timing of that phase of the mission is contingent upon further testing and design iteration.[4]

The following photo and diagram are taken from the Internet by searching "Plastic retention system 001/B" and illustrate the cork line and the parachute, respectively.

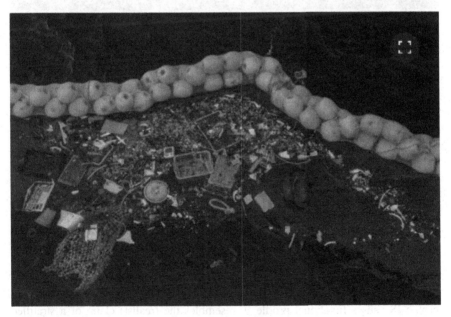

[o] **Plastic retention in System 001/B**

https://assets.theoceancleanup.com/app/uploads/2019/10/TheOceanCleanup_October2nd_Press_Briefing_System001B-24-640x425.jpg

---

[4]On Oct. 2, 2019, quoted from *https://theoceancleanup.com/updates/the-ocean-cleanup-success fully-catches-plastic-in-the-great-pacific-garbage-patch/* . Some days later, marine biologists reported that the spread photo showed (still?) much by-catch of sea animals (udfp).

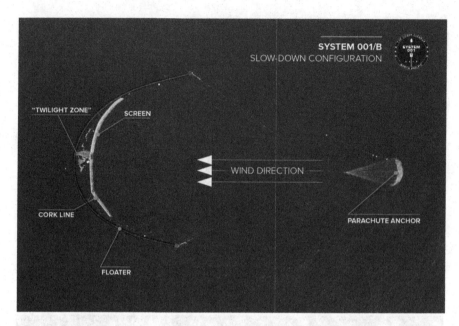

https://specials-images.forbesimg.com/dam/imageserve/5d95dd6dadeaab00067f8c89/960x0.png?
cropX1=-1&cropY1=-1&cropX2=-1&cropY2=-1&quality=75&fit=scale&
background=000000&uri=

Returning to the formal story, another option appears when, for some reason or other, OF(x) turns out to have some attractive aspects that DF does not have. This opens the way to adapting DF in the direction of the OF(x), which is of course the functional version of the notion of a concession that we met before; formally, replacing DF by DF* such that OF(x)ΔDF* ⊂ OF(x)ΔDF.

So far we assumed that OS(x) ≠ IS. In the other case, OS(x) = IS, the theory claims OF(x) = DF. If the latter claim is true, the theory is confirmed. Of course, this does not yet prove that the general claim of the theory is true. Hence, the prototype will have to be reproduced and tested again. However, if the claim OF(x) = DF turns out to be false, the theory is falsified. Assuming DF fixed, the task is of course to revise the theory by revising IS. Another option is again that of doing a functional concession, for OF(x) may be in some respects at least as attractive as DF.

Some similarities between design and nomic research are worth noting. To begin with, the (causal) claim of a design theory, recall: imposing intended structural profile IS causes functional profile F, resembles the (realist) claim of a stratified nomic theory, viz. that the theoretical entities and properties cause and explain such and such observational nomic facts. Typical examples of explanatory stratified causal theories are, e.g. the theories of Newton and Darwin, and utility theory.

Another important similarity (Kuipers 2001, Ch. 9) is the following. Design research aims at (theories about) products with certain desired functional features, and nomic research aims at theories about phenomena with certain already established observational features and hence it is desired that the theories at least respect or even explain them.

However, there are also important differences in this connection:

1. theoretical entities and properties are not observable,structural properties are, like functional properties, observable,
2. observational features are derivable from the stratified theory,functional properties are (claimed to be) caused by structural properties.

Hence, whereas stratified nomic theories explain observational features by non-observational ones, design theories are a kind of descriptive causal theories, similar to e.g. complex chemical reaction equations. Of course, if the causal relations are not transparent or not based on well-established theories, they ask for explanation of the working of the products.

Some other differences, which are elaborated in Kuipers (2001, Ch. 10), are:

1. a design target set is a free and changeable choice, whereas given a domain and a vocabulary, the nomic target set is fixed;
2. a design target set is known, whereas the nomic target set is not;
3. design research enables a straightforward (comparative) evaluation of a (new) prototype, in contrast to the indirect evaluation of a (new) theory;
4. to revise a prototype may be expensive for material reasons, to revise a theory is relatively inexpensive;
5. for a fixed domain and vocabulary, nomic truth approximation is, ideally speaking, free from external influences, whereas design research is basically open to such influences.

However this may be, as has been indicated, there are strong formal analogies, simply due to the fact that design research, like nomological research, can be characterized as aiming at a product which satisfies a certain target set of desired properties, while avoiding another set of undesired properties.

10.4   Generalized Target and Decreasing Symmetric Difference (DSD-)Condition

As suggested before, in the context of genuine scientific research aiming at nomic truth approximation T is the unknown, so can't be used. The situation is similar in the case of explicative research. We do not know in detail how the concept we are looking for looks like. Instead of saying that T is exactly the set of nomic possibilities or desired features (and hence U-T the set of nomic impossibilities or undesired features), we now deal with two more modest claims: that R is a set of known nomic possibilities or desired features and that S excludes some known nomic impossibilities or undesired features. That is, in such cases it is assumed to be possible to establish boundaries of T, notably subsets R and S such that we may assume that

Fig. 10.4 The generalized
symmetric difference
between X and the fitting
pair <R, S>: (X-S)∪(R-X),
i.e. the union of the two
@-areas

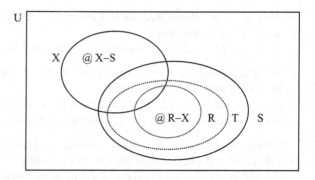

R ⊆ T ⊆ S, providing the *generalized target condition*. Let us call <R, S> a *fitting pair* if it satisfies the condition R ⊆ S and, for later use, a *T-fitting* pair if R ⊆ T ⊆ S. From now on <R, S> is supposed to be a fitting pair.

We define the *generalized symmetric difference* between X and a fitting ordered pair XΔ<R, S> as (X−S)∪(R−X), i.e. the union of the two @-areas in Fig. 10.4. We have inserted T as a dotted ellipse, such that <R, S> is T-fitting. It is easy to see that the generalized symmetric difference reduces to XΔT when R = T = S.

We now define the *generalized DSD-condition* as follows: Y is closer to the target <R, S> than X, or Y amounts to progress compared to X, iff YΔ< R, S > ⊂ XΔ< R, S>, that is, iff it satisfies the generalized DSD-condition. It is easy to check that this condition can be decomposed into:

> Y − S ⊆ X − S and R − Y ⊆ R
> − X and at least one of them must be a proper subset relation.

Figure 10.5 represents the situation that Y is closer to the target <R, S> than X as follows: shaded areas indicate empty sets and at least one of the two #-areas has to be non-empty. Like in Fig. 10.4, we have inserted T as a dotted ellipse, such that <R, S> is T-fitting.

However, now it may or may not be the case that there is some unknown final target T. In the first application, empirical progress, this will be the case, in the second, explicative progress, this assumption is at least problematic.

If the assumption makes sense, the following theorem is easy to prove and easy to see by comparing Fig. 10.2 and 10.5.

Success Theorem:
If Y is closer to target T than X and if <R, S> is a T-fitting pair (R ⊆ T ⊆ S) then Y is at least as close to the target <R, S> as X (whatever T is) and will become in the long run, assuming R increases and S decreases, closer to that target.

For the 'at least as close'-claim, i.e. Y−S ⊆ X−S and R−Y ⊆ R−X, note that the shaded areas in Fig. 10.5 are just parts of those in Fig. 10.2. For the 'long

Fig. 10.5 Y is closer to the
target <R, S > than X

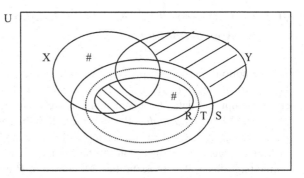

run'-claim, note that every relevant non-emptiness in Fig. 10.2 will sooner or later show up when R increases and S decreases.

Again there is a plausible way to liberalize the generalized condition to a quantitative one when U is finite. Instead of imposing the subset relation between the relevant generalized symmetric differences, we may impose the condition that the one symmetric difference is smaller than the other, formally: $|Y\Delta< R, S > |$ $< |X\Delta< R, S>|$. However, the corresponding success theorem is now not straight-forwardly valid. But it can be replaced by:

Expected success theorem
If Y is quantitatively closer to target T than X and if <R, S> is a T-fitting pair ($R \subseteq T \subseteq S$)
   then, assuming some plausible randomness conditions on the experiments, Y may be
   expected to be quantitatively at least as close to the target <R, S> as X (whatever T is)
   and that it will become in the long run, assuming R increases and S decreases, quantitatively closer to that target.

For an elaboration of this quantitative liberalization in the context of empirical progress and truth approximation (the topic of Sect. 10.5.1), see (Kuipers 2017, and 2019, Ch. 5). For the other concretizations in that context, refinement and stratifica-tion, see (Kuipers 2019, Chs. 6–7) or (Kuipers 2020a and 2020b, respectively).

10.5   Three Applications of the Generalized Conditions for Qualitative Progress

The generalized conceptual proposals enable us again to explicate two rather different kinds of progress in science, viz. empirical progress (and its relation to nomic truth approximation) (Sect. 10.5.1) and progress in concept explication (Sect. 10.5.2). But we will also briefly revisit progress in design research from the generalized perspective (Sect. 10.5.3).

10.5.1   Empirical Progress and Its Relation to Nomic Truth Approximation

For the notion of empirical progress we turn first to the representation of evidence, i.e. empirical data, for evidence will guide nomic truth approximation. In the nomic context empirical data at a given moment are typically asymmetric. They can be represented by a *data tuple* <R, S>, where R indicates the set of so far *realized possibilities*, e.g. the realized physical possibilities, and S ($\supseteq$ R) the (set of models of the) strongest law induced on the basis of R. Hence, cS indicates the induced nomic (e.g. physical) impossibilities. Of course, if R and S are correctly described and induced, respectively, then they constitute, whatever T is, a T-fitting pair: R $\subseteq$ T $\subseteq$ S. To be sure, the assumption of correct data is far from trivial. Regarding the inductively obtained S the correctness assumption is even much more demanding than that regarding R, which only demands correct case by case descriptions of the experimental results.

The following (basic) definition for exclusion theories X and Y, i.e. theories with the claim T $\subseteq$ X and T $\subseteq$ Y, respectively, is now plausible:

Definition:
Theory Y is (empirically) more successful relative to <R, S> than theory X iff:
Y−S $\subseteq$ X−S and R−Y $\subseteq$ R−X and at least one of the subset relations is proper, that is, Y satisfies the generalized DSD-condition relative to <R, S> in comparison with X.

Figure 10.5 now represents the situation that theory Y is (empirically) more successful relative to <R, S> than theory X. Note that the second condition (R−Y $\subseteq$ R−X) amounts to the claim that all established counterexamples of Y (R−Y) are counterexamples of X. The first condition (Y−S $\subseteq$ X−S) can be argued to amount to the claim that all established empirical laws explained by (in the minimal sense of being derivable from) X can be explained by Y. Briefly, any superset L of X$\cup$S, i.e. L such that X$\cup$S $\subseteq$ L, representing a consequence of both X and S, is a superset of Y$\cup$S.[5]

A real life example that fits this definition of empirical progress rather well concerns Einstein's special theory of relativity in comparison with six competing electrodynamic theories at the beginning of the twentieth century (Kuipers 2000, Ch. 6, and 2001, Ch. 8). The example is taken from (Panofsky and Phillips [1955] 1962, p. 282). Similarly, Einstein's theory of gravity is more successful than Newton's theory, notably in explaining several phenomena, partly known beforehand, partly predicted by Einstein.

---

[5]Recall that in Sect. 10.3.1 we mentioned, besides exclusion and inclusion theories, two-sided theories, consisting of a tuple <M, P> of subsets M (models) and P (postulates) of U, with M $\subseteq$ P, and an inclusion claim M $\subseteq$ T and an exclusion claim T $\subseteq$ P. From this perspective the notion of a generalized symmetric difference is one-sided. We get a sensible two-sided (applied) version by the union of (R−M)$\cup$(P−S) and (M−S)$\cup$(R−P). Note that they both equal the one-sided version when M = X = P.

In the present context the assumption of a not yet specified target, viz. the nomic truth, is relatively unproblematic. Hence, the success theorem now applies. It now amounts to the claim that 'closer to the nomic truth' entails 'being or at least becoming more successful (when new experiments are performed)'. This is crucial for the remaining methodological steps to complete the theory of nomic truth approximation (for details, see Kuipers 2019, Ch. 4). Assuming that a new theory is at a certain moment more successful than the old one, propose and test the following '*empirical progress hypothesis*': the new theory (is and) remains more successful in the basic sense than the old one. Note that such testing will lead to increasing R, due to new realized possibilities, and decreasing S, due to new accepted laws, which narrows down the set of not (yet) excluded possibilities. Assuming that after 'sufficient confirmation' this 'empirical progress hypothesis' is accepted (for the time being), argue *on the basis of a reconstruction of the way in which experiments have been done* and the success theorem, that the 'default explanation' for this case of empirical progress is the hypothesis that the new theory is closer to the nomic truth than the old one, *i.e., that this is a case of nomic truth approximation*. Note that this typically is an abductive reasoning.

In the typical situation of theoretical science, the theories will be stratified by the distinction between an observational language and a theoretical language. Of course, 'more successful' will then be expressed in observational terms and the success theorem then allows the abductive conclusion that nomic truth approximation has been obtained on the observational level. To (abductively) conclude that this is also a case of nomic truth approximation on the theoretical level some extra condition has to be satisfied, for theoretical nomic truth approximation does not unconditionally entail observational nomic truth approximation (see Kuipers 2019, Chs. 2 and 7, and 2020b). However, if we abductively conclude to observational nomic truth approximation we may go one step further and abductively conclude to theoretical nomic truth approximation in combination with the satisfaction of (the weakest version of) the extra condition, for that combination explains the observational nomic truth approximation and hence the empirical progress.

10.5.2 Progress in Explicative Research[6]

Scientists use concepts, principles and intuitions that are partly specific for their subject matter, but they also share part of them with colleagues working in different fields. Compare, for example, the chemical notion of the 'valence' of an atom with the notion of 'confirmation' of a hypothesis by certain evidence. An important task of 'the philosophy of (the special science of) chemistry' is the explication of the concept of 'valence'. Similarly, an important task of 'general philosophy of science' is the explication of the concept of 'confirmation'. In both cases it is evident that this

---

[6]This subsection is strongly based on (Kuipers 2007a).

only makes sense if one tries to do justice, as much as possible, to the actual use of these notions by scientists, without however following this use slavishly. That is, occasionally a philosopher may have good reasons for suggesting to scientists that they should deviate from a standard use. Frequently, this amounts to a plea for differentiation in order to stop debates at cross-purposes due to the conflation of different meanings. The distinction between 'probability' and 'truthlikeness' and that between 'confirmation' and 'conditional probability as degree of confirmation' are cases in point.

What has been said about concepts, also applies to principles and intuitions of scientists. Compare the subject specific 'principle of the conservation of energy' (see Kuipers 2019, Ch. 8) and the general intuition of 'diminishing returns from repeated tests' (see Kuipers 2007a). Both aren't crystal clear, they need explication; of course, beginning with the explication of the concepts involved.

Although the term 'explication' is not often used anymore by philosophers, albeit that there seems a revival of interest in the method, it is clear that when they discuss the meaning of concepts and propose or report specific definitions, characterizations, models, theories, accounts, conceptions, (rational) reconstructions or formalizations of them, they are practicing concept explication in a more or less explicit and rigorous way. Similarly, when philosophers propose or report specific analyses, accounts, reconstructions or formalizations of principles and intuitions, or dissolutions of paradoxes, explication is at stake. Both kinds of activity belong to the dominant ones among (systematic, constructive, analytical) philosophers, and not in the least among philosophers of science. However, explicitly calling these activities 'explication' is not very popular, let alone using the explication terminology in presenting results. There seem to be at least three reasons for the reluctance to use the word 'explication'. First, the word itself may be found a bit too pretentious. Second, making the application of the method explicit may not only lead to rather cumbersome texts, but also appear to be a difficult task. Finally, many philosophers do not like to be associated with the logical empiricists that introduced '(concept) explication' around 1950 as a technical term for this philosophical method, viz. Rudolf Carnap and Carl Hempel. To be sure, explication has nothing specifically to do with logical empiricism.

10.5.2.1   The Method of Concept Explication

As explained above, most philosophers apply the method of explication only informally and implicitly. Hence, the reconstruction of this method is itself a kind of meta-explication. Notably Rudolf Carnap (1950 and 1966) and Carl Hempel ([1950] 1966 and 1952), but also John Kemeny and Paul Oppenheim (1952) have articulated the method. Here we will freely describe and develop the method in their spirit. As suggested, it has been and can be applied in various degrees of rigor and explicitness.

The point of departure of concept explication in general philosophy of science is an informal, intuitive concept, the *explicandum*, which is frequently used by scientists in different fields. The aim is to define a concept, the *explicatum*, satisfying three general desiderata for concept formation and some specific ones. The three general desiderata are precision, fruitfulness and simplicity. The specific desiderata pertain to the similarity with the informal concept. This similarity is to be evaluated by two kinds of criteria:

1. The explicatum should apply to evident, undisputed *(types of)* examples of the informal concept and it should not apply to evident, undisputed *(types of)* *non-examples*.
2. The explicatum should fulfil (other) *conditions of adequacy* that have been derived from the informal concept, and occasionally it should violate *conditions of inadequacy*.

Evident non-examples are frequently generalizable to a condition of adequacy and evident examples to conditions of *in*adequacy. For this reason, the more usual combination of evident examples and conditions of adequacy is essentially enough.

Due to the general and specific desiderata, concept explication is an active enterprise, as opposed to the passive cognate of concept or meaning analysis (and its precursor 'logical analysis'). Following Carnap ([1950] 1966) and Hempel (1952), in the latter activity one merely tries to describe the uses of an informal concept, hence without any further desiderata. To be sure, meaning analysis may be very useful, if only to search for evident examples and non-examples and for finding further conditions of (in)adequacy for concept explication.

Let us briefly illustrate the specific desiderata, and its pitfalls, with the famous example of the explication of 'confirmation'. It is generally assumed, at least prima facie, that "this entity is a black raven" confirms the hypothesis that "all ravens are black", and hence provides an evident example of a confirmation. This is so-called Nicod's criterion. On the other hand, a white raven is a counterexample and hence an evident non-example. A black or non-black non-raven, e.g. a black or white tie, also seem evident non-examples. However, it also seems plausible that the explicatum of confirmation should satisfy the 'equivalence condition', that is, if certain evidence confirms a certain hypothesis then it confirms any of its logically equivalent versions. From this condition of adequacy it would follow that a non-black non-raven, e.g. a white tie, also confirms "all ravens are black", for it is, according to Nicod's criterion, an evident confirming instance of its logically equivalent version "all non-black objects are non-ravens". This is Hempel's so-called paradox of confirmation (Hempel [1945] 1965). It is clear that one can accept this consequence, assuming perhaps certain test conditions, or one can start to dispute either what is to be counted as an evident example or how plausible the equivalence condition in fact is as a condition of adequacy.

Following Hempel ([1945] 1965), scholars accepting the equivalence condition have also argued that the explicatum of confirmation should satisfy the '(special) consequence condition', that is, if certain evidence confirms a certain hypothesis then it confirms any of its consequences. However, it may also be argued that the opposite condition should be taken as a condition of adequacy, that is, if certain evidence confirms a certain hypothesis then it confirms any stronger hypothesis, the so-called 'converse consequence condition'. Although both conditions may have some intuitive appeal, as Hempel already pointed out, it is easily seen that imposing them together leads to absurd conclusions: any evidence confirms any hypothesis as soon as it confirms some hypothesis (if E confirms H, it confirms H&H∗, by the converse consequence condition, hence it confirms H∗ by the consequence condition). For Hempel this was a reason to (implicitly) classify the converse consequence condition as a condition of *in*adequacy. So far for the example of confirmation. For some clarification of the general desiderata, we refer to (Kuipers, 2007a).

10.5.2.2 Evaluation Report

The evaluation report of an explication makes clear to what extent a proposed explicatum satisfies the general and specific desiderata. As to the three general desiderata, an informal scoring will usually be the best that can be obtained. Of course, for the question whether progress has been made, see below, comparative scoring is more important than separate scoring.

Regarding the specific desiderata and starting with evident examples and non-examples, on the success side of the report we have the 'true positives' and the 'true negatives', that is, the evident examples fitting into the proposed explication and the evident non-examples excluded by it, respectively. On the problem side we get the 'false positives' and the 'false negatives', that is, the evident non-examples that nevertheless fit and the evident examples that don't fit, respectively. False positives show that the explication is too wide and false negatives that it is too narrow.

Besides clear-cut classifications of some evident (non-)examples as problems, there may well have been identified other 'evident' (non-)examples of which the question has arisen whether they are really evident. Such disputed (non-)examples may even lead to revision of the connotation of the concept, but they may also be seen as a consequence of initial idealization, requiring concretization of the explication in order to become (non-)examples.

Regarding the other type of specific desiderata, viz., conditions of (in)adequacy, a similar terminology is called for. Fulfilled conditions of adequacy and violated conditions of inadequacy are successes, and violated conditions of adequacy and fulfilled conditions of inadequacy are problems. Violated conditions of adequacy show that the explication is too wide and fulfilled conditions of inadequacy that it is too narrow. Again, conditions may be classified as questioned. Unfulfilled conditions of adequacy and fulfilled conditions of inadequacy may become disputed and lead to connotation revision, but such problems may again also be seen as a consequence of initial idealization, requiring concretization of the explication in order to become (un)fulfilled.

In sum, restricted to specific desiderata, an evaluation report of a proposed explication minimally specifies on the success side evident examples that are covered and conditions of adequacy that are fulfilled. On the problem side it at least lists evident examples that are not covered and conditions of adequacy that are not fulfilled. Finally, on both sides there may occur old and new (non-)examples and conditions of which the status still is or has become questioned.

Although it is intuitively appealing to continue to take evident non-examples and conditions of inadequacy explicitly into account in the evaluation report they can, as suggested before, implicitly be taken care of by appropriate conditions of adequacy, excluding the evident non-examples, and by appropriate evident examples, breaking the conditions of inadequacy, respectively.

10.5.2.3   Progress in Concept Explication

This brings us to the comparative evaluation of two explications of the same concept and the possibility of progress in concept explication or, briefly, *explicative progress*. Concept explication frequently leads to the conclusion that there are in fact two or more concepts that have to be distinguished, leading to a branching of the explication. However, when two explications of a concept are considered as rivals, the plausible question arises whether the one is better than the other. It is easy to see that there is a possibility to define 'strictly better', which assumes of course agreement about the relevant evaluation reports, both with respect to the relevant items as to their scores. The latter is the easiest to imagine by one person and, if relevant, from one and the same theoretical or programmatic perspective. Moreover, it explains why another person frequently disputes the claim by one scholar of having made progress on the basis of a disagreement about the separate evaluation reports.

Be this as it may, this does not prevent the following definition of explicative progress, neglecting evident non-examples and conditions of inadequacy, for the reasons indicated above.

Definition
E2 is a strictly better explication of a concept than E1 if and only if:
1. E2 satisfies the general desiderata at least as well as E1
2. E1 and E2 share all questioned examples and conditions of adequacy
3.1 E2 covers all evident examples covered by E1
3.2 E2 fulfils all conditions of adequacy fulfilled by E1
3.3 E2 covers some more evident examples and/or fulfils some more conditions of adequacy

It is easy to check that conditions 3.1–3.3 essentially correspond to the generalized DSD-condition of Sect. 10.4. For let U be the set of conceptual possibilities with which we are dealing. Let R indicate the set of conceptual possibilities that represent evident examples and let S indicate the set of conceptual possibilities that are not excluded by the conditions of adequacy. Finally, let X and Y represent the set of conceptual possibilities that satisfy explication E1 and E2, respectively.[7]

---

[7]From this correspondence it is clear that explications aiming at catching evident examples and fulfilling conditions of adequacy are a kind of exclusion theories, in which the conditions of adequacy function in fact as necessary conditions. Now it is interesting to note that the two conditions of our explication of (generalized) progress, the (generalized) target condition and the (generalized) DSD-condition, are not necessary conditions but function together as sufficient

In the present context the assumption of unique unknown target is problematic for there may be different 'solutions' to our conceptual puzzle. Hence, the success theorem cannot be straightforwardly applied. Looking for further evident examples and conditions of adequacy will have to narrow down the number of possible solutions and they may give rise to new reasons to deviate or to distinguish two or more specifications of the concept.

It is easy to imagine stronger, weaker and refined versions of various kinds of the above definition of explicative progress, for example, by requiring unintended successes, by merely counting numbers of covered examples and fulfilled conditions, and by assigning weights to them, respectively. Moreover, changes of questioned examples and conditions may be taken into account. In the terms of Rawls, further articulated by Thagard (1988), one may summarize the overall conclusion that one explication is better than another by claiming that the first better approaches a *reflective equilibrium* than the second. However, the most important point is that the ideal of explicative progress can be defined in a strict sense and hence that it can function as a regulative idea.

The notion of an evaluation report and the definition of explicative progress are modeled along the lines for evaluating theories and defining empirical progress in terms of counterexamples and explained empirical laws (Kuipers 2000, Chs. 5–6; Kuipers 2001, Chs. 7–8). Hence, the partial, formal analogy between empirical and explicative progress need not be surprising. Another reason why such an analogy may be expected is that in both cases we deal with progress in problem solving, albeit of different kinds.

As suggested before, a first explication may be a highly idealized way of catching cases and conditions, with the explicit intention to set up subsequent more realistic explications by accounting for cases and aspects that have first been neglected. This is a useful strategy, not only for concept explication, but also for concept formation in general. Moreover, it is largely analogous to what has been explicated for the empirical sciences as the strategy of idealization and concretization or factualization (Nowak 1980; Krajewski 1977); see also (Kuipers 2007b) for an explicit application to the explication of the intuitive notion of 'truth approximation'. Besides being a strictly better explication, the specific criterion for being a successful concretization of an idealized one is that the latter is an extreme special case of the former. For further examples of this kind, notably the earlier mentioned claim that "Jeffrey's explication of conditionalization is an improvement of the Bayesian one", we refer to (Kuipers 2007a).

---

condition. Hence, our initial and generalized explication are in fact a kind of inclusion theories (see Sect. 10.3.1): our claim is that transitions that satisfy these conditions are evident cases of progress. In this way we have discovered that there is another, inclusion, type of concept explication than the supposedly standard, exclusion, type dealing with conditions of adequacy. It would lead here too far to spell this inclusion type further out, but it is evident that the transition from the initial to the generalized explication of progress is itself a case of progress, for the latter just covers much more evident cases, without losing other ones.

10.5.3   Progress in Design Research Revisited

The refinements regarding design research, introduced in Sect. 10.3.2.2, can be realized in full by a generalized approach. The first point was the allowance of *possibly* relevant properties in U. This can optimally be realized by replacing the set of desired properties D by the tuple $<D^+, D^->$,[8] where D^+ indicates the definitely desired properties and D^- the definitely undesired properties, leading to a third set of not (yet) decided properties $U-(D^+\cup D^-)$. Note that D^+ corresponds to R in the generalized set-up and cD^- (the complement of D^-, $U-D^-$) to S. Progress can now be described in terms of the generalized DSD-difference: $(O(x)- cD^-)\cup(D^+-O(x)) = (O(x)\cap D^-)\cup(D^+-O(x))$.

This generalized set-up becomes particularly relevant when introducing the second refinement in Sect. 10.3.2.2, viz. the distinction between structural and functional properties. It may not only lead to a tripartition in the space of possibly relevant functional properties, but also in the space of possibly relevant structural properties. The consequence is that the presentation of the double generalized second refinement becomes rather complicated, but straightforward. So, we leave it as an exercise.

10.6   Concluding Remarks

We like to close with some suggestions for further research, notably case-studies and elaborating applications to e.g. technological, social and moral progress.

Although we have indicated some examples of progress in nomic, design and explicative research, it would be worth to start to look in detail at several examples to get an idea of how idealized the (generalized) model of progress is. This may be done in the area of scientific research, such as the three areas that were the focus of our applications. However, it is at least as interesting to do so outside scientific research. We will just make a few remarks on the ideas of technological and social progress.

Regarding technological progress, notably concerning products that come on the market, it is rather plausible, as already suggested at the end of Sect. 10.3.2.1, to submit that it will be possible to apply roughly the same means as for progress in design research. However, now it is even clearer that the desired properties cannot be assumed to be known in advance for in due course we learn what we ideally desire, for the time being, from for example a car for private use. But, given the present set of wishes of one person, one will agree that one's present car complies much better to one's wishes than the car of years ago. This may even fit the two conditions of Sect. 10.2. Of course, it is another question to what extent such technological

---

[8] See for this way of representation in the context of truth approximation (Kuipers 2014 and 2019, Ch. 15).

progress for one type of product for one person can be generalized to groups of people.

More complicated is the articulation of social progress, in one country or even worldwide, in line with our qualitative approach. It would be ideal if we could assume agreement on some minimum level of an ideal social state regarding all relevant respects. One might think of (almost) universally subscribed human rights. As books of Steven Pinker (2018) and Hans Rosling (2018) illustrate, there are many respects in which quantitative progress has been made. E.g. the percentage of children, girls in particular, that get at least elementary education has impressively increased in the last decades. Qualitative social progress can now be documented, e.g. in one country, in terms of those aspects which have obtained the minimum level and those that have not. But it is clear that a more refined approach, with quantitative means, is to be preferred.

A final challenge is whether our approach can be used for articulating the notion of moral progress. An interesting start is made by Jeremy Evans (2017). Some quotes will suggest that it is worth to further investigate the possibility to view it as an application of our general approach. Evans first notes that "Essentially every-one agrees that the outlawing of slavery, or the beginning of women's suffrage, or the defeat of Nazism constitute paradigmatic examples of moral progress in human history" (Evans 2017, 75). His leading ideas are the following:

> we will understand (global, population-level) moral progress in the most expansive terms, as the trajectory toward a human community that is more like the kind we ought to bring about. This reflects the fact that philosophical interest in moral progress is tied closely to the Enlightenment ideal that saw ethical theory as providing a foundation for constructing a better world for human beings to live together. The hope then is to identify some reliable, theoretically-neutral indicator for when populations are moving in the right direction (Evans 2017, 77).

This suggests, first, that there is in the background an idea of a perfect human community, in our terms, a kind of target T, and, second, that there is a close connection between social progress and moral progress: the former provides indicators for the latter or, in our terms, a kind of boundaries <R, S>. Regarding the indicators, Evans links up with "some theorists and policy organizations [that] are already taking a multi-criteria approach that combines objective and subjective measures of population welfare" (Evans 2017, 81), where the subjective measures refer to 'subjective well-being'. Evans' main claim is that "most normative traditions should concede a strong, if imperfect correlation between rising mean population welfare and moral progress" (Evans 2017, 84). However promising, there is not only still much to do by Evans and others, for example, his approach seems rather human-centered, but there also is still much to do to semi-formally reconstruct his approach to moral progress as an application of our general approach. The latter is certainly beyond the scope of this paper.

Acknowledgements This paper makes the informally described similarities between nomological, design and explicative research in Kuipers (2001, Ch. 9) formally explicit. I am very grateful to Gustavo Cevolani, Jeanne Peijnenburg and Luca Tambolo for their detailed comments on an earlier version.

References

Carnap, R. ([1950] 1963). On explication. In R. Carnap (Ed.), *Logical foundations of probability* (2nd ed., 1963, pp. 2–8). University of Chicago Press.

Carnap, R. (1966). Three kinds of concepts in science. In R. Carnap (Ed.), *An introduction to the philosophy of science* (pp. 51–61). Basic Books.

Cevolani, G., Crupi, V., & Festa, R. (2011). Verisimilitude and belief change for conjunctive theories. *Erkenntnis, 75*(2), 183–202.

Evans, J. (2017). A working definition of moral Progress. *Ethical Theory and Moral Practice, 20*(1), 75–92.

Hempel, C. G. ([1945] 1965). Studies in the logic of confirmation. In C. G. Hempel (Ed.), *Aspects of scientific explanation* (pp. 3–46). The Free Press. First published in *Mind*, 1945.

Hempel, C. G. ([1950] 1966). The empiricist criterion of meaning. In A. J. Ayer (Ed.), *Logical positivism* (pp. 108–126). The Free Press. (Originally published in 1950).

Hempel, C. G. (1952). *Fundamentals of concept formation*. University of Chicago Press.

Kemeny, J. G., & Oppenheim, P. (1952). Degrees of factual support. *Philosophy of Science, 19*, 305–330.

Krajewski, W. (1977). *Correspondence principle and growth of science*. Reidel.

Kuipers, T. (2000). *From instrumentalism to constructive realism*. Kluwer.

Kuipers, T. (2001). *Structures in science*. Kluwer.

Kuipers, T. (2007a). Introduction: Explication in philosophy of science. In T. Kuipers (Ed.), *Philosophy of science: Focal issues, handbook of the philosophy of science* (Vol. 1, pp. vii–xxiii). Elsevier.

Kuipers, T. (2007b). On two types of idealization and concretization. The case of truth approximation. In J. Brzezinski, A. Klawiter, T. A. F. Kuipers, K. Lastowski, K. Paprzycka, & P. Przybysz (Eds.), *The courage of doing philosophy: Essays dedicated to Leszek Nowak* (pp. 75–101). Rodopi.

Kuipers, T. (2013). Philosophy of design research. In H. Andersen, D. Dieks, W. J. Gonzalez, T. Uebel, & G. Wheeler (Eds.), *New challenges to philosophy of science* (pp. 457–466). Springer.

Kuipers, T. (2014). Dovetailing belief base revision with (basic) truth approximation. In E. Weber, D. Wouters, & J. Meheus (Eds.), *Logic, reasoning, and rationality. Proceedings-conference in Ghent, 2010* (pp. 77–93). Springer.

Kuipers, T. (2017). Quantitative nomic truth approximation by revising models and postulates. In M. Massimi, J. W. Romeijn, & G. Schurz (Eds.), *European studies in philosophy of science* (EPSA15 Selected Papers). Springer.

Kuipers, T. (2019). *Nomic truth approximation revisited, synthese library*. Springer.

Kuipers, T. (2020a). Refined nomic truth approximation by revising models and postulates. In Ch. J. Feldbacher-Escamilla, A. Gebharter, P. Brössel, & M. Werning (Eds.), *Logical perspectives on science and cognition – The philosophy of Gerhard Schurz*. Monographic issue, *Synthese, 197*(4), 1601–1625.

Kuipers, T. (2020b). Stratified Nomic realism. In W. J. Gonzalez (Ed.), *New approaches to scientific realism* (pp. 145–165). De Gruyter.

Kuipers, T., Vos, R., & Sie, H. (1992). Design research programs and the logic of their development. *Erkenntnis, 37*(1), 37–63.

Nowak, L. (1980). *The structure of idealization*. Reidel.

Panofsky, W., & Phillips, M. ([1955] 1962). Classical electricity and magnetism, 2 : Addison-Wesley. (1st ed., 1955).

Pinker, S. (2018). *Enlightenment now. The case for reason, science, humanism, and progress*. Viking.

Rosling, H. (2018). *Factfulness. Ten reasons we're wrong about the world and why things are better than you think*. Flatiron Books.

Thagard, P. (1988). *Computational philosophy of science*. The MIT Press.

Vos, R. (1991). *Drugs looking for diseases*. Kluwer.

Part V
Scientific Realism and the Instrumentalist Alternative

Chapter 11
Explicating Inference to the Best Explanation

Ilkka Niiniluoto

Abstract Inference to the best explanation (IBE) is a pattern of everyday and scientific reasoning, where a hypothesis is accepted if it gives a better explanation of the known evidence than any alternative hypothesis. This term was introduced by Gilbert Harman in 1965, and ever since IBE has been a central theme in the agenda of logic, artificial intelligence, and philosophy of science. However, a similar idea of a special kind of ampliative reasoning (besides deduction and induction) had been advocated by Charles S. Peirce already a century earlier: "hypothetical reasoning" proceeds backward from effects to causes, or "abduction" leads from surprising facts to explanatory theories. The key issues in the explication of IBE concern the definition of "the best explanation" and the possibility of analyzing the nature of explanatory reasoning by the Bayesian probabilistic approach or by the truthlikeness framework.

Keywords Abduction · Bayesianism · Confirmation · Explanation · Explanatory power · Induction · Likelihood · Peirce · Probability · Testability · Truthlikeness

11.1 Peirce's Abduction and Its Interpretations

The young Charles S. Peirce made a distinction between three modes of scientific reasoning already in his lectures in 1865–1866 and in the article "Deduction, Induction, and Hypothesis" in 1878. His starting point was the observation that a paradigm example of *deduction*, the Barbara syllogism of the first figure, can be inverted in two different ways. Using modern logical notation with ∴ as a general sign for inference and allowing singular terms, Barbara looks as follows:

I. Niiniluoto (✉)
Department of Philosophy, History, and Art Studies, University of Helsinki, Helsinki, Finland
e-mail: ilkka.niiniluoto@helsinki.fi

© The Author(s), under exclusive license to Springer Nature Switzerland AG 2022
W. J. Gonzalez (ed.), *Current Trends in Philosophy of Science*, Synthese Library
462, https://doi.org/10.1007/978-3-031-01315-7_11

$$(1) \quad (\forall x)(Fx \rightarrow Gx)$$
$$Fb$$
$$\therefore Gb.$$

Induction is the inference of the major premise (rule) from the minor premise (case) and the conclusion (result):

$$(2) \quad Fb$$
$$Gb$$
$$\therefore (\forall x)(Fx \rightarrow Gx).$$

Hypothesis is the inference of the minor premise from the major premise and the conclusion:

$$(3) \quad (\forall x)(Fx \rightarrow Gx)$$
$$Gb$$
$$\therefore Fb.$$

Thus, hypothesis leads from the rule and the result to the case (see *CP* 2.623). In other words, hypothesis is "an argument from consequence and consequent to antecedent" (*EP* 1:34–35).

According to Peirce, induction (2) is reasoning from particulars to the general law, and hypothesis (3) proceeds from effects to causes (*CP* 2.536), while a typical deduction (1) proceeds from causes to effects. Peirce also stated, already in 1865, that hypothesis is an *inference to an explanation*, and the resulting inference (1) is then an "explanatory syllogism" (*W* 1:267, 428, 440). Thus, hypothetical inference gives "scientific understanding" of a singular or general fact by providing the minor premise for its deductive explanation. Hence, Peirce's account of deduction covers what Carl G. Hempel in 1948 called the deductive-nomological (DN) explanation of singular facts and the DN explanation of laws (see Hempel, 1965). Peirce's probabilistic versions of (1), (2), and (3), developed in 1878–1883, where the general law is replaced with a statistical law, anticipate what Hempel in 1962 called the inductive-probabilistic (IP) explanation (see Niiniluoto, 2018a, 5–8).

In 1898 Peirce started to use the term *abduction* for his new account of hypothesis. The general form of "adopting a hypothesis for the sake of its explanation of known facts" or the "operation of adopting an explanatory hypothesis" is as follows (*CP* 5.189, *EP* 2:231):

$$(4) \quad \text{The surprising fact C is observed;}$$
But if A were true, C would be a matter of course.
Hence, there is reason to suspect that A is true.

This schema, which has become Peirce's best known or canonical formulation of abduction, indicates how a hypothesis can be "abductively conjectured" if it accounts "for the facts or some of them". Such abduction "only infers a *may-be*" from an actual fact (*CP* 8.238).

Schema (4) is obviously a generalization of the original pattern (3) of hypothetical inference: the emphasis that the fact C is surprising (and therefore in need of explanation) has been added, and there are no restrictions on the logical complexity of A. As here A may be a general theory, it might be said to express *theoretical abduction* (in contrast to singular abduction (3).[1] The idea of explanation is maintained in the second premise, but this is not any more explicitly associated with the relation of cause and effect.

Already in 1878 Peirce argued that hypotheses have to be put into "fair and unbiased" tests by comparing their predictions with observations (*CP* 2.634). In his papers and lectures in 1901–1903, Peirce defined induction in a new way as "the operation of testing a hypothesis by experiment" (*CP* 6.526), which is quite different from (2) or its modifications, whereas abduction is an "inferential step" which is "the first starting of a hypothesis and the entertaining of it, whether as a simple interrogation or with any degree of confidence" (*CP* 6.525). Here abduction and induction are successive steps in scientific inquiry: "Abduction is the process of forming an explanatory hypothesis", and "it is the only logical operation which introduces any new idea" (*EP* 2:216). After the abductive step in proposing the hypothesis A, the next step is to derive consequences from A by deduction, and then to put theory A into a severe observational or experimental test (*EP* 2:114). This account of testing follows the hypothetico-deductive (HD) model of science, but Peirce also allowed for cases where the test evidence is only a probabilistic consequence of the hypothesis.

When Peirce died in 1914, he was well-known as the founder of American pragmatism, but his seminal ideas about logic, semiotics, probability, and scientific inference became only gradually recognized with the publication of his *Collected Papers* (*CP*) in six volumes in 1931–1935 and two additional volumes in 1958. Especially his views about abduction were radical, since they differed from the two main positions in the mid-twentieth century philosophy of science: against the HD-method, Peirce allowed that there is an ampliative inference from our original problem situation to an explanatory hypothesis; against the inductivists, who thought that theoretical hypotheses are obtained from observations by inductive generalization, Peirce argued that his abduction is not identical with induction.

Peirce's abduction was also in tension with the sharp dichotomy between the context of discovery and the context of justification, as advocated by Hans Reichenbach and Karl Popper. Among philosophers of science, Norwood Russell Hanson (1958) suggested that abduction is a "logic of discovery", while Howard Smokler (1968) treated abduction as a method of confirmation. A stronger link to justification is advanced in Gilbert Harman's (1965) notion of "inference to the best

---

[1] For a useful classification of patterns of abduction, see Schurz (2008).

explanation" (IBE). An intermediate position is outlined by Larry Laudan (1980), whose abductive "logic of pursuit" aims to identify testworthy hypotheses.

Today many logicians and philosophers treat abduction as synonymous with IBE,[2] but some scholars argue that Peirce included only the generation of hypotheses (i.e., discovery and pursuit) in his account of abduction, so that the selection of the best explanation belongs to the later "inductive" stage of testing.[3] But even though Peirce's canonical schema (4) does not mention alternative explanations to A, he was aware that "the possible explanations of our facts may be strictly innumerable" (*EP* 2:107), and proposed a list of criteria or virtues for the comparison of hypotheses and choosing between them: explanatory power, testability, caution, breadth, and incomplexity (*CP* 7.220). In his later reflections in 1908–1910, he stated that an explanatory hypothesis may not only raise "a question meriting attention"[4] but also "justify us in seriously inclining belief in it" and in its acceptance, "as long as the phenomena be inexplicable otherwise" (*CP* 2.662, 2.469). On the whole, Peirce's dynamic abductive methodology gives reasons to doubt that the generation and the justification of hypotheses can always be clearly separated from each other (cf. Niiniluoto, 1999, 2018a). And it is evident that his influence is still prominent in current discussions of inference to the best explanation.

11.2 Inference to the Best Explanation

When Harman (1965) formulated his model of inference to the best explanation, he stated (without direct reference to Peirce) that IBE "corresponds approximately" to what others have called "abduction" or "hypothetical reasoning".

> In making this inference [to the best explanation] one infers, from the fact that a certain hypothesis would explain the evidence, to the truth of that hypothesis. In general, there will be several hypotheses which might explain the evidence, so one must be able to reject all such alternative hypotheses before one is warranted in making the inference. Thus, one infers, from the premise that a given hypothesis would provide a 'better' explanation for the evidence than would any other hypothesis, to the conclusion that the given hypothesis is true (Harman 1965, 89).

In this interpretation, abduction is not only an inference to one of the potential and testworthy explanations but to the *best* explanation, and the conclusion is claimed to be *true*. Harman's IBE thus recommends the inference from evidence E to the hypothesis H when H is in some sense a better explanation of E than its rivals:

---

[2] See e.g. Thagard (1978), van Fraassen (1989), Kuipers (1999, 2019), Lipton (2004), Schurz (2008), Bird (2010), Psillos (2011), and Rivadulla (2018).

[3] For a survey of these arguments, see Niiniluoto (2018a, 17–18).

[4] For the interrogative or strategic interpretation of abduction, see Hintikka (1998) and Shaffer (2019).

(IBE)   A hypothesis H may be inferred from evidence E when H is a better
          explanation of E than any other rival hypothesis.

Better explanations are characterized by Harman only with one sentence: they are
simpler, more plausible, less ad hoc, explain more, and so forth. According to IBE,
the best explanation H of evidence E is rationally acceptable on E:

(IBE′) If hypothesis H is the best explanation of evidence E, then conclude for the
          time being that H is true.

Inferences licensed by IBE or IBE′ are not deductively valid. In the spirit of Peirce's
fallibilism, acceptance as true is only tentative, since all factual assertions are
corrigible and revisable by further evidence.

In some everyday applications of IBE, there is a ready-made list of potential
explanations, and the task is to *select* the best of them (e.g., diseases and causes of
death in medical diagnosis). In other situations, typical of scientific reasoning, the
researcher has to *create* explanatory hypotheses with new theoretical concepts.[5]

Within IBE evidence E is known but surprising: given background knowledge B,
it is not known *why* E is the case, and the task of explanation H is to answer this
why-question (see Hempel, 1965). Depending on the relation of H and E, such
explanation may be deductive, inductive, or approximate (see Niiniluoto, 1999,
2018a, b). IBE is usually restricted to rational or scientific explanations, so that
appeal to miracles, God's will or evil demons is excluded.

The second sentence in the quotation from Harman seems to assume a specific
strong condition: the background information B together with evidence E excludes
all potential explanations but one. Here exclusion may be understood in the pragmatic sense that B and E give us good reasons to reject some of the available (so far
discovered) potential explanations. So as a special case of IBE, where a hypothesis
has no serious rivals, we have *inference to the only explanation*:

(IOE) A hypothesis H may be inferred from evidence E when H is the only
          available explanation of E.

In the limiting case, which Alexander Bird (2010) calls "eliminative abduction",
evidence eliminates or refutes by Modus Tollens all but one of the hypotheses. If the
relevant hypotheses constitute a partition, such eliminative abduction is a deductively valid argument.

Some researchers in logic and artificial intelligence (AI) have formulated
paraconsistent systems with an abductive rule of inference, taking its model from
Peirce's schema (3):

---

[5]For the distinction between selective and creative abduction, see Magnani (2001), 20.

$$A \rightarrow C$$
$$C$$
$$\therefore A.$$

This kind of rule is not an instance of IBE, since here A is not compared with any other conditions for C.[6] In Atocha Aliseda's (2006) logical approach, the successful search for explanations leads to the expansion or revision of our initial belief system. An alternative approach in the AI literature is to treat IBE as a form of probabilistic non-deductive reasoning:

C is a collection of data (facts, observations, givens)
H explains C (H would, if true, explain C)
No other hypothesis can explain C as well as H.
Therefore, H is probably true.

Here the criteria of the best explanation include: how good H is by itself, how decisively H surpasses the alternatives, and how thorough was the search for alternative explanations (see Josephson & Josephson, 1994).

Peter Lipton (1991, 58), has argued that IBE as a rule of acceptance should include a clause stating that the best explanation is "good enough":

> Given our data and our background beliefs, we infer what would, if true, provide the best of the competing explanations we can generate of those data (so long as the best is good enough for us to make any inference at all).

This point links IBE to what Herbert Simon calls "satisficing" as an alternative to optimizing: instead of dreaming that IBE in a single application gives us the best of all conceivable explanations, a scientist may adopt for the time being the first satisfactory hypothesis – and then be ready to correct it in the light of new evidence. But, whatever criterion we give for sufficiently promising explanations, sometimes none of the proposed explanations is good enough, and the most rational attitude is suspension of judgment.

Bas van Fraassen (1989) argues that IBE may lead to "the best of a bad lot". As IBE is in practice restricted to a set of historically given or formulated hypotheses, how could we know that the true hypothesis is among the so far proposed? A formal reply to this "bad lot objection" is to point out that IBE should be applied to a *partition* U, consisting of a set of mutually exclusive and jointly exhaustive hypotheses $H_1, \ldots, H_n$, so that their disjunction is a logical truth and, hence, precisely one of them is true.[7] In the cognitive problem of finding the true element of a partition U, the elements of U are the complete answers and the disjunctions of the elements of U

---

[6]Cf. the summary in Niiniluoto (2018a), Ch. 3.

[7]Levi (1967) argues convincingly that acceptance and rejection are relative to a partition. This comparative idea is common to Bayesianism and the Neyman-Pearson theory of statistics.

are partial answers.[8] In Jaakko Hintikka's inductive logic the choice of the partition is achieved by taking the relevant hypotheses to be constituents of a first-order language L, i.e., linguistic descriptions of alternative "possible worlds", and all generalizations in L can be expressed as disjunctions of constituents.[9] The simplest partition with two alternatives is of the form {H, ~H}. When the available alternative hypotheses H_i, i = 1, ..., n, are not exhaustive, the partition can be achieved by adding the "catch-all hypothesis" ~ (H_1 v ... v H_n) (see Niiniluoto, 1999). But such a catch-all hypothesis need not be an interesting hypothesis – or an explanation at all.[10] Accepting the catch-all hypothesis is then equivalent to the suspension of judgment between available explanations.

Finnur Dellsén (2021) suggests that Lipton's two-stage model of IBE, i.e., generation and selection, should be complemented with the third stage of "explanatory consolidation". Here accumulating evidence and repeated failures to formulate better alternatives warrant us in thinking that our best hypothesis is indeed better than all *possible* alternatives. But this kind of judgment is very bold in any historical situation. A Peircean fallibilist may rather insist that a typical feature of scientific progress is the strategic use of abduction as a tool of truth-seeking (cf. Hintikka, 1998; Niiniluoto, 2018a). Then, as Schupbach (2014) argues, IBE shares with deduction the capacity to "preserve good material content": if truth is among the relevant alternatives, then there is a good chance that IBE picks it out. But IBE is not discredited as a rule of inference, even if it sometimes happens to lead from false premises to a false conclusion.[11] Indeed, from the strategic perspective, even "a bad lot" may help us to improve our position by converging toward the truth via false but truthlike theories: as final truths have not been reached, there are always still better "unconceived alternatives" which are not yet known. Our task in science is to continue inquiry, to enrich our conceptual framework, to use abductive imagination to create new explanatory theories, to seek ever more test evidence by observation and experimentation, to revise our so far best hypotheses, and thereby to find new theories that are closer to the truth than the earlier ones. Abduction and IBE are not steps in two or three stages, but constitute a self-corrective cyclical back and forth process (see Tavory & Timmermans, 2014; Shaffer, 2019). From this perspective, there is no sharp division between the pursuit and tentative acceptance of explanatory theories (see Sect. 11.1).

---

[8] Dellsén (2016) argues that the rival hypotheses should be pairwise incompatible proposals for complete explanations, so that he would not allow disjunctive explanations.

[9] Constituents tell what kinds of individuals there are in the world. For a survey of inductive logic, see Niiniluoto (2011).

[10] See Douven (2017a). Cf. Dellsén's (2017) remark that in {H,~H} the negation ~H of an explanation H may fail to be an explanation.

[11] IBE is fallible also for the reason that evidence E may happen to be false. This aspect of IBE has not been systematically studied yet, even though Bird (2017) mentions Jeffrey conditionalization on uncertain evidence.

A fallibilist should also acknowledge that often our best theories include idealizations, so that they are known to be false.[12] However, idealized truthlike theories may be able to give approximate explanations, where the derived conclusion is close to the given explanandum.[13] In such cases the rule IBE in its standard form seems inappropriate, since the notion of truth does not have a comparative form "truer than". But, as Kuipers (1999) and Niiniluoto (1999) propose, instead of truth in the conclusion might be that the best explanation is "close to the truth". Using two explications of this notion, approximate truth and truthlikeness (verisimilitude), IBE can be replaced by the following principles:

(IBAE)   If the best available explanation H of evidence E is approximate, conclude for the time being that H is truthlike.

(IBAE′)   If the best available explanation H of evidence E is approximate, conclude for the time being that H is approximately true.

This kind of modification can be applied generally to any theoretical explanation (see Niiniluoto, 2018a, Ch. 8; Kuipers, 2019, Ch. 11):

(IBT)   If a theory H has so far proven to be the best among the available theories, then conclude for the time being that it is the closest to the truth of the available theories.

A comparative version of IBT, both in premises and conclusion, is the following:

(IBT^c)   If theory H′ is a better explanation of the available evidence E than theory H, then conclude for the time being that H′ is more truthlike than H.

The rules IBT and IBT^c avoid van Fraassen's bad lot argument, since they do not presuppose that we have access to the true theory. But, just as the original IBE, they still leave us the task of explicating what is meant by the "best" explanation.

---

[12] Dellsén (2017) uses this kind of example as a counterargument to Schupbach (2014), since here IBE does not preserve "inductive strength". But this is unfair, since if all alternative hypotheses H_1, ..., H_n are known to be false, it is clear that the best one of them cannot be expected to be probable, but its value has to be assessed by the notion of verisimilitude (see Niiniluoto, 1999).

[13] See Niiniluoto (2018b) for the treatment of idealized explanations.

11.3 Explanatory Power

The notion of "the best explanation" is of course ambiguous in several ways, since there is no agreement on the theoretical and explanatory virtues that are appreciated in science. In the spirit of scientific realism, Hempel (1965) demanded that an explanation should be *true*, but this property of a theory is not easily accessible – except in simple everyday applications. As truth has to be inferred indirectly and fallibly from available evidence and background information, it does not function as a criterion of theory choice in scientific inquiry, where we compare the merits of alternative potential explanations.[14] Peirce's own list of virtues included explanatory power, testability, caution, breath, and simplicity (see Sect. 11.1), and it has been elaborated especially with the notion of "consilience" or "unification" (see Thagard, 1978; Schurz, 2008; Niiniluoto, 2018a, Ch. 6.4). Bird (2017) makes a distinction between the *internal* and *external* explanatory virtues of a hypothesis H: the former concern H alone, the latter the relation of H to the relevant evidence E. When Lipton (1991) suggested that IBE should be understood as "Inference to the Loveliest Potential Explanation", where a *lovely* explanation is one which offers a high degree of "potential understanding", he seemed to be aiming at an external virtue. But, as pointed out by Hintikka (1968), it is also important to distinguish between local and global theorizing: in *local* theorizing, we are interested in finding an explanation of a particular piece of data or evidence E, while in *global* theorizing evidence E is only a stepping-stone to a more generally applicable theory.

The "explanationist" friends of abduction and IBE emphasize the role of explanations in scientific reasoning (see McCain & Poston, 2017). Therefore, the main virtue of theoretical hypotheses is their *explanatory power*, i.e., their ability to give information about the evidence (Hintikka, 1968; Niiniluoto & Tuomela, 1973) or to make evidence less surprising (Schupbach, 2017).[15] This notion, which is a key tool in the explication of the rule of IBE, usually presupposes that a probability measure P is defined over the sentences of the relevant scientific language. This measure satisfies Bayes's theorem, which states that posterior probability P(H/E) is proportional to prior probability P(H) and likelihood P(E/H):

$$(5) \quad P(H/E) = \frac{P(H)P(E/H)}{P(E)} = \frac{P(H)P(E/H)}{P(H)P(E/H) + P(\sim H)P(E/\sim H)}.$$

For a Bayesian, this conditional probability expresses a rational degree of belief in the truth of hypothesis H given an evidence statement E. According to Bayesian conditionalization, when E is learned and accepted as true, the prior credence P(H) is

---

[14] However, if truth value is chosen as an epistemic utility of a hypothesis, its expected value equals posterior probability (see Niiniluoto, 2018a, 112).

[15] There may be other virtues, like simplicity or parsimony, but for the realist they are secondary to explanatory power.

changed to P(H/E). A subjective Bayesian allows the researcher to choose her priors and likelihoods freely as long as probability axioms are satisfied, but a reasonable analysis of probabilistic explanation presupposes that the likelihood P(E/H) is based on objective lawlike propensities or statistical conditions of randomness via David Lewis' Principal Principle.[16]

The first measure of *explanatory power* of theory H with respect to evidence E was defined by Hempel in 1948 as

$$\mathrm{expl}_1(\mathrm{H}, \mathrm{E}) = P(\tilde{H}/\tilde{E})$$

(see Hempel, 1965). Here E may be a conjunction of several empirical statements. Hempel derived this measure as the ratio between the common information content of H and E (i.e., $1 - P(H \vee E)$) and the content of E (i.e., $1 - P(E)$). Popper (1959) defined explanatory power by the formula

$$\mathrm{expl2}(\mathrm{H}, \mathrm{E}) = \frac{P(E/H) - P(E)}{P(E/H) + P(E)} = \frac{P(H/E) - P(H)}{P(H/E) + P(H)}$$

and Hintikka (1968) by

$$\mathrm{expl3}(\mathrm{H}, \mathrm{E}) = \frac{P(E/H) - P(E)}{1 - P(E)}.$$

A related logarithmic measure was proposed by Håkan Törnebohm in 1966:

$$\mathrm{expl4}(\mathrm{H}, \mathrm{E}) = \frac{logP(E) - logP(E/H)}{logP(E)}$$

as a normalization of I. J. Good's 1960 definition logP(E/H) – logP(E). Jonah Schupbach and Jan Sprenger gave in 2011 an axiomatic derivation of the following measure:

$$\mathrm{expl5}(\mathrm{H}, \mathrm{E}) = \frac{P(H/E) - P(H/\tilde{E})}{P(H/E) + P(H/\tilde{E})}$$

(see Schupbach, 2017).[17] All of these measures share the same maximality condition:

---

[16] Jonathan Weisberg (2009) argues against the subjectivists that explanatory considerations should be used to fix objectively correct prior probabilities as well. Bird (2017) recommends objective evidence-based plausibilities which are not credences or personal degrees of belief.

[17] J. G. Greeno has defined the overall explanatory power of a theory with respect to a class **E** of exclusive observational statements as the reduction of uncertainty about **E** achieved by H. This is defined by the difference between the initial entropy of **E** and the relative entropy of **E** given H (see

$$(6) \quad expl(H,E) = 1 \text{ iff } H \text{ entails } E.$$

Thus, all deductive explanations of E by H are equally good, so that these measures can primarily be used for the comparison of non-deductive explanations. Most of these measures satisfy the comparative likelihood condition:

$$(7) \quad expl_i(H,E) < expl_i(K,E) \text{ iff } P(E/H) < P(E/K), \text{ for } i = 2, 3, 4, 5.$$

In other worlds, K is a better explanation of E than H if and only if the likelihood of K is larger than that of H. Further, for these measures

$$(8) \quad expl(H,E) > 0 \text{ iff } P(E/H) > P(E) \text{ iff } P(H/E) > P(H),$$

i.e., H has some explanatory power with respect to E if and only if H is *positively relevant* to E. Then $expl(H,E) = 0$ if and only if E and H are probabilistically independent. This amounts to the basic idea of Wesley Salmon's (1971) SR-model of probabilistic explanation.[18] Hempel's $expl_1$ behaves in a different way:

$$(9) \quad expl_1(H,E) < expl_1(K,E) \text{ iff } P(K)(1 - P(E/K)) < P(H)(1 - P(E/H)).$$

This measure thereby favors theories with high information content $cont(K) = 1 - P(K)$ and high likelihood $P(E/K)$, so that it combines global and local aspects of explanation. If $P(H) = P(K)$, then the comparison (9) reduces to the likelihood condition (7). If $P(E/H) = P(E/K) \neq 1$, then (9) reduces to $P(K) < P(H)$, i.e., the less probable of H and K is the better one. Further, the minimum value of $expl_1$ is 0, and

$$(10) \quad expl_1(H,E) > 0 \text{ iff } P(H \lor E) < 1 \text{ iff } \sim H \text{ does not entail } E.$$

Hempel himself applied the same formal definition as a measure of *systematic power*, when the capacity of a theory to give successful explanations and predictions are combined.

The measures of explanatory power help to define the notion of *best* explanation as well. If explanatory power is chosen as the epistemic utility, and the best hypotheses is the one with maximal expected utility, then IBE can be understood

---

Salmon, 1971). For example, **E** might be the class of state descriptions or constituents of a monadic predicate language (see Niiniluoto & Tuomela, 1973). This kind of measure, which is global with respect to the explanandum, has its maximal value if H entails one of the elements of **E**.

[18] Salmon relied on the frequency interpretation of probability instead of propensities.

as an acceptance rule within cognitive decision theory (see Niiniluoto, 2018a, Ch. 7). The choice of the local positive relevance measures $expl_i$, $i = 2, 3, 4, 5$, lead to R. A. Fisher's principle of *Maximum Likelihood*:

(11) Given evidence E, accept the theory H which maximizes the likelihood P (E/H).

For the case of deductive explanation, the rule (11) does not help much, since it cannot distinguish between potential explanations that make the likelihood P(E/H) equal to one. But even here a difference in the rival hypotheses may be detected by enlarging the description of the initial evidence E by new observations and tests. For example, suppose we are looking for an explanation of the death of a certain person, and there are available several possible causes H, H′, ... of her death. However, if the case is given a fuller description and examination, including facts about her life and the symptoms in her body, it may be that many of the potential hypotheses fail to give a causal explanation any more. In the limit, it may happen that only *one* of the potential hypotheses is left, and this is by IOE certainly the "best" explanation in this situation.

In global theorizing, we are trying to find a good hypothesis H which explains also other phenomena than the particular data E. One suggestion in this direction is to take the amount of the substantial *information content* of H, measured by cont (H) $= 1 - P(H)$, as the epistemic utility to be maximized. But this rule would have the undesirable feature that it favors a logically inconsistent theory H (with $P(H) = 0$ and cont(H) $= 1$). It does not help at all to replace cont(H) by the relative content cont(H/E) $= 1 - P(H/E)$ of H given E (see Hintikka, 1968). Hempel's measure $expl_1(H,E) = P(\sim H/\sim E)$ is maximized by minimizing $P(H)(1 - P(E/H))$, so that it favors hypotheses with high content and high likelihood, but it also receives its maximal value for a contradiction H. Hence, the rule IBE would lead to absurd consequences if the "best explanation" is simply characterized as the theory which has the maximal degree of information content or Hempelian systematic power. But, as Hintikka (1968) notes, the situation changes decisively if information content and systematic power are treated as *truth-dependent* epistemic utilities. For example, if the utility of accepting H is cont(H) when H is true and -cont($\sim$H) when H is false, then the expected utility rule recommends to accept the hypothesis with the highest value of the relevance measure:

(12) Given evidence E, accept the theory H which maximizes $P(H/E) - P(H)$.

The same result is obtained by choosing the two utilities as $expl_1(H,E)$ and $-expl_1(\sim H,E)$. According to this rule (12), as $P(H/E) - P(H) = P(H/E) + cont(H) - 1$, the best hypothesis should have both high posterior probability given evidence and high information content. As posterior probability is equal to the expected truth-value of H given E, the rule (12) is also obtained by choosing truth and information as the basic utilities. Thus, (12) is a way of balancing the demands of truth and information (systematic power) in scientific inquiry.

Following Levi's (1967) model of acceptance relative to a partition U, the rule (12) recommends the rejection of elements H_i of U with $P(H_i/E) < P(H_i)$, and the acceptance of the disjunction of the others. In other words, all non-explanatory hypotheses in U are ruled out, and the disjunction of all explanatory hypotheses is accepted as the strongest on the basis of evidence.[19] A similar idea is suggested by Dellsén (2017), whose rule ARI_k of "abductively robust inference" considers the consequences of the disjunction of k most lovely hypotheses that explain the evidence E. The choice $k = 1$ gives the rule IBE. He illustrates this proposal with four most plausible alternative explanations of the origin of life on Earth, which all share the assumption that life emerged from a genetic replicator.

If hypothesis H is logically stronger than hypothesis H', i.e., H entails H', then we have $P(H) \leq P(H')$ and $cont(H) \geq cont(H')$. If, in addition, both H and H' entail a contingent evidence statement E, then we have $P(H/E) - P(H) \leq P(H'/E) - P(H')$. In this deductive case, the rule (12) recommends that a logically weaker hypothesis is always better than a stronger one. However, in the probabilistic case, where the hypotheses do not entail the evidence, this need not be the case.

Bayesian approaches to abduction often assume that IBE should recommend the acceptance of the most probable explanation:

(13)   Given evidence E, accept the explanation H of E such that H has the maximal posterior probability $P(H/E)$ on E.

As self-explanations are excluded, in (13) H cannot be E (or contain E as its conjunct), even though $P(E/E)$ has the maximal value 1. The condition that H has to explain E also excludes trivial tautologies as candidates for H.[20] For theories which deductively explain E, it follows from Bayes's Theorem that rule (13) recommends the choice of the theory H that has the highest prior probability $P(H)$, i.e., H is initially the most plausible of the rival explanations. With the same assumptions, this H also maximizes the difference $P(H/E) - P(H)$, so that (12) and (13) are equivalent in this case. However, as we already noted, if rule (13) is applied to inductive explanation, (12) and (13) may lead to different results: the theory with maximal likelihood need not be the one with the highest posterior probability, if it has a very low prior probability. If the rival explanations have the same prior probability, rules (12) and (13) give the same result.

---

[19] See Niiniluoto (2018a), 112.

[20] However, the lottery paradox is usually taken to imply that high probability cannot be sufficient for acceptance. Still, many inductive acceptance rules require as a necessary condition that one of the elements of a partition has the probability larger than ½.

For Popper, prior improbability is the main virtue of a scientific theory, since "science aims at a high informative content, well backed by evidence" (Popper, 1959, 399). By (8), this condition is satisfied by Hempel's measure $expl_1$, but not by Popper's own $expl_2$. Especially in the context of global scientific theorizing, it is indeed important to explicate the Popperian intuition about the significance of independent testability and unifying power (see Schurz, 2008; Niiniluoto, 2018a). But, in spite of Popper's critique of "inductivism", strong explanation is compatible with the demand of high *posterior* probability (cf. Niiniluoto & Tuomela, 1973). We noted that for deductive explanations, the Bayesian rules (12) and (13) favor hypotheses with high initial probability. For example, it has been suggested that the *plausibility* P(H) of H depends on its simplicity or its coherence with background theories (Salmon, 2001). This requirement that P(H) is high seems most natural in the context of everyday reasoning about alternative complete explanations, where the plausibility P(H) is measured by relative frequencies. If I see in the morning that grass is wet, this could be explained by two conditions: it was raining last night or the sprinkler was on, but the former is more common and thus acceptable as the more probable explanation.

In Sect. 11.2 we noted that a criterion is needed for saying when an explanatory hypothesis is "good enough" for acceptance. The answer can be sought from the quantitative measures of explanatory power: a hypothesis H is sufficiently good if its degree of explanatory power expl(H,E) with respect to evidence E exceeds a threshold value.[21] Depending on the choice of expl, and the rules (12), (13), and (14), this gives a constraint on the likelihood P(E/H), the difference P(H/E) – P(H), or the posterior probability P(H/E). The Bayesians have usually agreed that high posterior probability P(H/E) is a necessary condition of acceptance – and this condition also guarantees that H is more probable on E than its rival ~H. Gerhard Schurz (2008) defends the plausible condition that an acceptable hypothesis should explain at least two independent facts, but a Bayesian can explicate this principle with probabilistic measures of confirmation.[22]

11.4 IBE and Bayesian Confirmation

Inference to the best explanation is a form of ampliative inference, and it is important to ask how such non-deductive reasoning could be justified. For Harman (1965), IBE is more basic than induction, since he claims that inductive generalization is a special case of IBE, while there are applications of IBE which cannot be formulated as

---

[21] See Niiniluoto (2018a, 118). Dellsén (2021) doubts that the attempt to specify a loveliness-threshold is too arbitrary. It is clear that such a threshold has rely on a context-dependent scientific judgment.

[22] For the argument due to Wayne Myrvold, see Niiniluoto (2018a), Ch. 6.4. Schurz's common cause abduction resembles the principle of criminal courts that the verdict is "beyond reasonable doubt" only if corroborated by two independent witnesses (cf. Amaya, 2009).

enumerative induction (e.g., detective stories, reasoning about criminal cases, sub-atomic particles, and mental experiences of other people). The latter claim is in agreement with Peirce, but the former conflicts Peirce's distinction between induction and abduction (cf. Niiniluoto, 2018a, 16–17). Stathis Psillos (2002) has defended the independence of IBE by employing the notion of defeasibility. He appeals to coherence conditions and contextual background knowledge to show that IBE "has resources to show when a potential undercutting defeater can be neutralized", where a defeater could be a hitherto unthought but better explanation.[23] He further shows how arguments for IBE may be rule-circular in a non-vicious way. Even if IBE is ampliative, and cannot be reduced to a conclusive deductive argument, "one cannot seriously question the fact that a hypothesis stands out as the best explanation of the evidence offers defeasible reason to warrantedly accept this hypothesis" (Psillos, 2009, 190). The lively debate about the power and justification of IBE is continued in the collection *Best Explanations* (2017), where the fundamentality of explanatory reasoning is defended by the editors Kevin McCain and Ted Poston as well as Ali Hasan, Ruth Weintraub, and Elizabeth Fricker. The autonomy of IBE is attacked by Richard Fumerton, who argues for its reduction to enumerative induction.

It is useful to remember that some philosophers have explicitly challenged the "explanationist" view that explanatory virtues are relevant to confirmation: according to Salmon (2001), "confirmation is logically independent of explanation". Bas van Fraassen (1989) makes a distinction between "confirmational" and "informational virtues" of theories, arguing that explanatory power belongs to the latter type. Dov Gabbay and John Woods (2005) argue that abduction is "ignorance-preserving". In the same way, it is sometimes claimed that *accommodation* (i.e., the adoption of an explanatory hypothesis to account for already known data) has to be sharply distinguished from *prediction* in the sense that accommodation does not have any power of confirmation.[24] A different view was defended already by Howard Smokler (1968), who distinguished between *inductive* and *abductive confirmation*. The former, typical of enumerative induction, satisfies the qualitative conditions of *entailment* (if evidence E entails hypothesis H, then E confirms H) and *special consequence* (if E confirms H, and K is entailed by H, then E confirms K). The latter, typical of situations where a theory is supported by its power to explain and predict surprising phenomena, satisfies the principles of *converse entailment* (if H entails non-tautological E, then E confirms H) and *converse consequence* (if K entails H and E confirms H, then E confirms K).

Smokler's division between two kinds of confirmation is motivated by the observation that the principles SC and CC are incompatible and likewise CE and

---

[23] For coherence conditions as the key to IBE, see Thagard (1989), Amaya (2009), and Poston (2014).

[24] See the discussion in Lipton (2004).

SC are incompatible.[25] It follows that the attempt to extend inference to the best explanation to "indirect IBE", where K is acceptable on the basis of evidence E if K is entailed by the best explanation H of E,[26] is problematic, since it combines the abductive CE with the inductive SC.

Even though Peirce as a frequentist was sharply critical of the classical Bayesian theory of probabilistic inference (see e.g. *EP* 2:215), it can be shown that Bayesianism provides a useful framework for studying abduction and induction as forms of ampliative reasoning. Smokler's insight can be at least partly reconstructed by the quantitative probabilistic approach without reducing IBE to induction.

Lipton (1991) distinguished "likely" explanations, which take into account the overall credibility or warrant of the explanation, and "lovely" explanations, which provide potential understanding about the explanandum. A likely explanation is probably true relative to the total evidence. As a reply to "Voltaire's objection" about the correlation of loveliness and truth, he argued that at least on some conditions the loveliest explanation is also the likeliest potential explanation. In the discussion, likeliness has been associated with the prior probability $P(H)$ or the posterior probability $P(H/E)$, whereas loveliness might correspond to explanations with a high value of $P(E/H)$.[27] Lipton (2004), in the second edition of his 1991 book, extended his account to promote "compatibilism" or "friendship" between Bayesianism and IBE, but a sharper treatment is obtained by quantitative measures of explanation and confirmation.

There are two main probabilistic definitions of confirmation:

(HP)   *High Probability Criterion*: E HP-confirms H iff $P(H/E)$ is sufficiently large.

(PR)   *Positive Relevance Criterion*: E PR-confirms H iff $P(H/E) > P(H)$.

PR is equivalent to conditions $P(E/H) > P(E)$ and $P(H/E) > P(H/\sim E)$. Further, E *disconfirms* H if $P(H/E) < P(H)$, while E is *irrelevant* to H if $P(H/E) = P(H)$. Both HP and PR can be relativized to some background knowledge B. It immediately follows that what Smokler called "enumerative" principles are satisfied by the HP criterion. On the other hand, his "abductive" principle of converse entailment CE is satisfied if

---

[25] These results are due to Hempel and Mary Hesse. See Niiniluoto and Tuomela (1973) and Roche (2017).

[26] This was proposed already by Harman (1965). See also Lipton (2004), 63–64, and Dellsén (2017).

[27] Cf. Niiniluoto (1999) and Okasha (2000). Leah Henderson (2014) proposes a two-level treatment, where a theory H achieves a better explanation when it explains the evidence E more with core claims and less with auxiliary hypotheses. When the likelihood $P(E/H)$ is calculated as the average over the likelihoods $P(E/H\&F_i)$ of specific models F_i, and the prior weights for F_i are uniform, then a better theoretical explanation has a larger likelihood.

confirmation is defined by the PR criterion. However, positive relevance does not generally satisfy the controversial converse consequence condition CC (see Niiniluoto & Tuomela, 1973).

If H entails E, we have $P(E/H) = 1$. Thus, by Bayes's Theorem (5), $P(H/E) = P(H)/P(E)$. Hence,

(14)   If H logically entails E, and if $P(H) > 0$ and $P(E) < 1$, then $P(H/E) > P(H)$.

This result about PR-confirmation is completely general in the sense that it is valid for all epistemic probability measures P and for all non-zero prior probabilities $P(H) > 0$. It allows H to be a strong and informative theory with theoretical concepts.

Andrés Rivadulla (2018) argues that, in spite of the increase of probability in (14), Bayesianism achieves no genuine confirmation, since for two rival theories H and K the prior ratio $P(H)/P(K)$ and the posterior ratio $P(H/E)/P(K/E)$ are equal to each other. It is correct that this equality holds if both H and K entail the evidence E. But, as we have repeatedly observed, the situation is different for the case where E is only a probabilistic consequence of the theory. As positive relevance PR is a symmetrical relation, it is sufficient for the confirmation of H by E that H is positively relevant to E. If inductive explanation is defined by the positive relevance condition, i.e., H inductively explains E if and only if $P(E/H) > P(E)$, then

(15)   If H is a positively relevant inductive explanation of E, then E PR-confirms H.

Combining these results (14) and (15), we have

(16)   If H deductively or inductively explains E, then E PR-confirms H.

Thus, the Bayesian approach immediately justifies the idea that *explanatory success is confirmatory or credence-increasing*. But by the same arguments we can establish a similar conclusion for prediction:

(17)   If H deductively or PR-inductively predicts E, then E PR-confirms H.

The result (16) express *the Basic Theorem of Abductive Confirmation* (see Niiniluoto, 1999). This simple application of Bayes's Theorem is sufficient to show that Peirce's schema (4) of abductive reasoning is truth-conducive in the positive but weak sense that explanatory success increases the probability that the explanatory theory is true.

A natural and useful way of refining the Basic Theorem is to introduce quantitative degrees of explanatory power and degrees of confirmation, with the aim of showing comparatively that *better explanations receive stronger confirmation* (see Niiniluoto, 2018a, Ch. 6.3):

(18)   If K is a better explanation of E than H, then E confirms K more than H.

To prove this result, we have to correlate the definitions of *explanatory power* expl (H,E) of H with respect to E and the *degree of confirmation* conf(H,E) of H by E. Popper (1959) attempted to achieve this directly by defining his measure of corroboration as a function of explanatory power $expl_2$.[28] The result (18) is also immediately obtained by the statistical school which defines empirical support by the likelihood function (see Bandyopadhyay & Forster, 2011). The simplest definition of *degree of confirmation*, related to the High Probability criterion, is posterior probability:

$$\text{conf}_1(H/E) = P(H/E).$$

Rudolf Carnap proposed in 1962 the difference measure, which is related to the Positive Relevance criterion:

$$\text{conf2}(H/E) = P(H/E) - P(H) = \frac{P(H)}{P(E)}(P(E/H) - P(E)).$$

Its variants include the ratio measure

$$\text{conf3}(H/E) = \frac{P(E/H)}{P(H)} = \frac{P(E/H)}{P(E)}.$$

and logarithmic differences

$$\text{conf}_4(H/E) = \log P(H/E) - \log P(H).$$

$$\text{conf}_5(H/E) = \log P(H/E) - \log P(H/\tilde{E}).$$

All of these measures conf_i, $i = 2, 3, 4, 5$, have a positive value if and only if E is positively relevant to H. They also satisfy the requirement that confirmation increases when the evidence is more surprising or improbable.

Relating these measures of explanatory power and confirmation gives as results the following cases of (18). If the explanation is chosen by the high posterior probability rule (13), then trivially we have the maximal value of conf_1. If H explains E better than K in the sense of expl_i, $i = 2, 3, 4, 5$, then $\text{conf}_3(H/E) > \text{conf}_3(H/E)$ and $\text{conf}_4(H/E) > \text{conf}_4(H/E)$. If H explains E better than K in the sense of the expected

---

[28] For criticism, see Niiniluoto and Tuomela (1973, 120–122).

value of $expl_1$, then $conf_2(H/E) > conf_2(K/E)$. For many purposes, $conf_2(H/E)$ gives similar results as the posterior probability measure $conf_1(H/E)$. In particular, if evidence E is entailed by both H and K, then $conf_1(H/E) > conf_1(K/E)$ iff $conf_2(H/E) > conf_2(K/E)$ iff $P(H) > P(K)$. More generally, we see that higher initial plausibility and higher explanatory likelihood together are sufficient for higher confirmation: if $P(K) \geq P(H)$ and $P(E/K) \geq P(E/H) > P(E)$, then $conf_1(K/E) \geq conf_1(H/E)$ and $conf_2(K/E) \geq conf_2(H/E)$.

We have seen that theoretical hypotheses H with $P(H) > 0$ may receive support from their empirical successes, but so far we have shown only comparative results about confirmation. Is it possible to guarantee that their posterior probability $P(H/E)$ is high? The application of Bayes's formula (5) to calculate the absolute value of $P(H/E)$ presupposes that the prior probability $P(E)$ of evidence E is known, and this is equal to the weighted average of the likelihoods $P(E/H_i)$ on all alternative hypotheses H_i.[29] Due to the difficulty in evaluating $P(E)$, Dellsén (2018) interprets IBE as a heuristic guide to the comparative probability values of available hypotheses.[30] By (14) successful deductive explanation of E by H increases the probability of H on E, even when the likelihood $P(E/{\sim}H)$ is unknown. In those cases where this likelihood is estimated to be sufficiently small, so that H is the only viable explanation of E, the increase in the probability $P(H/E)$ will be significant.[31] In the limiting case of IOE, where $P(E/{\sim}H) = 0$, we have $P(H/E) = 1$ independently of the prior probability $P(H) > 0$.

Let us illustrate the calculation of posterior probability by an example, where a person is tested for the corona infection (H) and evidence E is a positive test result. Assume that the test is completely reliable in the sense that $P(E/H) = 1$, but 5% of positive outcomes are mistaken, so that $P(E/{\sim}H) = 0.05$. Assuming that 1% of the population has infection, so that by this base rate $P(H) = 0.01$ and $P({\sim}H) = 0.99$, we have by (5)

$$P(H/E) = \frac{0.01}{0.01 + 0.99 \times 0.05} \eqsim \frac{1}{1+5} = \frac{1}{6}.$$

This value is much larger than the prior 0.01, so that by the conditions of (11) (likelihood) and (12) (increase of probability) the infection hypothesis H is strongly supported by E as the best explanation. But at the same time this posterior probability is low, in fact much less than 0.5, so that $P(H/E) < P({\sim}H/E)$ and H is not acceptable on E by (13). But why is it the case that we are inclined to approve the

---

[29] In a comparative statistical Bayes-test between hypotheses H_0 and H, the probability $P(E)$ cancels out, since $P(H_0/E) > P(H/E)$ iff $P(H_0)P(E/H_0) > P(H)P(E/H)$ iff $P(E/H_0)/P(E/H) > P(H)/P(H_0)$.

[30] Dellsén is also cautious in suggesting that the most probable explanation serves as a working hypothesis to be tested further, which agrees with the pursuit interpretation of Peirce's abduction.

[31] In Hintikka's inductive logic, there is one constituent which asymptotically receives the posterior probability one with increasing observational evidence. Among constituents compatible with evidence E, this hypothesis H is also the most parsimonious (assumes the least number of kinds of individuals) and the least probable, so that it maximizes both $P(H/E)$ and $P(H/E) - P(H)$ (see Niiniluoto, 2011).

results and act on the basis of such medical tests? One possible answer is that the choice of the population is ambiguous (should we use the country, city, or district to fix the base rate?), so that one might instead apply the Principle of Indifference and choose the priors as $P(H) = P(\sim H) = 0.5$.[32] Then the posterior probability is very high:

$$P(H/E) = \frac{0.5}{0.5 + 0.5 \times 0.05} = \frac{1.00}{1.05} = \frac{20}{21}.$$

Another remark is that if the evidence consists of two independent positive tests E_1 and E_2, then the posterior probability is again high:

$$P(H/E_1 \& E_2) = \frac{0.01}{0.01 + 0.99 \times 0.05 \times 0.05} \approx \frac{1}{1 + 0.25} = \frac{4}{5}.$$

As H is the best explanation of the two agreeing test outcomes, the Bayesian acceptability of H on $E_1 \& E_2$ agrees with Schurz's (2008) principle of Common Cause Abduction.

Combining the results (16) and (17), we see that the total confirmation of a hypothesis comes from its success in explanation and prediction. In addition to the initial problem situation with evidence E, there may be new evidence E' from surprising observations and tests. The prior plausibility of a hypothesis is now its probability $P(H/E)$ relative to the initial evidence E, which depends on the ability of H to explain E. The confirmation of H due to the new evidence E' relative to E is then definable by $conf_2(H,E'/E) = P(H/E' \& E) - P(H/E)$. The joint confirmation of H by E and E' can be defined by $conf_2(H/E \& E') = P(H/E \& E') - P(H)$. Therefore, it seems that we need not assume an essential difference between accommodation and prediction as methods of confirmation. Similarly, the measures of explanatory power can be systematically reinterpreted so that they cover both explanatory and predictive relations between hypothesis and evidence.

Even though explanation and prediction behave in a symmetrical fashion with respect to the probabilistic confirmation of rival hypotheses, this is problematic in the case of acceptance. A hypothesis that explains our initial data, and is thereby confirmed by it to some extent, may still be ad hoc. To remove this doubt, the hypothesis should be independently testable, i.e., it should either explain some old evidence or be successful in new serious tests. Therefore, one may argue that IBE as an acceptance rule should contain an additional condition stating that the "best" hypothesis is one with both explanatory and predictive power. Nevertheless, the postulate of dark matter is an important contemporary example of the power of explanatory considerations even without successful prediction. The majority of

---

[32] This is indeed the most common behavior that Tversky and Kahneman find in their famous psychological experiments (see Niiniluoto, 1981). Henderson (2017b) argues that there is reason identify the prior probability $P(H)$ as the base rate with respect to population U only if the theory H is randomly selected from U, and this condition is only rarely satisfied.

physicist accept the existence of dark matter as an abductive solution to the problem of "missing masses", since it explains anomalous observations about galaxies, even though dark matter by definition is invisible (does not interact with electromagnetic radiation), and experimental attempts to verify or test its existence have failed. The alternative explanation without dark matter would require the revision of Newtonian or relativistic dynamics, which is a difficult step for the physicists, but probably there are still unconceived explanations to be investigated in the future.

11.5 Modified Bayesian Approaches

Bas van Fraassen (1989) argued that IBE and Bayesianism are incompatible. Our results in Sects. 11.3 and 11.4 show that on the contrary IBE has a natural formulation in two Bayesian variants: acceptance (full belief) and confirmation (partial belief). The results (14) and (18) illustrate what Henderson (2017a) calls "emergent compatibilism": a reasonable formulation of IBE emerges automatically from the Bayesian model without the need to give extra boosting by explanatory considerations to prior probabilities and likelihoods.

Besides the bad lot objection, van Fraassen (1989) formulated two other critiques of IBE. Even though they are not fatal to the basic idea of explanatory reasoning, they have inspired the development of new modified Bayesian explications of IBE.

One on van Fraassen's arguments claims that the Bayesian account of abduction leads to incoherence: if explanatory success is credited with adding an extra bonus to the posterior probability, this procedure allows the construction of a dynamic Dutch Book against the scientist. As a straightforward reply, we may note that the point of the Bayesian approach is not to add separate bonus points, but to show that explanatory virtues enter into the assessment of posterior probability via the normal elements of Bayes's Theorem, viz. prior probability and likelihood (see Kvanvig, 1994; Niiniluoto, 1999). No such special assumption was needed to prove the basic theorem (18) about the connection between explanation and confirmation. But another response has been developed by Igor Douven (2017b), who is not worried about dynamic incoherence. His suggestion is to modify Bayesian conditionalization, as based on the probabilistic update formula (5) relative to the partition of hypotheses H_i, $i = 1, \ldots, n$, with a function $f(H_i, E)$ which assigns a bonus point to the hypothesis that explains E best and nothing else to other hypotheses:

$$(19) \quad P(Hi/E) = \frac{P(H_i)P(E/H_i) + f(H_i, E)}{\sum (P(H_j)P(E/H_j) + f(H_j, E))}.$$

Douven shows that the rule (19), when applying Popper's explanatory power $expl_2$ and Good's measure $expl_4$ in the definition of the bonus f, improves the Bayesian update: it is on average faster and more accurate than the Bayes rule in identifying the true bias of a coin. A further modification of (19) credits each hypothesis with a bonus which is proportional to its explanatory power. Douven's approach opens the experimental comparison of the merits of a variety of modified models of IBE.[33]

Other examples of the boosting mechanism add a statement about explanation to the probability function. Roche and Sober (2013) argue that "explanatoriness is evidentially irrelevant". They defend this conclusion by claiming that P (H/E&X) = P(H/E), when X states "were H and E true, H would explain E". Note, however, that our basic theorem (16) is expressed as a conditional "if X then P(H/E) > P(H)". It is not clear that X can occur as an argument in a probability function, or a statement in probability space. But if this makes sense, then theorem (16) states that P(H/E&X) > P(H/X), i.e., E is positively relevant H given X, instead of the equation of Roche and Sober. A related argument, with the opposite conclusion, is given by Tesic et al. (forthcoming). They show that P(H/E&X) > P(H/E), if X states "H would be the best explanation of E, were E to obtain", H and X are probabilistically independent, P(E/H,X) ≥ P(E/H,~X), and P(E/~H,X) > P(E/~H, ~X). They see this theorem as a defense of IBE within the purely Bayesian framework, since explanatory success with respect to new evidence E gives an additional boost to the confirmation of hypothesis H.

van Fraassen's third criticism notes that the result (14) about confirmation is blocked if the prior probability P(H) is zero. This is the most challenging of his objections to IBE, since the reply needs new logical tools – the notions of approximate truth and truthlikeness – beyond the probability calculus. As noted by Niiniluoto (1999), there are two cases in which such a zero probability is natural: (a) H is a *sharp* hypothesis, i.e., H belongs to an infinite set of rival hypotheses **H**, e.g. the set of real numbers; (b) H is an *idealized* hypothesis, i.e., the class of rival hypotheses (including H) is defined relative some counterfactual assumption. The case (a) can be solved with the notion of *probable approximate truth*, which can be increased by evidence even though P(H) = 0 (see Niiniluoto, 2018a, 139–140). This notion of ap-confirmation satisfies the special consequence principle. For the case (b), it is useful to define the *expected verisimilitude* ver(H/E) of a hypothesis H given evidence E by

$$(20) \quad ver(H/E) = \sum P(C_i/E)Tr(H, C_i),$$

---

[33] Climenhaga (2017) argues that Douven's rule leads to synchronic probabilistic incoherence. Climehaga also claims that the rules IBE and IBT may lead to inconsistent conclusions with the same evidence. This issue deserves to be investigated further, since it involves delicate questions about the relativity of acceptance to a partition (cf. Levi, 1967; Niiniluoto, 1976).

where the sum goes over all i $= 1, \ldots, $ n. Here $C_1, \ldots, C_n$ are the alternative constituents (complete states of affairs) expressible in some language, and $Tr(H,C_i)$ is what the degree of truthlikeness of theory H would be if C_i were the true state. The value ver(H/E) generally differs from posterior probability P(H/E), and we may have ver(H/E) > 0 even when P(H) = 0 or P(H/E) = 0. When new evidence E' is obtained, the value ver(H/E&E') is defined by updating the probabilities $P(C_i/E\&E')$ of constituents by Bayesian conditionalization. Expected verisimilitude ver can be used to assess the success of the modified rules IBAE and IBT (see Sect. 11.3) to establish a fallible link between explanatory success and truthlikeness. With the ver-function we can also formulate a rule of acceptance:

(21)   Given evidence E, accept the hypothesis H with maximal value ver(H/E).

which is a direct generalization of the principle (13) of maximizing posterior probability. Here it is important to investigate the conditions under which the following kinds of principles are valid:

(22)   If H' is a better approximate explanation of E than H, then ver(H'/E) > ver (H/E).

(23)   If H approximately explains E, and H may incompatible with E, then the expected verisimilitude of H given E is high.

(24)   If H approximately explains E, then E increases the expected verisimilitude of H.

Here (24) is generalization of the Basic Theorem (17) for explanation and probabilistic PR-confirmation. If these principles hold, then (relative) explanatory success gives us a rational warrant for claims about (relative) truthlikeness.[34] Thereby they also allow the formalization of the explanationist defense of scientific realism, known as the "ultimate" or "no miracles" argument for realism:

(NMA)   The empirical success of science would be a miracle, unless the best scientific theories are true, approximately true or truthlike.

---

[34] See Niiniluoto (2018a, 143–145) for illustrations in special cases. Further work in this direction is needed.

Therefore, the status of scientific realism is dependent on our ability to explicate the modified rules IBAE, IBAE', IBT, and IBT^c within the truthlikeness framework.[35] NMA can also be defended by arguing in the spirit of IOE that realism is the only viable explanation of the success of science (See Niiniluoto, 2018a, 161–163).

References

Aliseda, A. (2006). *Abductive reasoning: Logical investigations into discovery and explanation*. Springer.

Amaya, A. (2009). Inference to the best legal explanation. In H. Kaptain, H. Prakken, & B. Verteij (Eds.), *Legal evidence and proof: Statistics, stories, logic* (pp. 135–159). Ashgate.

Bandyopadhyay, P. S., & Forster, M. (Eds.). (2011). *Philosophy of statistics*. Elsevier.

Bird, A. (2010). Eliminative abduction – Examples from medicine. *Studies in History and Philosophy of Science, 41*(4), 345–352.

Bird, A. (2017). Inference to the best explanation, Bayesianism, and knowledge. In K. McCain & T. Poston (Eds.), *Best explanations: New essays on inference to the best explanation* (pp. 97–120). Oxford University Press.

Climenhaga, N. (2017). Inference to the best explanation made incoherent. *Journal of Philosophy, 114*(5), 251–273.

Dellsén, F. (2016). Explanatory rivals and the ultimate argument. *Theoria, 82*(3), 210–237.

Dellsén, F. (2017). Abductively robust inference. *Analysis, 77*(1), 20–29.

Dellsén, F. (2018). The heuristic conception of inference to the best explanation. *Philosophical Studies, 175*(7), 1745–1766.

Dellsén, F. (2021). Explanatory consolidation: From 'best' to 'good enough'. *Philosophy and Phenomenological Research, 103*(1), 155–177.

Douven, I. (2017a). Abduction. In E. Zalta (Ed.), *Stanford encyclopedia of philosophy*. Stanford University. http://plato.stanford.edu/archives/spr2011/entries/abduction/

Douven, I. (2017b). Inference to the best explanation: What is it? And why should we care? In K. McCain & T. Poston (Eds.), *Best explanations: New essays on inference to the best explanation* (pp. 7–24). Oxford University Press.

Gabbay, D., & Woods, J. (2005). *The reach of abduction: Insight and trial*. Elsevier.

Hanson, N. R. (1958). *Patterns of discovery*. Cambridge University Press.

Harman, G. (1965). The inference to the best explanation. *The Philosophical Review, 74*(1), 88–95.

Hempel, C. G. (1965). *Aspects of scientific explanation*. The Free Press.

Henderson, L. (2014). Bayesianism and inference to the best explanation. *The British Journal for the Philosophy of Science, 65*(4), 687–715.

---

[35] See Niiniluoto (2018a), Ch. 9. Henderson (2017b) gives a Bayesian formalization of NMA by the assumptions

P(T is empirically successful/T is approximately true) is high
P(T is empirically successful/T is not approximately true) is very small,

which imply by Bayes's theorem that P(T is approximately true/T is empirically successful) is close to one. This straightforward application of the notion of probable approximate truth is not unproblematic, since also the empirical success of approximately true theories is typically approximate. In approximate explanation the idealized theory H and the original observational evidence E are incompatible, so that $P(E/H) = 0$. Therefore, the precise analysis of NMA should involve counterfactual probabilities and estimation of verisimilitude (see Niiniluoto, 2021).

Henderson, L. (2017a). The no-miracles argument and the base rate fallacy. *Synthese, 194*(4), 1295–1302.

Henderson, L. (2017b). Bayesianism and IBE: The case of individual vs. group selection. In K. McCain & T. Poston (Eds.), *Best explanations: New essays on inference to the best explanation* (pp. 248–261). Oxford University Press.

Hintikka, J. (1968). The varieties of information and scientific explanation. In B. van Rootselaar & J. F. Staal (Eds.), *Logic, methodology, and philosophy of science III* (pp. 151–171). North-Holland.

Hintikka, J. (1998). What is abduction? The fundamental problem of contemporary epistemology. *Transactions of the Charles S. Peirce Society, 34*, 503–533.

Josephson, J., & Josephson, S. (Eds.). (1994). *Abductive inference.* Cambridge University Press.

Kuipers, T. (1999). Abduction aiming at empirical progress or even truth approximation leading to a challenge of computational modelling. *Foundations of Science, 4*, 307–323.

Kuipers, T. (2019). *Nomic truth approximation revisited.* Springer.

Kvanvig, J. L. (1994). A critique of van Fraassen's voluntaristic epistemology. *Synthese, 98*(2), 325–348.

Laudan, L. (1980). Why was the logic of discovery abandoned? In T. Nickles (Ed.), *Scientific discovery, logic, and rationality* (pp. 173–183). D. Reidel.

Levi, I. (1967). *Gambling with truth.* Alfred A. Knopf.

Lipton, P. (1991). *Inference to the best explanation* (first ed.). Routledge.

Lipton, P. (2004). *Inference to the best explanation.* Routledge. (First edition in 1991.)

Magnani, L. (2001). *Abduction, reason, and science: Processes of discovery and explanation.* Kluwer and Plenum.

McCain, K., & Poston, T. (Eds.). (2017). *Best explanations: New essays on inference to the best explanation.* Oxford University Press.

Niiniluoto, I. (1976). Inquiries, problems, and questions: Remarks on local induction. In R. Bogdan (Ed.), *Local induction* (pp. 263–296). Reidel.

Niiniluoto, I. (1981). L. J. Cohen versus Bayesianism. *The Behavioral and Brain Sciences, 4*(3), 349.

Niiniluoto, I. (1999). Defending abduction. *Philosophy of Science (Proceedings), 66*, S436–S451.

Niiniluoto, I. (2011). The development of the Hintikka program. In D. Gabbay, S. Hartmann, & J. Woods (Eds.), *Handbook of the history of logic 10: Inductive logic* (pp. 311–359). North-Holland.

Niiniluoto, I. (2018a). *Truth-seeking by abduction.* Springer.

Niiniluoto, I. (2018b). Explanation by idealized theories. *Kairos. Journal of Philosophy & Science, 20*(1), 43–63. https://doi.org/10.2478/kjps-2018-003

Niiniluoto, I. (2021). Vassend on verisimilitude and counterfactual probability. *Philosophy of Science, 88*(3), 554–561.

Niiniluoto, I., & Tuomela, R. (1973). *Theoretical concepts and hypothetico-inductive inference.* D. Reidel.

Okasha, S. (2000). Van Fraassen's critique of inference to the best explanation. *Studies in History and Philosophy of Science, 31*(4), 691–710.

Peirce, C. S. (1931–35, 1958). *Collected papers* 1–6 (C. Hartshorne & P. Weiss, Eds.), 7–8 (A. Burks, Ed.). Harvard University Press. (*CP*).

Peirce, C. S. (1982–2010). *Writings of Charles S. Peirce: A chronological edition*, Vol. 1–6, 8 (M. Fisch et al. Eds.). Indiana University Press. (*W*).

Peirce, C. S. (1992). *The essential Peirce vol. 1 (1867–1893)* (N. Houser & C. Kloesel, Eds.). Indiana University Press. (*EP* 1).

Peirce, C. S. (1998). *The essential Peirce vol. 2 (1893–1913)* (N. Houser & C. Kloesel, Eds.). Indiana University Press. (*EP* 2).

Popper, K. R. (1959). *The logic of scientific discovery.* Hutchinson.

Poston, T. (2014). *Reason and explanation: A defense of explanatory coherentism.* Palgrave Macmillan.

Psillos, S. (2002). Simply the best: A case for abduction. In A. Kakas & F. Sadri (Eds.), *Computational logic* (pp. 605–625). Springer.

Psillos, S. (2009). *Knowing the structure of nature: Essays on realism and explanation.* Palgrave Macmillan.

Psillos, S. (2011). An explorer upon untrodden ground: Peirce on abduction. In D. Gabbay, S. Hartmann, & J. Woods (Eds.), *Handbook of the history of logic. Vol. 10: Inductive logic* (pp. 117–151). North-Holland.

Rivadulla, A. (2018). Abduction, Bayesianism and best explanations in physics. *Culturas científicas, 1*(1), 63–75.

Roche, W. (2017). Explanation, confirmation, and Hempel's paradox. In K. McCain & T. Poston (Eds.), *Best explanations: New essays on inference to the best explanation* (pp. 219–241). Oxford University Press.

Roche, W., & Sober, E. (2013). Explanatoriness is evidentially irrelevant; or, inference to the best explanation meets Bayesian confirmation theory. *Analysis, 73*(4), 659–668.

Salmon, W. (1971). *Statistical explanation and statistical relevance.* University of Pittsburgh Press.

Salmon, W. (2001). Explanation and confirmation: A Bayesian critique of inference to the best explanation. In G. Hon & S. S. Rakover (Eds.), *Explanation: Theoretical approaches and applications* (pp. 61–91). Kluwer.

Schupbach, J. N. (2014). Is the bad lot objection just misguided? *Erkenntnis, 79*(1), 55–64.

Schupbach, J. N. (2017). Inference to the best explanation, cleaned up and made respectable. In K. McCain & T. Poston (Eds.), *Best explanations: New essays on inference to the best explanation* (pp. 38–61). Oxford University Press.

Schurz, G. (2008). Patterns of abduction. *Synthese, 164*(2), 201–234.

Shaffer, M. (2019). The availability heuristic and inference to the best explanation. *Logos & Episteme, 10*(4), 409–432.

Smokler, H. (1968). Conflicting conceptions of confirmation. *The Journal of Philosophy, 65*, 300–312.

Tavory, I., & Timmermans, S. (2014). *Abductive analysis: Theorizing qualitative research.* The University of Chicago Press.

Tesic, M., Eva, B., & Hartmann, S. (forthcoming). *Confirmation by explanation: A Bayesian justification of IBE.* www.academia.edu/43018036/

Thagard, P. (1978). The best explanation: Criteria for theory choice. *The Journal of Philosophy, 75*(2), 76–92.

Thagard, P. (1989). Explanatory coherence. *Behavioral and Brain Sciences, 12*(3), 435–502.

van Fraassen, B. (1989). *Laws and symmetry.* Oxford University Press.

Weisberg, J. (2009). Locating IBE in the Bayesian framework. *Synthese, 167*(1), 125–144.

Chapter 12
Re-inflating the Realism-Instrumentalism Controversy

Stathis Psillos

Abstract Compatibilist instrumentalism, aka irenic instrumentalism, exemplified in the work of Rudolf Carnap and Ernest Nagel, tended to underestimate the ontic differences between realism and instrumentalism making the claim that the two positions are merely different modes of speech (Nagel) of that they differ only qua languages (Carnap). Compatibilism as such did not last for a long time but it's taken two new leases of life. The first is in the very influential paper by Stein in the early 1990s. The second is in Kyle Stanford's recent work. Between them, the arguments by Stein and Stanford aim to deflate the realist-instrumentalism controversy and to make instrumentalism palatable as a view of science. The aim of this paper is to show that compatibilism fails. Instrumentalism and realism are genuine rivals. Hence, the difference between realism and instrumentalism (even when they are taken to be 'sophisticated') is deep and philosophically significant.

Keywords Realism · Instrumentalism · Carnap · Nagel · Stein · Stanford

12.1 Introduction

Ever since instrumentalism, based on Craig's theorem, was almost summarily dismissed by a series of arguments that favoured scientific realism, in the 1960s, denialism – the view that the so-called theoretical entities do not exist and that the only substantive dispute between realism and instrumentalism concerns whether theoretical terms have meaning or not – was split into two positions: agnosticism (exemplified in Bas van Fraassen's constructive empiricism) and compatibilism, called irenic instrumentalism, by Wilfrid Sellars (1965), (exemplified in the work of Rudolf Carnap and Ernest Nagel). The latter approach tended to underestimate the ontic differences between realism and instrumentalism making the claim that the two

S. Psillos (✉)
Department of History and Philosophy of Science, National and Kapodistrian University of Athens, Athens, Greece
e-mail: psillos@phs.uoa.gr

© The Author(s), under exclusive license to Springer Nature Switzerland AG 2022
W. J. Gonzalez (ed.), *Current Trends in Philosophy of Science*, Synthese Library 462, https://doi.org/10.1007/978-3-031-01315-7_12

positions are merely different modes of speech (Nagel) of that they differ only qua languages (Carnap).

Compatibilism as such did not last for a long time but it's taken two new leases of life. The first is in the very influential paper by Stein in the early 1990s. In this Stein argued that "between a cogent and enlightened 'realism' and a sophisticated 'instrumentalism' there is no significant difference – no difference that makes a difference" (1989, 61). The second is in Kyle Stanford's recent work. After trying to give a new lease of life to instrumentalism by advancing the unconceived alternatives argument in his (2006), Stanford has been pushing the line of a new compatibilism between 'historicist' instrumentalism and those versions of realism, which, as he put it, have granted to instrumentalism "that we are in the midst of an ongoing historical process in which our theoretical conceptions of nature will continue to change just as profoundly and fundamentally as they have in the past" (2015a, 875). Between them, the arguments by Stein and Stanford aim to deflate the realist-instrumentalism controversy and to make instrumentalism palatable as a view of science. At the same time, however, instrumentalism as a genuine rival to realism is revamped by a younger generation of philosophers, notably Brad Wray (2018) and Darrell Rowbottom (2019). And while Wray ventures for agnosticism (see my 2020), Rowbottom goes for (a new) denialism (see Psillos & Zorzato, 2020).

The aim of this paper is to show that compatibilism fails. Instrumentalism and realism are genuine rivals. Hence, the difference between realism and instrumentalism (even when they are taken to be 'sophisticated') is deep and philosophically significant.

Here is the road map. Taking a cue from Rowbottom's claim that instrumentalism is a *movement*, it will be shown in Sect. 12.2 that the saving-the-phenomena tradition, which Pierre Duhem thought was instrumentalist, was not really part of the movement. Then in Sect. 12.3 I will identify two core commitments of the movement, viz., denialism and non-eliminativism. In Sect. 12.4 I will argue that there is a tension between these two commitments and in Sect. 12.5, it will be shown that Craig's theorem was an aborted attempt to relieve this tension by abandoning non-eliminativism. Sections 12.6, 12.7, 12.8 and 12.9 will examine the main forms and arguments for compatibilism: Carnap's, Nagel's, Stein's and Stanford's.

12.2   To Save the Appearances

Rowbottom (2018), one of the few outspoken current instrumentalists, takes it that instrumentalism treats science as an instrument; he also adds that instrumentalism takes it that science is about the observable (or the *phenomena*); scientific theories help us "orient" ourselves in the world around us. That's a very broad understanding of a "movement". What makes it "a form of anti-realism"? Surely, both of the points above are consistent with realism. The view that theories are useful instruments for prediction and control is actually a consequence of realism: if theories do succeed in uncovering the unseen world, then surely they succeed in what they assert about the phenomena. In fact, the realist would argue that the latter success is fully due to the former. What then makes the instrumentalist movement *anti-realist*? It's precisely something added on to the two claims made above, and this more is typically captured by a 'nothing but' clause: theories are *nothing but* instruments for prediction and control. (Rowbottom puts "primarily" before "instruments", but the force is the same.) This *nothing-but-ness* is what realism denies; hence the incompatibility.

The *nothing but* is usually qualified by expressions such as 'in the final analysis' or 'the cash value is nothing but. . .'. But the key point should be clear: the *credo* of the instrumentalist movement is that any kinds of aspirations we might have that science goes beyond the phenomena is unwarranted and/or superfluous. This kind of block to science's cognitive aspirations is either because of ontological denialism (e.g., there is nothing beyond the appearances) or because of epistemological prudentialism (e.g., even if there is something beyond appearances, for instance, their unseen causes, either we are unable to find out what they are or it does not matter for doing science well that we try to map the unseen causes).

Instrumentalism is indeed a movement. Both Mach and Duhem flirted (to say the least) with it. But to be part of the movement you have to adopt the fundamental credo. This credo is captured nicely in the anonymous preface of the famous *De Revolutionibus Orbium Coelestium*, in 1543, the posthumously published masterpiece of Copernicus, where it is written:

> For it is the duty of an astronomer to compose the history of the celestial motions through careful and expert study. Then he must conceive and devise the causes of these motions or hypotheses about them. Since he cannot in any way attain to the true causes, he will adopt whatever suppositions enable the motions to be computed correctly from the principles of geometry for the future as well as for the past. The present author has performed both these duties excellently. For these hypotheses need not be true nor even probable. On the contrary, if they provide a calculus consistent with the observations, that alone is enough. (. . .) For this art, it is quite clear, is completely and absolutely ignorant of the causes of the apparent nonuniform motions. And if any causes are devised by the imagination, as indeed very many are, they are not put forward to convince anyone that are true, but merely to provide a reliable basis for computation ([1543] 1992, xix).

As it turned out, the preface was written not by Copernicus himself but by Andreas Osiander, a Lutheran theologian. The intention was obvious: to avoid the wrath of the church. But leaving this aside for a minute, the key *argument* for instrumentalism is simply that the astronomer cannot find the truth about the causes of motion. Now, in this particular context, the truth is "divinely revealed"; hence neither the astronomer nor the philosopher will be able to attain it by evidence plus sound reasoning. But why should we take this argument seriously in a post-revelation context?

Pierre Duhem in his famous essay *ΣΩZEIN TA ΦAINOMENA* [*To Save the Phenomena*] (1908) secularised the argument by embedding into a long homonymous tradition which goes back to Plato. Before Copernicus, the dominant astronomical theory was Claudius Ptolemy's (ca. 85–ca.165). He had assumed, pretty much like Aristotle and Plato, a geocentric model of the universe. To save the appearances of planetary motions, Ptolemy had devised a system of deferents and epicycles. There were alternative mathematical models of the motion of the planets (e.g., one based on a moving eccentric circle), but Ptolemy thought that since all these models were saving the appearances, they were good enough. The issue of their physical reality was not raised (though at least some medieval philosophers understood these models realistically). Astronomical hypotheses were "*certa instrumenta*", as Erasmus Reinhold put it in 1551, for the construction of astronomical tables. Copernicus's heliocentric system was achieving this goal more efficiently than Ptolemy's, and this accounted, according to Duhem, for the fact that it started to win the attention of astronomers in the saving-the-phenomena tradition.

So, for Duhem astronomy aimed merely at calculating the various astronomical phenomena; astronomical hypotheses were (by default) fictions and not realities, whose "whole purpose" was to save the phenomena; hence, it's only natural that astronomers use different hypotheses to do their business ([1908] 1969, 22). Then, Osiander's plea for instrumentalism was just the culmination of a long instrumentalist tradition.

It should be noted however that the key instrumentalist argument that Osiander put forward, viz. (roughly) that truth cannot be known, is not part and parcel of the saving-the-phenomena tradition. Hence this tradition was not outright instrumentalist. Rather it was a mixed bag. There are a number of important issues that ideally should be discussed here. One of them is the relation between physics and astronomy in the ancient Greek tradition and beyond. It was physics which was looking for causes (the essence and the power) of things, whereas astronomy (being in effect applied geometry) relied on principles of physics (e.g., that orbits are circular) but aimed at capturing the motions of the heavenly bodies and not, at least in the first instance, their causes. This is something that had to do with the nature of astronomy qua science: it relied heavily on mathematical (i.e., geometrical) models and there can be (in fact there are) more than one of them to save the appearances. Mathematics was taken to be an instrument that was transforming astronomy into an exact science, leading it to exact predictions with ever-increasing levels of accuracy. As Samuel Sambursky put it, in his classic study of ancient Greek science "Mathematics had transformed astronomy into what Iamblichus called a 'prognostic science'" (1962, 58). Still, it doesn't follow that the causes are unknown or unknowable. Nor does it follow that astronomy was an instrumentalist science. This kind of view, fostered by Duhem's reading of the saving-the-phenomena tradition, was challenged by GER Lloyd in his (1978).

A very interesting case was Geminus's account of Posidonius's *Meteorologica*, which is reported by Simplicius, Aristotle's commentator, in his Commentary on Physics, Book 2 (cf. 2014a). According to Geminus, despite the fact that physics and astronomy have different aims, they study the same entities (e.g., "that the sun is a sizeable body, that the earth is spherical") albeit from a different perspective and using different methodologies. Whereas physics treats of heavenly bodies qua substances and examines their powers, "the astronomer argues from the properties of their shapes and sizes, or from quantity of movement and the time that corresponds to it" (292, 7–8). So, astronomy, qua mathematical science, deals only with the geometrical properties of bodies and the quantitative aspects of local motion. These are abstractions from the actual properties of bodies. Hence, unlike physics, which "often touches on causes" (292, 10) astronomy is not in the business of causal explanation. This happens at three levels.

First, astronomy doesn't "pay any great attention to causes" (292, 12), that is, to actual physical bodies, since the objects of astronomy are abstractions. That's clearly the case when the astronomer represents "the earth or the heavenly bodies as spherical" that is as perfect spheres.

Second, astronomy "does not even attempt to find the cause" (292, 12), For instance, astronomy is not concerned with *why* an eclipse happens but with *when*.

Third, occasionally astronomy relies on assuming "certain orbits by whose presence appearances will be saved" (292, 15). That's the territory of epicycles and eccentrics. This level is non-causal because the assumptions made (the orbits chosen) are constrained by the demand of saving the phenomena. In this sense, any assumption will do, provided that it saves the phenomena.

But that's only half of the story. The other half, so to speak, is in what Geminus, as reported by Simplicius, goes on to say, viz., that *the very idea* of saving the phenomena aims to produce hypotheses such that "the treatment of the planets is squared with the accepted method of causal explanation" (292, 19–20). That is, the various hypotheses should be in accord with the principles of physics so that a causal explanation of the motions of the planets is possible. In particular, this causal explanation should be based on principles of accepted physics, viz., that the motions of heavenly bodies are uniform, regular and always in the same direction. So, it turns out that the saving-the-phenomena tradition is quite far from being instrumentalist, pure and simple. Astronomy was taken not to aim, in and of itself, at causal explanations of the motions of heavenly bodies, since (a) it deals with abstractions and (b) it's not concerned with the reason why X happens, but with when it happens.

In achieving (b) astronomy relies on hypotheses which, though they do not offer causal explanations themselves, they should be constrained by principles which are causal-explanatory, viz., that the motions of heavenly bodies are uniform, regular and always in the same direction. Now, it turns out that there are more than one hypotheses which save the appearances consistently with the causal-explanatory principles of physics. Astronomy has to state and examine all of them; it's not part of its business to pick one of them, simply because of (b) above. Hence, astronomers will aim to examine all relevant mathematical models.

The problem of saving the appearances by relying on correct principles of physics was accentuated by the fact that the motions of the planets presented various irregularities, such as the retrograde motion of Mercury and Venus. These were apparent motions which had to be explained away by reference to real motions – real in the sense that they satisfied the principles noted above, viz. uniformity, regularity and uni-directionality. Which of the possible real motions were actual was not the business of the astronomer to find out. Geminus notes this very well when he points out that astronomers should "enquire closely in what ways it is possible for these appearances to be produced" (292, 18–19), and hence to save the appearances, whilst at the same time the treatment of the planets offered should be "squared with the

accepted method of causal explanation". This dual task, as it were, renders astronomy peculiar. On the one hand, it should investigate all possible geometrical
arrangements of heavenly bodies which save the appearances. Hence Geminus
reports the heliocentric model of Heraclides of Pontus adding that it's not the
concern of astronomers to find out which heavenly body, by nature, is at rest and
which by, nature, moves. Rather they should look for configurations of motion and
rest among the planets, the sun and the earth consistent with the appearances. On
the other hand, however, astronomers should rely on "basic" principles that identify
the real motion and properties of the heavenly bodies, viz., that "the dance of the
heavenly bodies is simple, regular and ordered" (292, 27).

Now, Duhem treats Geminus's views as instrumentalist. He writes that
Posidonius as reported by Geminus appeals to equivalent hypotheses "in order to
drive home the astronomer's inability to grasp the true nature of the heavenly
motions ([1908] 1969, 11). But, there is no such point in Geminus's account. In
fact, if we look at Simplicius's own views, we realise that this tradition was not
instrumentalist. Simplicius was an Aristotelian. No wonder then that he quotes
Aristotle with approval when he says that "it would be absurd for the natural scientist
to know nothing of the essential properties of natural bodies" (292, 33).

So, what's the point of the saving-the-phenomena tradition? For Simplicius, the
key to understanding this view lies in the irregularities there are in the motions of the
heavenly bodies. These are merely appearances in the sense that though the motions
of the planets are regular and uniform they *appear* to us differently. In his Commentary of Aristotle's *On the Heavens* (2.10–2.14) (cf. 2014b), Simplicius notes:
"The true account does not accept that they [the planets] stand still or move
backward, or that there is in addition or subtraction in the numbers of their motions
[i.e., their speed changes] even if they are observed to move in this way, nor does it
admit hypotheses of this kind; rather it demonstrates that the heavenly motions are
simple, circular, uniform, and ordered, using as evidence their substance"
(488, 10–15). Astronomers, he adds, were unable to "grasp with precision how
what occurs in the heaven is only the appearance of their condition and not the
truth"; hence, they "were content to find out on what hypotheses the phenomena
concerning the stars which are said to wander could be preserved by means of
uniform, ordered, circular motions" (488, 17–19).

In citing this passage, Duhem takes it that Simplicius considers the various
astronomical hypotheses to be fictions and not realities ([1908] 1969, 22), And
yet, the only fictions in Simplicius's account are the supposed irregularities in the
motions of the planets. The contrast Simplicius draws is between the fictitious (i.e.,
apparent) irregularities and the true circular and uniform motions.

What's the point of all this? It tuns out that this textbook case of an 'instrumentalist movement' is not really (or purely) instrumentalist. Astronomical theories were
never 'nothing but' mere instruments dealing in fictions. They were not offering
causal explanations of the planetary motions and yet the geometrical models were
constrained by causal principles. Hence the fact that astronomical methods, bound
up as they were by geometrical models, could not map out the true motions of the
planets implies noting about an inherent inability to find out the true motions. After

all, we know that the very issue of 'the system of the world' was settled by dynamical considerations after Newton showed that the centre of gravity of the solar system was close to the sun.

What this case shows, I think, is that the case for compatibilism is not brought out by historical facts. On the contrary, the instrumentalist credo, captured nicely by the Osiander quote above, sets severe limits to any claim of compatibility. As noted already, the 'nothing but' view is based on an argument, viz., (roughly) that truth cannot be known, that no realist would endorse and which we have every reason to believe it's incorrect.

12.3 Varieties of Instrumentalism

What is instrumentalism? Here are some ways to describe it:

> The instrumentalistic fictionalist (or "instrumentalist" as we shall call him) does not propose to eliminate the non-E [i.e, the theoretical] portion of science, but simply to treat it differently from the E-portion [i.e., the observational]. Unlike the latter, which he holds significant and therefore true or false, confirmable or disconfirmable, and possibly expressive of belief, he treats the non-E portion as machinery, not to be qualified in any of these ways, though useful in the scientist's work. (...) Though, like all fictionalists, the instrumentalist holds that a uniformly significant E is not capable of absorbing all of science as a *de facto* body of doctrine, he is able to make a weaker affirmation, i.e,, he can say that E as he construes it *is* capable of expressing all *genuine assertions or beliefs* contained in the body of science (Scheffler, 1963, 186).

> [Instrumentalists] are those who claim that scientific theories are not true accounts of the world; theories are not so much true as *useful*. Since the task of scientific theories is to organise the data of our experience in such a way that predictions about, and eventual control of, the future are possible, theories are simply *instruments*; theoretical objects are by and large convenient fictions (Lambert & Brittan, 1987, 149).

> [S]cientific instrumentalism doubts that theoretical entities (. . .) literally exist, or at least that we can never know of their existence. Rather we are only justified in assuming the existence of *observable* entities or properties. Theoretical concepts possess only an instrumental function or the purpose of a 'most economical representation' of empirical knowledge, as Ernst Mach put it (Schurz, 2014, 291).

Formulations such as the above are quite typical. They are not equivalent with each other but this need not detain us here. What matters is that instrumentalism is a rival of scientific realism; and hence a kind of anti-realism. Two are the main tenets.

The first is *denialism*. Instrumentalism denies the existence of a certain kind of entities. It denies those entities which are required for the truth of a theory and they do not belong to an elite set of entities which satisfy a certain characteristic, typically observability. Hence, instrumentalism takes it that there are entities which are real and are, or could be, given to us in experience. Let's call them OK-entities. But instrumentalists also recognise that the putative content of scientific theories exceeds whatever can be asserted of the OK-entities. Of this excess content, they claim that it is a *mere* instrument, albeit useful, which facilitates predictions about, and classification of, the OK-entities.

Hence, the second tenet of instrumentalism is *non-eliminativism*: theoretical discourse is not eliminable and hence its content is not fully captured in a language which refers only to OK-entities. Given denialism, the excess content of theories cannot be about not-OK entities; hence, it's about (useful) fictions.

Let's call the advocate of both tenets, fictionalist instrumentalist, or simply fictionalist. Not all self-proclaimed instrumentalists are denialists; nor non-eliminativists. Hence, not all instrumentalists are fictionalists. Take John Dewey, for instance, who invented the term "Instrumentalism". He took it to be a view akin to pragmatism; a view about the "proper" objects of science and their relations to the things of ordinary experience. For Dewey the objects of science are tools or means for supplying connections among otherwise disconnected appearances or experiences. He has that vivid image of a series of mountains which are submerged except for their tops. These tops, which correspond to appearances, would be totally unrelated and disconnected, if it was not accepted that there are submerged – hence unseen – mountain parts underneath the visible peaks. In a similar fashion, science, says Dewey, posits theoretical objects whose constant relationships explain the order of appearances. These instrumentalities, as Dewey called the objects that scientific theory posits, are more 'real' than the objects of common sense only in the sense that they fulfil "the function of instituting connections" among the latter: they weave together otherwise disconnected objects of appearances into "a consecutive history" (1958, 139). Dewey's instrumentalism denies that there is anything more to the reality of the "conceptual objects" of science.

However, Dewey-instrumentalism is not denialist. It does not deny the reality of the objects of science. Rather it puts a gloss on what it is for them to be real. Dewey-instrumentalism is a kind of *contextualism*: reality is not an absolute category firmly attributed to some entities and firmly denied to others. Rather, what counts as real depends on the context. Speaking in an auto-biographical way, he noted: "That the table *as* a perceived table is an object of knowledge in one context as truly as the physical atoms, molecules, etc. are in another situational context and with reference to another *problem* is a position I have given considerable space to developing" (1939, 537). The context and the problem are determined, at least partly, by the things one does with an entity and by the role an entity plays within a system – for instance, one can put books on a table but one cannot put books on a swarm of molecules. One may well question the motivation for this view, since it is not clear, to say the least – how contexts are separated. But here, in any case, is a variant of instrumentalism that does not reduce theories to calculating devices and does not deny that (in a sense, at least) explanatory posits are real.

Take Philipp Frank, who's been one of the very few outspoken instrumentalists. In his (1932), he started his treatment of causality by stating that science is an instrument and that scientific formulas are not propositions about the real world. Science, he argued, aimed at prediction and theories are merely tools for this aim – in particular, symbolic tools that do not (aim to) represent the real world. Theories, for

Frank, are neither true nor false; they have only instrumental value that is cashed in terms of predictions of future observations on the basis of present ones. As he characteristically put it, a theory can be called 'true' or 'false' in the sense that "'a badly sharpened knife is a false instrument for cutting" (1932, 19) and not in the sense in which statements about ordinary objects of experience are false (or true).

Let us call this *non-cognitivist instrumentalism*. It is non-eliminativist but only in the sense that theories (the not-E portion of theories, as Scheffler has put it) lacks assertoric content and is only a means to facilitate moving from claims about observables to predictions about observables. In this sense, theoretical premises are merely inference-tickets. And it is clearly denialist since it takes it that all this theoretical discourse comes ontically to nothing, since there is nothing to be about. As Frank put it, a scientific theory is not a *replica of the real world*.

Non-cognitivist instrumentalism is not taken seriously any more. The reason is simply that, judged by its fruits, it does a very poor job even when it comes to explaining how prediction and control are possible. One important argument against non-cognitivism (of the sort espoused by Frank) is that it is a reconstruction of science that turns a perfectly meaningful practice – where there is communication and understanding – into a meaningless manipulation of symbols underlied by problematic and context-dependent rules that connect some of the symbols with experience (and hence give them some partial meaning). More importantly, however, non-cognitivist instrumentalism offers a causally disconnected image of the world. Given its way of reading the theoretical superstructure of theories, it's not even possible that the observable phenomena are connected to each other via processes and mechanisms that involve unobservable entities. The problem here is not that causal laws become very complicated by being meant to connect only observable magnitudes. Rather, the problem is that causal laws become arbitrary, since they are *after all* formulated by means of symbols such that, were they to be taken literally would refer to unobservable entities, but, being not-taken literally, there is simply no fact of the matter as to whether these or other (incompatible) causal laws describe the workings of nature, provided that they are all co-ordinated in the appropriate manner with observables.

12.4   A Tension for Fictionalism

The instrumentalist "movement" is too diverse. I argued that the core position involves denialism and non-eliminativism, aka fictionalism.

Fictionalism is a view with a long pedigree. It was defended for the first time by Hans Vaihinger in 1911. It is the view that some entities whose existence is implied by the truth of a theory are not real, but useful fictions. Hence, on the fictionalist approach, scientific theories which are prima facie committed to the existence of unobersvable entities are false, simply because there are no such entities for the theories to be committed. On this view, to say that one accepts the proposition that *p as if* it were true is to say that *p* is false but that it is useful to accept whatever p asserts as a fiction. Introducing this view in *The Philosophy of As If*, Vaihinger noted that what is meant by saying that matter consists of atoms is that matter must be treated *as if* it consisted of atoms. But what does it mean that "matter must be treated as if it consisted of atoms?" As he said: "It can only mean that empirically given matter must be treated as it would be treated *if* it consisted of atoms" (1911, 93). Though it is false that matter has atomic structure, Vaihinger argued that the as-if operator implies a decision to maintain formally the assumption that matter has atomic structure *as a useful fiction*. Hence, we may willingly accept falsehoods or fictions if this is useful for practical purposes or if we thereby avoid conceptual perplexities. We then act *as if* they were true or real. It should be noted that Vaihinger's fictions are not *just* false assumptions – they are *knowingly* false and, in their stronger version, impossible to be true. Vaihinger distinguished between fictions and hypotheses – the latter can be true or false and it is an open issue what they are; the former *cannot* be true.

In contemporary philosophy, fictionalism was famously introduced by Hartry Field as an alternative to mathematical realism. On this view, there are no numbers (or other mathematical entities), but mathematics is still useful being a conservative extension of mathematics-free (that is nominalistic) scientific theories: it facilitates deductions; all that is required for this is that a mathematical theory M be consistent.

Now, it should be stressed that fictionalism allows for an important distinction to be drawn between what the theory says that the world is like and what the world is *really* like. It leaves a gap between the claim (a) that *T* is false and the claim (b) that everything in experience is *as if T* were actually true. It is fully consistent to argue, according to fictionalism, there are no electrons and yet everything in experience is *as if* there were actually electrons. This combination of claims is logically consistent precisely because fictionalism takes theories at face-value: it does not re-interpret them; nor does it claim that they refer to entities other than those implied by a literal understanding of them. In other words, fictionalism is a non-eliminativist position: theoretical entities, were they to exist at all, would be irreducible to observable entities; but (the fictionalist says) they do *not* exist.

Putnam, who was a major critic of fictionalism in the 1970s, offered some credit to fictionalism, and rightly so, precisely because fictionalism allows for a distinction between a theory's being true and everything in experience being such that the theory is actually false. This leaves it entirely open that the theory is empirically adequate and yet false. Why then should fictionalism be rejected? Because it does not make sense to have a *merely* fictionalist stance towards a theory that has been accepted and employed in the explanation and prediction of observable phenomena. The fictionalist would typically read the theory literally, would treat the theoretical concepts as indispensable and would accept a theory 'for scientific purposes' but would refrain from commitment to the reality of the entities implied by the theory since she would take it that theory – though perhaps empirically adequate – is *false*. But what possibly could show to a fictionalist that the theory is true? Putnam (1971) takes it that the fictionalist would demand a deductive proof of the theory and rightly objects that if this were the golden standard for acceptance as true, no non-trivial observational statements would be accepted as true either. The fictionalist would end up with scepticism. So Putnam challenges the fictionalist to draw and motivate a robust distinction between rationally accepting a theory T (but treating its supposed entities as useful fictions) and rationally accepting that T is true. As Putnam (1971, 354) put it, if one rationally accepts a theory for scientific purposes, "what further reasons could one want before one regarded it as rational to *believe* a theory?" His answer was that these reasons are good enough! Being a fallibilist, Putnam adds that belief in a theory is belief in that the theory is "an approximation to the truth which can probably be bettered".

In light of this, we can see that there is a tension between denialism and non-eliminativism. If t-terms (which taken literally putatively refer to not-OK entities) cannot be dispensed with when a theory is presented (non-eliminativism) and if, on top of this, they do play a useful role in prediction and classification, is it *reasonable* to adopt denialism? In other words, if theories are indispensable for achieving what instrumentalists are after, viz. classifying and predicting facts about the world of experience, why should we deny the reality of the alleged not-OK entities whose existence is implied by the theories?

The non-eliminativist tenet implies that its advocate takes it that the theory has excess content which, on the face of it, is about the not-OK entities. But denialism takes it that there are no not-OK entities for theory to be about. But if reasons to accept a theory are *ipso facto* reasons to believe it is true, non-eliminativism is not great bedfellow for denialism.

12.5   Craig to the Rescue

Can fictionalism be improved upon? Naturally, if it could be shown that t-terms are dispensable, that they can be eliminated without loss of OK-content, then fictionalism would avoid the foregoing tension. Even if theories *appear* to have excess content over their OK-content, this is illusory since, ultimately the vehicle for talking about the alleged not-OK entities, ie., theoretical vocabulary, is dispensable. And that's the reason why Craig's theorem fell like manna from heaven to all aspiring instrumentalists. This theorem was taken as proof that a certain theory T which implies – if literally understood and true – the reality of certain entities can be replaced by another theory T' which does not imply commitment to the reality of the 'suspicious entities'.

In his Ph.D Thesis in 1951, logician William Craig constructed a general method according to which given any first-order theory T and given any effectively specified sub-vocabulary O of T, we can construct another theory T' whose theorems are exactly those theorems of T which contain no constants other than those in the sub-vocabulary O. But as Hempel noted and Craig (1956) himself proved, the application to philosophy of science of Craig's theorem yields the following powerful result: given any first-order theory T and any effectively specified observational sub-vocabulary V_O of T, one can construct another theory T' whose theorems are exactly those theorems of T which contain no terms other than those already in the sub-vocabulary V_O. What came to be known as Craig's Theorem is the following: for any scientific theory T, T is replaceable by another (axiomatisable) theory Craig (T), consisting of all and only the theorems of T which are formulated in terms of the observational vocabulary V_O. Craig showed how to construct these axioms of the new theory Craig(T). There will be an infinite set of axioms (no matter how simple the set of axioms of the original theory T is), but there is an effective procedure that specifies them. The new theory Craig(T) which replaces the original theory T is 'functionally equivalent' to T, in that all observational consequences of T also follow from Craig(T): the latter establishes all those deductive connections between observation sentences that the initial theory T establishes.

The gist of Craig's theorem is that a theory is a conservative extension of the deductive systematisation of its observational consequences. As Hempel put it in a note on Craig's theorem in 1954: "The method then yields a theory in terms of observables only which has, in Craig's (or Feigl's) terms, the same 'cash value' at the original theory with its constructs" (Feigl Archive, University of Pittsburgh. All rights reserved. Doc. HF 09-38-04). In a Conference on Physicalism in Princeton, April 1954, where Carnap was present, too, Hempel presented Craig's result as a dilemma: "There is also a theorem of Craig proved in an unpublished doctoral thesis which is damaging to the idea that a theory is necessary at all <u>or</u> to the idea that explanation is deduction" (Carnap Archive, University of Pittsburgh. All rights reserved. Doc. 090-62-06, p.

And yet, this promising route to fictionalism by means of Craig-eliminability was short-lived. One key reason was that Craig(T) as a replacement of the original theory T fell short of the very instrumentalist standards for a scientific theory. This was pointed out by Nelson Goodman – who was no friend of realism – in his review of Craig's (1956) in JSL in 1957. Goodman's point was that Craig(T) lacks a key virtue that theories should possess even by instrumentalist standards: that's not truth but *integration* – aka economy of thought. But Craig(T) will have an infinity of axioms, no matter how simple the original theory T might be. Hence, Goodman says:

> Evidently preservation of all theorems in non-suspect [i.e., observational] language is far from enough. What more, then, is required [for a good theory]? For one thing, that the replacing system [i.e., Craig(T)] have an appreciable degree of deductive coherence or economy. The chief purpose of proof in a philosophical system is less to convince of truth than to integrate. If every theorem has its own postulate, no integration is achieved (1957, 318).

Putnam (1965) mounted a formidable attack on the philosophical significance of Craig's theorem arguing that a) theoretical terms are meaningful, taking their meaning from the theories in which they feature and b) scientists employ terms like 'electron', 'virus', 'spacetime curvature' and so on – and advance relevant theories – because they wish to *talk about* electrons, viruses, the curvature of spacetime and so on; that is scientists want to find out about the *unobservable* world. Theoretical terms provide scientists with the necessary linguistic tools for talking about things they want to talk about. Besides, the very idea of Craig-eliminability requires the clear separation of the theoretical vocabulary from the observational one but this is wishful thinking since there is no such sharp distinction. Rather, this distinction is vague and context-dependent.

If abandoning non-eliminativism is not open to aspiring fictionalists, would abandoning denialism be? Strictly speaking one cannot be a fictionalist without denialism. Yet, more loosely speaking, one can combine the thought that not-OK, i.e., theoretical, entities are useful fictions with agnosticism, as opposed to denialism. On this combination of views, one need not deny the reality of the entities assumed by a literal reading of a theory – but instead be *agnostic*. This is the position articulated by van Fraassen (1980). On this view, the failures of fictionalism (or denialist instrumentalism) does not make scientific realism the only rational option. An agnostic variety of instrumentalism is not, *ipso facto*, ruled out: one can always remain agnostic as to the truth-value of the particular theoretical descriptions of the world offered by a theory, even though one uses the theory for all scientific purposes. Notably, van Fraassen capitalised on the very fact that Putnam credited to fictionalism, viz., that there is a gap between empirical adequacy and truth, and argued that this very gap allows for a position such that "theories could agree in empirical content and differ in truth-value" (1980, 36). This position, once occupied, would have scientific realism going for a "leap of faith" to close the gap between empirical adequacy and truth. And as van Fraassen stressed: "The decision to leap is subject to rational scrutiny, but not *dictated* by reason and evidence" (1980, 37). But given that I've dealt with van Fraassen extensively elsewhere (e.g., my 2012), I will say no more here.

12.6   Compatibilism I

Is denialist instrumentalism dead then? The answer that became possible in the 1960s was that if it is, so is realism. Why? Because the two views are compatible with each other, after all. In its very, influential *The Structure of Science*, in 1961, Ernest Nagel famously argued that realism and instrumentalism are merely *different languages* about theories and the choice between them is only a choice of the preferred mode of speech. Here is how he put it:

> It is therefore difficult to escape the conclusion that when the two apparently opposing views on the cognitive status of theories [realism and instrumentalism] are each stated with some circumspection, each can assimilate into its formulations not only the facts concerning the primary subject matter explored by experimental inquiry but also all the relevant facts concerning the logic and procedure of science. In brief, the opposition between these views is a conflict over preferred modes of speech (1961, 152).

Nagel takes instrumentalism to be the view that theoretical propositions are, in effect, inference tickets, which licence transitions from observational premises, which are truth-evaluable to observational conclusions. Being disguised rules of inference, theoretical claims lack truth-values. Hence, they are not in the business of assertion. Here is a striking passage: "a theory is held to be a rule or a principle for analysing and symbolically representing certain materials of gross experience, and at the same time an instrument in a technique for inferring observation statements from other such statements" (1961, 129). This passage is striking because it aligns Nagel with Frank's non-cognitivism. But then the argument against non-cognitivism, viz., that it makes a mockery of a perfectly meaningful practice, holds against Nagel too. But still one may wonder: how can it be that realism and instrumentalism are compatible?

Nagel's argument capitalises on the difference between treating a statement as a rule or as a premise. Take, for example the statement 'All As are B'. On the face of it, it looks like it's a universal generalisation. It can be a premise in the following argument: a is A; All As are B; therefore a is B. However, it can be the case that we can move directly from 'a is A' to 'a is B", via the material rule of inference: from 'x is A' infer 'x is B'. This rule is material and not formal in the sense that it implicates a specific subject-matter. For instance, from 'copper is a metal and 'copper is heated', we can infer 'copper expands' using the rule: from 'x is metal' and 'x is heated' infer 'x expands'. Now, Nagel takes this distinction between premises and rules of inference to imply that "a given statement may function as a premise in one context but may in effect be used as a leading principle in another context, and vice versa" (1961, 138). And from this he infers that "there is on the whole only a verbal difference between asking whether a theory is satisfactory (as a technique of inference) and asking whether a theory is true (as a premise)" (1961, 139).

This 'irenic instrumentalism', as Sellars (1965) put it, faces two important difficulties. The first is that a theory, on the face of it, is a putative *map* of reality and not a set of principles for *constructing maps*. This is acknowledged by Nagel (1961, 139) but its implications are not fully appreciated. The theory has a part (the E-portion, as Scheffler put it) which is a map of reality, according to instrumentalists. This part offers the premises upon which the rules operate and also contains the conclusions of the rules. The question then is: how is the distinction between rules and premises to be drawn? Why, in other words, assume that some of the statements of the theory have a fact-reporting role, while others do not? This problem is far from trivial, since it amounts to the problem of whether there is a sharp line between observational statements (for which the issue of factual truth is open) and theoretical statements (for which there is no issue of factual truth). And we all know that there is no sharp line to be drawn.

The second difficulty relates to Nagel's idea that the theory should be "satisfactory" as a rule of inference. As Sellars pointed out, what else could 'satisfactory' mean other than truth-preserving? Suppose that observational statement O1 entails O2, via statements T1&T2&T3. The truth of T1&T2&T3 explains why the inference O1 |- O2 is truth preserving. Lacking this explanation, that O1→O2 is true is totally ungrounded.

This difficulty becomes a serious problem if we think of the role theories play in the confirmation of observational statements, the idea being that theories are often necessary for the establishment of inductive connections between seemingly unrelated observational statements. Here is Putnam's (1963) own example. Consider the prediction H: 'When two subcritical masses of U_{235} are slammed together to form a supercritical mass, there will be a nuclear explosion'. H could be re-written in an observational language – that is without the term 'Uranium $_{235}$' – as O_1: 'When two particular rocks are slammed together, an explosion will happen'. Consider now the available evidence, namely O_2: 'Up to now, when two rocks were put together nothing happened'. Given this, it follows that $prob(O_1/O_2)$ is very low, if it can be determined at all. But consider the posterior probability of O_1 given the past evidence and the whole of atomic theory T which entails that the uranium rocks would explode if critical mass were attained quickly enough. It is obvious that prob $(O_1/O_2\&T)$ is now determined and much greater than $prob(O_1/O_2)$.

All this speaks directly against instrumentalism. But there is still the following worry: how could it *possibly* be that realism and instrumentalism are compatible, given that realism affirms the existence of t-entities while instrumentalism denies it? Nagel claims that the very idea of the physical existence or reality of entities is vague. He takes it that "real" in an "honorific" term which is used "to express a value judgment and to attribute a 'superior' status to the things asserted to be real". And he adds: "it would be desirable to ban the use of the word altogether" (1961, 151). But the point is not to find a way for the instrumentalist to understand 'being real' such that they can assert that e.g., atoms are *real*. Rather the point is that the reality attributed by an instrumentalist to atoms is bound to be always 'second-class' vis-à-vis the reality attributed to tables and pointer readings. And that's exactly the point Nagel *missed* when he stressed:

> to assert that in this sense atoms exist is to claim that available empirical evidence is sufficient to establish the adequacy of the theory as a leading principle [i.e., rule of inference] for an extensive domain of inquiry. But (...) this is in effect only verbally different from saying that the theory is so well confirmed by the evidence that the theory can be tentatively accepted as true (Nagel, 1961, 151).

12.7   Compatibilism II

A few years after the publication of *The Structure of Science*, Carnap advocated compatibilism in his *Philosophical Foundations of Physics*, referring with approval to Nagel. He said:

> It is obvious that there is a difference between the meanings of the instrumentalist and the realist ways of speaking. My own view, which I shall not elaborate here, is that the conflict between the two approaches is essentially linguistic. It is a question of which way of speaking is to be preferred under a given set of circumstances. To say that a theory is a reliable instrument-that is, that the predictions of observable events that it yields will be confirmed-is essentially the same as saying that the theory is true and that the theoretical, unobservable entities it speaks about exist. Thus, there is no incompatibility between the thesis of the instrumentalist and that of the realist. At least, there is no incompatibility so long as the former avoids such negative assertions as, '... but the theory does not consist of sentences which are either true or false, and the atoms, electrons, and the like do not really exist' (1966, 256).

And in a footnote he added: "An illuminating discussion of the two or three points of view on this controversy is given by Ernest Nagel, The Structure of Science

(New York: Harcourt, Brace & World, 1961), Chapter 6, 'The Cognitive Status of Theories'."

Carnap's point is striking, since as he emphatically notes, realism and instrumentalism are compatible (they can both be true at the same time) provided instrumentalism abandons Denialism! Insofar as instrumentalist avoids negative claims of the kind 't-entities do not exist' or 't-propositions are not assertions', then instrumentalism is compatible with realism. But can there be such an instrumentalist?

A few years later in the paperback edition of the 1966 book, Carnap replaced the previous paragraph with the following and dropped the reference to Nagel. The paragraph now reads as follows:

> It is obvious that there is a difference between the meanings of the instrumentalist and the realist ways of speaking. My own view, which I shall not elaborate here, is that the conflict between the two approaches is essentially this. I believe that the question should not be discussed in the form: "Are theoretical entities real?" but rather in the form" "Shall we prefer a language of physics (and of science in general) that contains theoretical terms, or a language without such terms?" For this point of view the question becomes one of preference and practical decision (1974, 256).

The footnote now reads: "In my view greater clarity often results if discussions of whether certain entities are real are replaced by discussions of preference of language forms. This view is defended in detail in my *Empiricism, Semantics and Ontology* (1950)".

The change is rather deep and interesting. Carnap moved from the strong compatibilist claim that there is no need to choose among Realism and Instrumentalism because 'they say the same thing' to the claim that commitment to entities is a matter of a choice of language (that is a choice to use of t-terms or not); and hence it's a practical issue in line with the views in his *Empiricism, Semantics and Ontology* (1950). So do or don't Realism and Instrumentalism say the same thing? And given that the famous (1950) paper predates both Nagel's account as well as Carnap's strong compatibility view, what accounts for the change in the two editions and what does this change imply?

Between 1950 and 1966, Carnap had re-invented the Ramsey-sentence approach to scientific theories. I have told this story elsewhere in great detail (see my 1999, chapter 3). So I will be brief. Not only did Carnap know of Craig's theorem, but he also extended Craig's results to "type theory, (involving introducing theoretical terms as auxiliary constants standing for existentially generalised functional variables in 'long' sentences containing only observational terms as true constants)". (As it is reported in the protocol of the Los Angeles Conference in August 1955 – Feigl Archive 04–172–02:14). In the same protocol, Carnap is also reported to have shown that '[a]n observational theory can be formed which will have the same deductive observational content as any given theory using non-observational terms. (Namely, *by existentially generalising non-observation terms*)' (ibid., 19).

What's special about Ramsey-sentences? What are the advantages over Craig's method? Three points are relevant:

- RS is finitely axiomatizable.
- RS retains the logical structure of the original theory.
- RS lays a claim of satisfaction by certain entities (those in the range of the existentially bound variables).

As Carnap put it: "the Ramsey sentence has precisely the same explanatory and predictive power as the original system of postulates" (1966, 252). And it achieves this in a way that is tempting for an empiricist since it delivers us from metaphysics. There are no t-terms and predicates in the RS of a theory; hence the issue of their reference evaporates. Still, however, the RS implies existential commitments to "whatever it is in the external world that is symbolized by" the t-terms. *The vehicle of non-observational commitment is the existential quantifier.* Here is a long quotation characteristic of Carnap's enthusiasm:

> In Ramsey's way of talking about the external world, a term such as 'electron' vanishes. This does not in any way imply that electrons vanish, or, more precisely, that whatever it is in the external world that is symbolized by the word 'electron' vanishes. The Ramsey sentence continues to assert, through its existential quantifiers, that there is something in the external world that has all those properties that physicists assign to the electron. It does not question the existence-the 'reality'-of this something. It merely proposes a different way of talking about that something. The troublesome question it avoids is not, 'Do electrons exist' but, 'What is the exact meaning of the term 'electron'?' In Ramsey's way of speaking about the world, this question does not arise. It is no longer necessary to inquire about the meaning of 'electron', because the term itself does not appear in Ramsey's language (1966, 252).

Now, it's not surprising that the Ramsey-sentence RS(T) of theory T is consistent with T. RS(T) logically follows from T; it cannot be the case that RS(T) is false and T true. So is the point that RS-instrumentalism is compatible with realism? Is the Ramsey way instrumentalism? Grover Maxwell was quick to point out that it is not. Commenting on Carnap's use of the Ramsey sentence, Maxwell wrote to Carnap: "I disagree that thinking theoretical entities 'in the Ramsey way' should be associated with instrumentalism' (Maxwell to Carnap, 24 June 1966; Carnap Archive 027–33–29). Carnap replied on 9 December 1967 conceding the point and putting the blame to carelessness and confusion on his part:

> You are quite right in the one critical remark you make, that the Ramsey way should not be associated with instrumentalism. In an earlier version of the manuscript I had distinguished three instead of two views on the question of the reality of entities, by splitting off instrumentalism into two forms, a negativistic one and a neutral one which I identified with the Ramsey way. Then a reader of the manuscript pointed out that the distinctions were not in agreement with the customary terminology; in particular that the term 'instrumentalism' is always used in the negativistic sense. Then I made a radical change, distinguishing only two points of view. This I did in great haste and so I mixed things up. For a future edition of the book I have decided on a reformulation which you see on the enclosed sheet (Carnap Archive 027–33–28).

If Carnap can be confused, anyone else can be. Certainly confusion did play a role. On the face of it, the Ramsey way is neither realist, nor instrumentalism; yet it appears to be a halfway house. But there is a deeper reason why Carnap took the Ramsey way to be instrumentalist.

In his reply to Hempel in the Schilpp volume, Carnap says:

> the Ramsey-sentence does indeed refer to theoretical entities by the use of abstract variables". But he immediately adds that "these entities are not unobservable physical objects like atoms, electrons, etc., but rather (*at least in the form of the theoretical language which I have chosen in [MCTC] §VII*] purely logico-mathematical entities, e.g. natural numbers, classes of such, classes of classes, etc. (1963, 963).

On this reading, the Ramsey sentence says that "the observable events in the world are such that *there are numbers*, classes of such, etc., which are correlated with the events in a prescribed way and which have among themselves certain relations; and this assertion is clearly a factual statement about the world' (ibid.).

Here again I invite the reader to look at my (2007) for the details of Carnap's views about the language of science and his claim that every legitimate scientific concept is extensionally equivalent with a certain mathematical function. For our purposes, it's enough to say that it is this particular reading of RS(T) which makes it consistent with instrumentalism. To sum up, RS(T) does not deny the existence of unobservable realisers of the theory; it's not denialist; yet it's fully consistence with the claim that the realisers of the theory is not physical stuff (i.e., physical unobservables) but instead mathematical entities. And to this extent, the Ramsey way can be seen as a form of instrumentalism. It's a *kind* of instrumentalism since RS(T) can be satisfied without having any commerce with those parts of the physical world, if there are any, that are not given to us I experience. But it's a *strange* kind of instrumentalism, since the satisfiers of RS(T), being mathematicalia, are not given to us in experience.

Carnap's neo-Pythagoreanism, as Sellars (1965, 177) put it, might secure some kind of instrumentalist reading of Ramsey sentences but it creates more problems than it solves. A rather severe one is that, on the face it, saddles instrumentalism with a rampant commitment to abstracta. So an instrumentalist might be accused of swallowing a camel and straining at a gnat, the camel being absracta and the gnat being physical unobservable entities. I think that's why Carnap in the end abandoned the strong Nagel compatibility and opted for the weaker claim that the choice between instrumentalism and realist is a matter of choice of language, where picking a language is a matter of practical expedience. Hence the shift from Nagel's *Structure* to his own *ESO*.

12.8   Compatibilism III

In a seminal piece that appeared in 1989 Stein aimed to save compatibilism by noting that "between a cogent and enlightened 'realism' and a sophisticated 'instrumentalism' there is no significant difference – no difference that makes a difference" (1989, 61). But what is this 'sophisticated instrumentalism'? Let's call 'philosophical type

X' the view, or the philosopher who holds the view, that these is no difference between sophisticated instrumentalism and enlightened realism. The basis for X is the claim that theories are instruments. Yet, it's not the case that they are nothing but instruments. According to Stein, a 'sophisticated instrumentalist' should extend further their conception of what the theories are instruments *for*. Going beyond the view that a theory is a *mere* instrument for calculating the outcomes of experiments, X must also encompass the claims that a theory is an instrument for *representing* the phenomena of nature (1989, 50) and for *guiding* the inquiry (1989, 52).

Let us briefly see the concessions realists should make to be X. Stein's key complaint is that the unreconstructed realist is committed to a semantic theory of the relation between theory and the world such that t-terms have factual reference. To be more precise, Stein seems to be driven by an epistemic worry, which he puts as follows: 'How can you *know* that things are as you say they are? If the claimed "reference" of the theory is something beyond its correctness and adequacy in representing phenomena' (1989, 50). The implication is clear: any realist claim of truth and factual reference of a theory that goes beyond whatever is involved in rendering a theory a correct and adequate representation of the phenomena is a difference that makes no difference.

A whole paper could be written on the issue of the correct and adequate representation of the phenomena. Hence, I will restrict myself to a general argument, which starts with a question: what more, or other, than a correct and accurate representation of the phenomena should or would a realist demand? The obvious answer is: nothing. When a realist takes theories to be, roughly put, maps of the world, the aim is a correct and accurate representation of the world and hence of the phenomena (narrowly understood to refer to *observable* phenomena). This map won't be a mirror of reality, for reasons that have to do with the fact that representation involves idealisations, abstractions and perspectives. Still, it can be more or less accurate and more or less correct if it is world-involving; that is, if it identifies relevant causal factors and, more broadly, ways in which the phenomena are brought about. In the case discussed in the second section, for instance, most (all?) of astronomers knew the limitations of astronomical representations of the planetary motion and yet they were trying to make them world-involving by basing them on what they had taken to be correct general principles.

If the realist doesn't have to concede much to become Stein's Philosophical Type X, how about the instrumentalist? For one, an instrumentalist would have to abandon denialism. If the correct and accurate representation of the phenomena requires commitment to theoretical entities, then anyone of philosophical type X would not be a denialist. For another, if instrumentalism takes 'representation' of the phenomena to involve only OK-entities, or to be such that whatever goes beyond the OK-entities are mere fictions, then theories end up being *nothing but* instruments.

If going for a notion of representation does not help the compatibilist cause how about the extra requirement that theories should play a role as resources to inquiry? This requirement has been stressed by Richard Boyd as offering a methodological advantage to realism. Stein's position is the following:

Either the methodological principles in question, and the fact of their success thus far in the history of inquiry, are susceptible of clear formulation in terms of the relations of theories to phenomena (and this, I think, is what Boyd intends) – in this case, the instrumentalist ought to adopt those principles, on the grounds of the evidence for their instrumental success; or they are not susceptible of such formulation – but then, where is the argument? In short, the instrumentalist's own principles entail no prohibitions: what the realist can do, the instrumentalist can do also (1989, 52).

Here again the keyword is "the relations of theories to phenomena". For Boyd and other realists like myself, the success of a theory is best explained by the theory's being approximately true. So theories are related to the phenomena by offering an approximately true account of them. Stein does not recognise approximate truth among the attributes that theories must have to serve as resources of inquiry. Instead he talks about the logical structure of the theories and their relation to phenomena and notes that "the attributes our theories must possess for them to serve as resources for inquiry are all attributes that concern the logical structure of the theories" (1989, 53).

What exactly is the logical structure of the theory and what are its attributes? Stein does not say anything directly on it. He does talk about mathematical structures, and gestures (uncharacteristically vaguely) that improved understanding consists in "structural deepening". But it's not clear at all how he understands the relation between mathematical structures and the physical world. What he is clear about is his insistence on the futility of "hypostatizing" (unobservable) entities. Now, to use Stein's ironic comment "it would be possible to do a lengthy dialectical number on this; but in brief": the long debates over structural realism have taught us that (a) it's hard to understand structures without entities and (b) hypostatising structures would not be less of a sin for an instrumentalist than hypostatising entities simply because they would end up with abstracta.

In any case, it might be that the logical structure of a theory is captured by the Ramsey-sentence RS(T) of a theory T. The chief relevant attributes of the Ramsey-sentence approach are (1) that T and RS(T) have the same logical-deductive structure and (2) they have exactly the same observational consequences. At the same time, RS(T) dispenses with the hypostatization of entities, since it replaces all theoretical predicates with existentially bound variables.

But if *that* is the 'sophisticated instrumentalism' that Stein has in mind, then there is not much progress since Carnap's compatibilism. To sum up: Stein claims that there is what I called a philosophical type X which going beyond the nothing-but-ness of instrumentalism by incorporating the two following roles at science:

– The theories represent the phenomena of nature.
– The theories are guides to inquiry.

According to Stein's main thesis, this philosophical type X is all we need to accept in order to understand science and its relation to the world, to reality. All the rest that the realist may add (genuine reference of the theoretical terms hence hypostatisation of the unobservable entities) is a difference that makes no difference. In other words, an instrumentalist can do whatever a realist does and because of that the claimed difference between them does not exist. But it turns out that the sought after philosophical type X is, at best, out old fellow: Ramsey-sentences.

12.9   Compatibilism IV

It might be surprising that a leading current advocate of instrumentalism has been flirting with compatibilism. Indeed, Kyle Stanford has recently published an article titled: "'Atoms exist' is probably true, and other facts that shouldn't comfort scientific realists" (2015b). In this he says quite emphatically that the anti-realist need not and should not deny any particular existential commitment to unobservables issued by scientific theories. As he put it: "we should decline to saddle the scientific realist's historicist opponent with the belief that there are no such things as genes or atoms or that the terms 'gene' and 'atom' do not refer" (2015b, 411).

To motivate, and see the limits of, this irenic stance, let us define two philosophical types: let's call them R and I. Both types, presumably, have learned the lesson from theory-change in the history of science. Both types acknowledge that a number of current scientific beliefs will be overthrown in the future. Both types acknowledge that a number of particular claims made by current science, and the concomitant existential commitments, are part and parcel of any future science. Perhaps surprisingly, type I takes it that atoms and genes and other such stuff are real and that at least some of the claims made about them are true or roughly so. And yet, type I emphasises the discontinuities in theory-change, whereas type R the continuities. Is there a big difference between R and I? Or is the difference really a matter of where the emphasis is placed?

Note that type I is not an outright denialist. So type I is not a fictionalist. More interestingly, type I does not endorse the views of past compatibilists. So, if we think of type I as instrumentalist, what kind of instrumentalism is this? Before I answer this question, let me say something about scientific anti-realism in general.

Traditionally, scientific anti-realism has been an all-or-nothing stance towards theories. Better put, it's been a dichotomous position. The content of a theory is split into two principled parts – let's call them the OK-part and the not-OK-part. The dichotomy (OK/not-OK) has an epistemic motivation. The OK-part is about things (or that part of the world) we can have epistemic access to; the not-OK part is 'about' those things that were they existed they would cognitively impenetrable by us. The criteria of cognitive penetrability, and hence, of epistemic accessibility, differ. It might be perception and hence observability; or something more sophisticated. But whatever the line of distinction is, commitment is an all-or-nothing matter (see my 2009). All not-OK entities are suspicious: unreal, fictions etc. Instrumentalism is a species of antirealism; it too is dichotomous. Once the distinction between OK and not-OK parts is drawn, denialism is an attitude towards the not-OK part.

So back to the question above: what kind of instrumentalism is type I? Strictly speaking, it's not instrumentalism. There is no denialism; nor any sharp dichotomy between the OK and not-OK parts of the theory. Besides, there is no way to tell ahead of time what unobservable (putative) entities will be accepted as real. Hence, the list of entities that will be accepted as real is open-ended. Stanford stresses:

> Even when we have found one of a theory's central existential commitments overturned in the course of further inquiry, many otherwise similar commitments have remained in place, so even in cases in which a successful scientific theory does have central existential commitments that will ultimately be judged to be false and/or central theoretical terms that will ultimately be judged non-referential, we should not imagine that we can specify in advance which existential commitments and/or theoretical terms these will be (2015b, 402).

Insofar as type I is instrumentalist, it's wait-and-see instrumentalism, to use John Earman's (1978) apt expression. Any putative entity that, were it to exist, it would be talked about in the not-OK part of the theory according to traditional instrumentalism, might be end up being talked about in the OK-part.

This last point is critical: type I, *aka* wait-and-see instrumentalism, leaves it up to science and the evidence-based consensus-forming mechanisms, to tell us what there is. There is no longer a principled philosophical argument to the effect that some kind of entities (e.g., unobservable entities) are useful fictions and the like. In effect, only persistence in time, i.e., withstanding major revolutions and breaks, can tell us what there is. In this sense, wait-and-see instrumentalism is compatible with type R, viz., realism. Actually, wait-and-see instrumentalism is more or less identical with an extra cautious realism.

12.10   A Coda

So what's the point of the realism-instrumentalism controversy? Stanford draws on the debate between catastrophists and uniformitarianism in geology and argues that the debate is about large-scale patterns in theory-change in the history of science. Instrumentalists claim that "the central commitments of future theoretical orthodoxy will (or would) ultimately be separated from those of the present by differences as fundamental, profound, far-reaching, and unpredictable as those that separate our own theories from their historical predecessors" (2015b, 415). While realists take it that at least the most central and fundamental claims of our most successful scientific theories as firmly established in ways that may be supplemented or modified but are quite unlikely to be overturned in the course of further inquiry" (2015b, 413).

If *that's* the difference, then there is not much progress made vis-à-vis type I or wait-and-see instrumentalism. Though showing this would require a careful look at the history of theory-change so far, there are two considerations that strongly support it.

The first consideration is that there are no clear-cut criteria about what's more likely to happen in the future, that is: radical wholesale changes or extensive continuity with some piecemeal changes? The historical record doesn't speak with the voice of an angel; so what patterns have prevailed so far is a matter of interpretation. We can certainly say that the history of theory-change is a history of discontinuities and continuities. Overall, I would say that there is a pattern of retention; and in particular a pattern in which more and more unobservable entities are accepted as real (see my 2021). Others might disagree. But that's precisely the point. In more or less any future time t, the situation will be the same as it is now: the past will be a history of discontinuities and continuities. Faced with *this* history, future realists and instrumentalists will still argue about the prevailing patterns. This would give an air of futility to Stanford's suggestion about the point of the realism-instrumentalism controversy.

The second consideration is that history is made by human subjects – in this case, scientists. So whether or not changes in the future science will be *fundamental, profound, far-reaching*, and *unpredictable* depends on how science will be practised in the future. If scientists use past experience as a guide, they can learn to do better science, e.g., to look for more and varied evidence, to formulate and test alternatives etc. And if this the case, we have every reason to believe that as science grows the changes will be less and less *fundamental, profound, far-reaching*, and *unpredictable*.

Where does all this leave us? Compatibilism fails: instrumentalism and realism are rival views of science; the differences are deep and interesting. And yet, insofar as instrumentalism is a cautious wait-and-see attitude towards current science, emphasising the discontinuities, the raptures, in theory-change, insofar as instrumentalism is captured by type I above, it becomes a view that scientific realists can coexist with.

Acknowledgements   Work which led to this paper has been presented in the Philosophy of Science reading group in the University of Athens. Many thanks to the Usual Suspects for comments; and especially to Maria Panagiotatou for extensive discussions of Howard Stein's paper 'Yes, But. . .". Thanks are also due to Wenceslao J. Gonzalez and an anonymous reader for encouragement. I want to dedicate this paper to the memory of Richard Boyd who taught me how to be a realist.

References

Carnap, R. (1950). Empiricism, semantics and ontology. *Revue Intérnationale de Philosophie, 4*, 20–40.
Carnap, R. (1963). Replies and systematic expositions. In P. Schilpp (Ed.), *The philosophy of Rudolf Carnap* (pp. 859–1013). Open Court.
Carnap, R. (1966). *Philosophical foundations of physics*. Basic Books.
Carnap, R. (1974). *An introduction to the philosophy of science*. Dover Publications.
Copernicus, Nicolaus. ([1543] 1992). On the revolutions (Translation and Commentary by Edward Rosen). The Johns Hopkins University Press, 1992.
Craig, W. (1956). Replacements of auxiliary assumptions. *Philosophical Review, 65*, 38–55.
Dewey, J. (1925). *Experience and nature*. Open Court Publishing Company/Dover Publications 1958.
Dewey, J. (1939). Experience, knowledge and value: A rejoinder. In P. A. Schilpp & L. E. Hahn (Eds.), *The philosophy of John Dewey* (pp. 517–608). Open Court.

Duhem, P. (1908). *ΣΩΖΕΙΝ ΤΑ ΦΑΙΝΟΜΕΝΑ. Essai sur la notion de théorie physique de Platon à Galilée*. Paris: Librairie scientifique A. Hermann. Translation: *To Save the Phenomena: An essay on the idea of physical theory from Plato to Galileo*. The University of Chicago Press, 1969.

Earman, J. (1978). Fairy tales vs an ongoing story: Ramsey's neglected argument for scientific realism. *Philosophical Studies, 33*, 195–202.

Frank, P. (1932). *The law of causality and its limits* (M. Neurath & R. S. Cohen, Trans.). Kluwer.

Goodman, N. (1957). Review: William Craig, replacement of auxiliary expressions. *Journal of Symbolic Logic, 22*, 317–318.

Lambert, K., & Brittan, G. G. (1987). *An introduction to the philosophy of science* (3rd ed.). Ridgeview Publishing Company.

Lloyd, G. E. R. (1978). Saving the appearances. *Classical Quarterly, 28*, 202–217.

Nagel, E. (1961). *The structure of science*. Harcourt, Brace & World.

Psillos, S. (1999). *Scientific realism: How science tracks truth*. Routledge.

Psillos, S. (2007). Carnap and incommensurability. *Philosophical Inquiry, 30*, 135–156.

Psillos, S. (2009). *Knowing the structure of nature*. Palgrave-MacMillan.

Psillos, S. (2012). One cannot be just a little bit realist: Putnam and van Fraassen. In J. R. Brown (Ed.), *Philosophy of science: The key thinkers* (pp. 183–206). Continuum Books.

Psillos, S. (2020). Resisting scientific anti-realism: Review of K. Brad Wray. Resisting scientific realism. *Metascience, 29*, 17–24.

Psillos, S. (2021). From the evidence of history to the history of evidence: Descartes, Newton and beyond. In T. Lyons & P. Vickers (Eds.), *Contemporary scientific realism and the challenge from the history of science* (pp. 70–98). Oxford University Press.

Psillos, S., & Zorzato, L. (2020). Against cognitive instrumentalism. *International Studies in Philosophy of Science, 33*, 247–257.

Putnam, H. (1963). "Degree of confirmation" and inductive logic. In P. Schilpp (Ed.), *The philosophy of Rudolf Carnap* (pp. 761–783). Open Court.

Putnam, H. (1965). Craig's theorem. *Journal of Philosophy, 62*, 251–260.

Putnam, H. (1971). *Philosophy of Logic*. G Allen & Unwin Ltd.

Rowbottom, D. P. (2019). *The instrument of science: Scientific anti-realism revitalised*. Routledge.

Rowbottom, D. P. (2018). In J. Saatsi (Ed.), *The Routledge handbook of scientific realism* (pp. 84–95). Routledge.

Sambursky, S. (1962). *The physical world of late antiquity*. Routledge and Kegan Paul.

Scheffler, I. (1963). *The anatomy of inquiry*. Bobbs-Merrill.

Schurz, G. (2014). *Philosophy of science: A unified approach*. Routledge.

Sellars, W. (1965). Scientific realism or irenic instrumentalism: A critique of Nagel and Feyerabend on theoretical explanation. In R. C. M. Wartofsky (Ed.), *Boston studies in the philosophy of science* (Vol. II, pp. 171–204). Humanities Press.

Simplicius. (2014a) On Aristotle Physics 2 (Trans. Barrie Fleet). Bloomsbury.

Simplicius. (2014b). On Aristotle On the heavens 3.7–4.6 (I. Mueller, Trans.). Bloomsbury.

Stanford, P. K. (2015a). Catastrophism, uniformitarianism, and a scientific realism debate that makes a difference. *Philosophy of Science, 82*, 867–878.

Stanford, P. K. (2015b). "Atoms exist" is probably true, and other facts that should not comfort scientific realists. *Journal of Philosophy, 112*, 397–416.

Stein, H. (1989). Yes, but ... some skeptical remarks on realism and anti-realism. *Dialectica, 43*, 47–65.

Vaihinger, H. 1911. *The philosophy of 'As if'* (C. K. Ogden, Trans.). Kegan Paul, Trench, Trubner and Co Ltd (1935)

van Fraassen, B. C. (1980). *The scientific image*. Clarendon Press.

Wray, K. B. (2018). *Resisting scientific realism*. Cambridge University Press.

Name Index

A
Ackermann, R., 165, 188
Ackland, R., 114, 140
Aitken, C.G.G., 156, 160, 162–163
Alberti, L.B., 171–172, 174
Alcolea, J., 11–12, 43
Aliseda, A., 240, 258
Allen, M.R., 62
Allen, S.K., 61
Allmer, J., 119, 139
Alonso, A.M., 11
Altman, D.G., 42
Amaya, A., 248f–249, 258
Andersen, H., 141, 231
Anderson, C., 82, 98
Anderson, T., 162
Angell, M., 39, 42
Arbesman, S., 86f, 95, 98
Arnold, V.I., 173, 188
Arrojo, M.J., 107f, 114, 116, 142
Askitas, N., 119, 139
Asquith, P., 163
Aubrey, W., 79

B
Bacon, F., 66, 70
Bandyopadhyay, P.S., 252, 258
Barabási, A.L., 107, 139, 143
Barber, G., 92, 99
Barnes, P., 92, 100
Bassens, A., 100
Battaglia, P., 101
Batterman, R., 93f, 99
Bauer, J.M., 143

Beebe, J.R., 2f, 11
Bell, A., 162
Bell, J., 153
Benedetti, F., 22f, 41
Bengio, Y., 73, 78, 81, 94, 99
Berners-Lee, T., 108, 111, 115, 117, 121f–124f,
    129f, 134f, 137, 139, 142
Bex, V., 61
Beyleveld, G., 100
Biedermann, A., 156–157, 159–160, 162–163
Bilsland, E., 79
Bird, A., 8, 11, 238f–239, 241f, 243–244f, 258
Bishop, C.M., 72, 78, 95f, 99
Bishop, R.C., 123f, 139
Bodziak, W., 157, 162
Bogdan, R., 259
Bohr, N., 213
Boschung, J., 61
Boumans, M., 61
Boutron, I., 18, 42
Bowler, P.J., 125, 139
Boyd, R., 279–280, 283
Bozza, S., 160, 163
Bradley, R., v, vii, 3, 45, 47, 52f, 61
Bradley, S., 52f
Bradshaw, G., 100
Bragge, J., 143
Brahe, T., 173
Breiman, L., 72, 78
Bridgland, A., 78
Brittan, G.G., 267, 284
Brockman, J., 88, 91, 99
Brooks, R., 85f, 99
Brössel, P., 231
Broussard, M., 85f, 99

© The Author(s), under exclusive license to Springer Nature Switzerland AG 2022
W. J. Gonzalez (ed.), *Current Trends in Philosophy of Science*, Synthese Library
462, https://doi.org/10.1007/978-3-031-01315-7

Brown, J.R., 284
Brynjolfsson, E., 113f, 140
Brzezinski, J., 231
Buchan, D.W.A., 74, 78
Buchanan, B.G., 68–69, 78–79
Buckner, C., 81, 89, 90f, 92, 99
Bueno, O., 11
Burks, A., 259
Burrell, J., 92, 99
Buyya, R., 125, 144

C
Campaner, R., 153, 162
Cao, L., 118, 129, 132, 140
Carlsen, M., 88
Carnap, R., 66, 67, 70, 78, 142, 224–225, 231,
        252, 261–262, 272, 275–278, 280,
        283–284
Cartwright, N., 152, 162, 193, 206
Cerf, V.G., 142
Cevolani, G., 213, 230–231
Champod, C., 156, 159, 162
Chang, H., 152, 162
Chen, K., 140
Chen, P., 85f, 100
Chilautsky, R.L., 69, 78
Clark, D.D., 108–109, 123–124f, 127–128, 140,
        142
Clark, S., 156
Climenhaga, N., 256f, 258
Clune, J., 101
Cohen, M.R., 150f, 162
Cohen, R.S., 142, 284
Collins, C., 42
Collins, J.M., 49f, 61
Colombo, M., 99
Cooke, R.M., 60f–61
Cooper, G., 94f, 100
Copernicus, N., 172–174, 181, 263–264, 284
Corrado, G.S., 140
Courville, A., 78
Cowls, J., 143
Craig, W., 261–262, 271–272, 276, 283–284
Creel, K., 91f, 94, 99
Crew, H., 188
Cristianini, N., 72, 78
Crossan, S., 78
Crupi, V., 231
Culp, S., 193, 206
Currie, G., 142

D
Danks, D., 101
Daston, L., 87, 99
Davidson, D., 89f, 99
Davis, E., 85f, 94, 100
Davis, R., 68, 78
Dawid, P.A., 156f, 158f, 162
Dayan, P., 79
de Berk, L., 156
de Clare, M., 79
de Finetti, B., 70
de Grave, K., 79
de Regt, H., 95f, 99, 152, 162
de Salvio, A., 188
Dean, J., 116, 140
Dellsén, F., 2f, 11, 241–242f, 247–248f, 250f,
        253, 258
Deng, J., 73, 78
Dennett, D., 90, 90f, 94, 99
Descartes, R., 98
Devin, M., 140
Dewey, J., 95, 99, 152, 200, 206, 268, 283
Diacu, F., 174, 188
Dick, S., 87–89f, 99
Dickson, M., 153
Dieks, D., 141, 162, 231
Dietz, S., 45f
Domingos, P., 94, 99
Dong, W., 78
Dorigo, T., 88f, 99
Douven, I., 241f, 255–256f, 258
Downing, T.E., 62
Drake, S., 169, 188
Dreyfus, H.L., 68, 78, 165, 188
Du, H., 62
Duhem, P., 262–264, 266, 284
Dupré, J., 193, 206
Dürer, A., 170, 172–174
Dürrenmatt, F., 198, 206
Dutton, W.H., 104, 140, 143

E
Earman, J., 282, 284
Ebers, G.C., 42
Edmond, G., 157
Eigner, K., 99
Einstein, A., 42, 85, 198, 208, 222
Elgin, C., 201, 206
Erickson, P., 87, 89f, 99
Estany, A., 12, 141

Estellat, C., 42
Eubanks, V., 85f, 99
Eva, B., 260
Evans, J., 230–231
Evans, R., 74, 78
Everitt, B.S., 155, 162
Evett, I.W., 157, 162

F
Fei-Fei, L., 78
Feigenbaum, E.A., 68–69, 78
Feigl, H., 272, 276
Feldbacher-Escamilla, Ch. J., 231
Feldmann, A., 106f, 140
Ferguson, D., 42
Festa, R., 231
Fiebrink, R., 79
Finkelstein, G., 97, 100
Fisch, M., 259
Floridi, L., 106f, 113f, 117f–118f, 123f, 126,
        133, 135f, 138, 140
Forfang, E., 42
Forster, M., 252, 258
Fouquet, M., 143
Fourier, J., 186
Frank, P., 268–269, 274, 284
Frege, G., 167
Fricker, E., 249
Frigg, R., v, vii, 3, 10, 45, 47f, 50f, 61–62, 84f,
        98
Fritchman, D., 92, 100

G
Gabbay, D., 249, 258–260
Galavotti, M.C., vi–vii, 6–7, 10, 140, 147, 153,
        156, 158f, 162
Galilei, G., 168–169, 178, 180, 184–185, 188
Galton, F., 83
Garbolino, P., 160, 162–163
Garcia Bouza, A., 11
Garson, J., 81, 99
Gebharter, A., 231
Gebru, T., 92, 100
Gelman, A., 154, 162
Geminus of Rhodes, 264
Gensollen, M., 12, 141
Giere, R.N., 150, 153
Gigerenzer, G., 95f, 100
Gillies, D.A., v, vii, 5, 10–11, 65, 67, 69–71,
        78, 85f, 87f–88f, 94f, 98, 100, 139,
        152–153, 162

Gillies, M., v, vii, 5, 65, 67, 79, 85f, 87f–88f, 98
Glass, K.C., 42
Glymour, C., 94, 100, 101
Glymour, M., 101
Gobet, F., 82, 100
Gonzalez, R.A., 143
Gonzalez, W.J., iii, iv, v, vii, 1–7, 10–12, 15,
        43, 45, 65, 81, 87f, 89f, 98, 100,
        103–105, 107f, 111f–118f, 120, 123,
        125–126, 129–131, 133–134f,
        136f–137, 140–142, 147, 162, 165, 188,
        191, 207, 231, 235, 261, 283
Goodfellow, I., 72, 78
Goodman, N., 272, 284
Gould, S.J., 197, 206
Graham, G., 123, 142
Green, T.F.G., 74, 78
Greengard, M., 77–78
Greeno, J.G., 244f
Greenstein, S., 108f, 111f, 142
Greenwald, K., 101
Grimes, D.A., 16f–20, 24, 26–28, 30–32, 34,
        37–38, 40, 42
Gryb, S., 45
Guestrin, C., 79
Guillan, A., 11
Gunning, D., 93f, 99

H
Haahr, M.T., 42
Hacking, I., 163, 165, 188
Hahn, L.E., 283
Hall, W., 139
Hamilton, A., 79
Hanseth, O., 116f–117f, 129, 142
Hanson, N.R., 149, 237
Harman, G., 8, 12, 235, 237–239, 248, 250f,
        258
Hartmann, S., 162, 259–260
Hartshorne, C., 259
Hasan, A., 249
Hassabis, D., 78, 101
Hastie, R., 160, 163
Hauser, F., 120f
Heal, G., 53f, 61
Heelan, P., 165, 188
Helgeson, C., 62
Hellmann, G., 153
Hempel, C.G., 70, 149, 162, 224–226, 231,
        236, 239, 243–246, 248, 250f, 258, 260,
        271–272
Henderson, L., 250f, 254f–255, 258–259

Hendler, J., 106f, 114–115, 126f, 139, 142–143
Hennig, C., 155
Hill, B., 52f, 61
Hintikka, J., 238f, 241, 243–244, 246, 253f,
    259
Hinton, G.E., 79, 81
Hochstatter, C., 143
Hodges, A., 127f, 142
Holmes, P., 174, 188
Hon, G., 260
Hooke, R., 180
Hooker, C., 129f, 142
Houser, N., 259
Howick, J., 20
Howson, C., 21, 42
Hrobjartsson, A., 18–19, 42
Hull, D., 142
Humphreys, P., 91f, 93, 100
Hunter, K., 79
Hutchinson, B., 92, 100

I

Ihde, D., 165–166, 168, 188
Ippoliti, E., 85f, 100–101
Isaacson, W., 122f, 142

J

Jackson, P., 68, 78
James, W., 152, 201, 205–206
Jensen, F.V., 72, 78
Jewson, S., 61
Jones, D.T., 74, 78
Josephson, J., 240, 259
Josephson, S., 240, 259
Jumper, J., 78

K

Kahn, R.E., 142
Kahneman, D., 254f
Kakas, A., 260
Kang, C., 132, 142
Kaptain, H., 258
Kasparov, G.K., 68, 71, 77–78
Kavukcuoglu, K., 78
Kaye, D., 160
Kellert, S.H., 2, 12, 153, 163
Kelly, K., 88
Kemeny, K., 224, 231
Kepler, J., 74, 171f, 173
Kevles, D.J., 197, 206

Kindermans, P.-J., 94f, 100
King, R.D., 78
Kirkpatrick, J., 78
Kitcher, Ph., vi–vii, 7–8, 10, 12, 191, 193–195,
    199–200, 206
Klawiter, A., 231
Kleinberg, J., 91
Kleinrock, L., 142
Kloesel, C., 259
Knight, W., 85f–86f, 100
Knutti, R., 50f, 61
Koza, J., 82f, 100
Kozyrkov, C., 96, 100
Krajewski, W., 228, 231
Krohn, J., 85, 92, 100
Kuhn, Th.S., 10f, 125f, 150, 191, 192, 194, 201,
    206
Kuipers, T.A.F., vi–vii, 8, 10, 208, 211,
    213–214, 218–219, 221–224, 226, 228,
    229f–231, 238f, 242, 259
Küng, L., 116f, 142
Kurzweil, R., 95
Kvanvig, J.L., 254, 259
Kvasz, L., vi–vii, 7, 11, 165–166, 172f, 176,
    183, 188

L

Lakatos, I., 10F, 136, 142, 149
Lambert, K., 267, 284
Landau, S., 162
Langley, P., 87, 100
Lastowski, K., 231
Latzer, M., 143
Laudan, L., 10f, 20f, 92, 100, 191, 194, 200,
    206, 238, 259
Le, Q.V., 140
Lecun, Y., 82
Lee, K., 101
Leese, M., 162
Leiner, B.M., 108f–109, 142
Lempert, R., 160, 163
Lenat, D., 86f
Leonelli, S., 99
Levi, I., 240f, 247, 256f, 259
Lewis, D., 244
Li, K., 78
Li, L.J., 78
Ligertwood, A., 157–158, 163
Lindley, D., 160
Lipton, P., 238f, 240–241, 243, 249f–250,
    259
Lipton, Z., 91f, 94, 100

Liu, F., 115f, 142, 144
Livengood, J., 45
Lloyd, G.E.R., 265, 284
Locke, J., 15, 42
Longino, H.E., 12, 153
Lopez, A., 62
Lynch, D.C., 142
Lynch, M., 85f, 100
Lyons, T., 284
Lyytinen, K., 116f–117f, 129, 142

M
Mach, E., 263, 267
Machamer, P., 162
Mackenzie, D., 83, 94f, 101
Magnani, L., 239f, 259
Makridakis, S., 115, 142
Manso, G., 101
Mao, M.Z., 140
Marcus, G., 85f, 94, 100
Marinacci, M., 50f, 61
Martini, C., 61
Mason, J., 143
Máté, A., 188
Mathias, P., 143
Maxwell, G., 185–186, 277
McAfee, A., 113f, 140
McCain, K., 243, 249, 258–260
McGuinness, D.F., 188
McMullin, E., 186, 188
McShea, D.W., 124, 142
Meehl, P., 95f, 100
Meeker, M., 104f, 143
Mercoier, H., 89f, 90f, 100
Meyer, E.T., 110f, 143
Michalski, R.S., 69, 78
Michel, G., 61
Midgley, P.M., 61
Millner, A., 53f, 61
Mitchell, M., 92, 100
Mitchell, S., 152, 163
Mnih, V., 84, 100
Moher, H., 42
Monga, R., 140
Montague, P.R., 79
Morgan, M.G., 60f–61
Morgan, T., 196
Muggleton, S., 69, 71, 74, 76, 78
Mullainathan, S., 91
Musgrave, A., 142
Musk, E., 94–95

N
Nagel, E., 150, 152, 162–163, 261–262,
    273–276, 278, 284
Nauels, A., 61
Nelson, A., 78, 272
Neuber, M., 150f, 163
Neurath, M., 148, 284
New, M., 62
Newell, A., 87, 100
Newman, M., 127f, 143
Newton, I., 97, 169, 172–174, 185–186, 193,
    208, 218, 222, 255, 267
Ney, P.G., 18, 42
Ng, A.Y., 140
Nguyen, A., 85f, 101
Nickles, Th., v, vii, 1, 5–6, 10, 81, 85f, 95f, 101
Nicolis, C., 131f, 143
Nicolis, G., 131f, 143
Niehörster, F., 62
Nielsen, P.H., 88
Nielsen, M., 82, 101
Niiniluoto, I., vi–vii, 4f, 8–9, 11–12, 105, 114f,
    143, 235–236, 238–243, 245f, 247f–259
Noble, S., 85f, 101
Norton, J.D., 153, 163
Noseworthy, J.H., 17f, 42
Nowak, L., 228, 231

O
O'Hara, K., 108, 139
Oliver, S.G., 79
O'Neil, C., 76, 78, 85f, 101
Ono, R., 104f, 114, 127, 132, 144
Oppenheim, P., 225, 231
Orgs, G., 79
Ornes, S., 117f, 143
Osiander, A., 263–264, 267

P
Page, K.L., 127, 143
Panagiotatou, M., 283
Panofsky, W., 222
Paprzycka, K., 231
Park, J.H., 117f, 143
Parker, W.S., 47f, 50f, 61
Pasquale, F., 85f, 101
Pearl, J., 83, 94, 101
Pears, D.F., 188
Pearson, K., 83
Peijnenburg, J., 230f

Peirce, C.S., 8–9, 152, 235–239, 241, 243,
    249–251, 253f, 259–260
Penedones, H., 78
Pennington, N., 160, 163
Petersen, S., 78
Phillips, M., 222, 231
Philp, T., 45, 51, 61
Picard, R.G., 142
Piirainen, K.A., 143
Pinker, S., 230–231
Pipek, V., 143
Plato, 263
Plattner, G.K., 61
Pockley, S.M., 188
Polanyi, M., 86f, 93f, 101
Popper, K.R., 10f, 20, 66–67, 70, 75, 77–78,
    149, 163, 237, 244, 248, 252, 256, 259
Posidonius, 264, 266
Postel, J., 142
Poston, T., 243, 249, 258–260
Potochnik, A., 186, 188
Prabakaran, M., 118f, 144
Prakken, H., 258
Presenti, J., 86
Priestley, M., 112f, 143
Przybysz, P., 231
Psillos, S., vi–vii, 9, 11, 238f, 249, 260–262,
    284
Ptolemy, C., 174, 260–264
Putnam, H., 270–274, 284

Q
Qin, C., 78
Qin, D., 61
Quinlan, J.R., 69, 72, 78–79

R
Rahnama, M., 61
Rajeswari, K., 118f, 144
Raji, I.D., 92, 100
Rakover, S.S., 260
Ramon, J., 79
Ramsey, F.P., 70, 276–278, 280–281, 284
Randall, D., 138f, 143
Ranzato, M.A., 140
Rasmussen, C.E., 79
Ravaud, P., 42
Rawls, H., 228
Rédei, M., 188
Reichenbach, H., 140, 148–149, 163, 237
Rendtorff, J.D., 141

Rescher, N., 2f, 8f, 12, 119f, 121f–123f, 125f,
    137, 140, 143
Rey, J., 11
Ribeiro, M., 76, 79
Rice, C., 93f, 99
Richardson, D., 79
Rivadulla, A., 238f, 251, 260
Roberston, C., 61
Roberts, L.G., 142
Roberts, R., 42
Robertson, B., 157, 163
Robinson, J.A., 88
Roche, W., 250f, 256, 260
Rodosek, G.D., 143
Rohde, M., 143
Romeijn, J.W., 231
Rosen, E., 283
Rosling, H., 230–231
Roussos, J., v, vii, 3, 45, 61
Rowbottom, D.P., 262–263, 284
Rudin, C., 76, 79
Rumelhart, D.E., 73, 79
Ruse, M., 142
Russell, S., 89, 101
Rutherford, E., 213

S
Saatsi, J., 284
Sabbatelli, T., 50f, 61
Sackett, D.L, 17f, 19–20, 24, 26, 30–38, 40, 42
Sadri, F., 260
Salmon, W.C., 70, 245, 245f, 248–249, 260
Scazzieri, R., 140, 162
Scheffler, I., 267, 269, 274, 284
Scheibe, E., 163
Scheines, R., 101
Schilpp, P.A., 278
Schmidt, K., 143
Schneider, S., 95f, 101
Schönwälder, J., 106f, 143
Schredelseker, K., 120f, 143
Schroeder, R., 143
Schultz, W., 74, 79
Schultze, S.J., 109, 143
Schulz, K.F., 16f–18f, 19–20, 24, 26–28,
    30–32, 34, 37–38, 40, 42
Schum, D., 162
Schupbach, J.N., 241, 242f, 243–244, 260
Schurz, G., 231, 237f–238f, 243, 248, 254, 260,
    267, 284
Sedol, L., 72, 89
Sejnowski, T., 84f, 101

Sellars, W., 203, 206–261, 274, 278, 284
Sen, A., 130f, 143
Senior, A.W., 78, 140
Senn, S.J., 18f, 28–29, 29f, 42
Sent, E.M., 153
Shadbolt, N., 108, 143
Shaffer, M., 238f, 241, 260
Shapiro, S., 42
Shawe-Taylor, J., 72, 78
Shome, N., 49f, 61
Shortliffe, E.H., 68–69, 78–79
Sie, H., 231
Sifre, L.,78
Silver, D., 78, 89, 101
Simon, H.A., 3f, 12, 87–88, 100, 111f, 114f,
    115, 117–118, 125f, 135, 143–144, 240
Simonyan, K., 78
Simplicius, 264–266, 284
Singh, S., 79
Siow, E., 118f, 144
Sirawaraporn, W., 79
Sjegedy, C., 85f, 101
Skerrir, R., 45
Skinner, B.F., 85f, 91, 101
Slat, B., 208, 216
Sluckin, T.J., 143
Smith, L.A., 50f, 61–62
Smokler, H., 237, 249–250, 260
Sober, E., 256, 260
Socher, R., 78
Soldatova, L.N., 79
Somers, J., 85f, 101
Sommerfeld, A., 213
Sparkes, A., 79
Spelke, E., 101
Spensor, C., 42
Sperber, D., 89f–90f
Spirtes, P., 94f, 101
Spitzer, E., 92, 100
Sprevak, M., 99
Staal, J.F., 259
Stadler, F., 148f, 163, 188
Stahl, D., 162
Stainforth, D.A., 50f, 52f, 61–62
Stanford, K., 9, 12, 261–262, 281–284
Steel, K., 47f, 61
Stein, H., 9, 12, 261–262, 278–280, 283–284
Stevens, G., 143
Stocker, T.F., 61
Stoeltzner, M., 162
Stoney, D.A., 162

Strevens, M., 93f, 101, 126f, 144
Strogatz, S., 98, 101
Suckling, E.B., 62
Sudbury, A., 94f, 100
Sullivan, E., 91f, 93f, 101
Suppes, P., 7, 140, 147, 150–153, 161–163
Sweeney, P., 85f, 101
Swenson-Wright, J., 162

T
Tambolo, L., 230
Taroni, F., 156, 159–163
Tarski, A., 163
Tavory, I., 241, 260
Tenenbaum, J., 101
Tesic, M., 256
Thagard, P., 125f, 228, 231, 238f, 243, 249f,
    260
Thompson, E., 60–62
Thompson, N., 86f, 101
Thompson, W., 157–158, 163
Tian, Y., 92, 101
Tignor, M., 61
Tillers, P., 159–160, 163
Timmermans, S., 241, 260
Tiropanis, T., 105f, 107–108f, 115–116, 132f,
    139, 143–144.
Tishby, N., 94f, 101
Todd, P., 95, 100
Torricelli, E., 177–178
Towse, R., 142
Trachok, M., 88f
Tredger, E.R., 62
Tribe, L., 159–160, 163
Trout, J.D., 95f, 99, 101
Tuboly, A.T., 150, 163
Tucker, P., 140
Tuomela, R., 243, 245f, 248, 250f–252f, 259
Turing, A., 67, 79, 88, 90f, 101, 127f, 142
Tversky, A., 254f
Twining, W., 162
Tybjerg, K., 162

U
Uebel, T., 11, 141, 231
Uhlig, S., 112f, 129, 144
Ullman, T., 95, 101
Uncles, M.D., 127f, 143
Urbach, P.M., 16f, 21, 42

V

Vaihinger, H., 269–270, 284
van Fraassen, B.C., 2f, 9, 10f, 192–193, 206,
    238f, 240, 249, 255–256, 259–261, 273,
    284
Van Rootselaar, B., 259
Vandervoort, M.K., 42
Varghese, B., 125, 144
Vasserman, L., 92, 100
Vasuki, K., 115f, 116, 118f, 144
Vertei, B., 258
Vickers, P., 284
Vignaux, G.A., 157, 163
Vincent, P., 78
Vos, R., 214, 231

W

Wachter-Boettcher, S., 85f, 102
Walsh, K., 49f, 61
Walsh, T., 86f, 89, 102
Wang, F., 76, 79
Ward, J.A., 74, 79
Waring, D., 42
Wartofsky, R.C.M., 284
Washington, R., 62
Waters, C.K., 12, 153, 163
Waters, T., 50f
Weber, E., 231
Weber, M., 12, 163
Weinberger, D., 85f, 88, 102
Weintraub, R., 249
Weisberg, J., 260
Weiss, P., 259
Weitzner, D.J., 139
Werndl, C., 61
Werning, M., 231
Westcott, J.H., 143
Wheeler, G., 11, 141, 231
Whitt, R.S., 109, 143
Williams, C.K.I., 72, 79
Williams, K., 74, 79

Williams, R.J., 79
Wilson, P., 61
Wilson, T., 90f, 93f, 102
Winsberg, E., 91f, 102v
Woods, J., 249, 258–260
Woodward, J., 94f, 102
Worrall, J., v, vii, 1, 3, 10, 10f–12, 15, 20f,
    42–43, 142
Wray, K.B., 262, 284
Wright, A., 107, 144
Wu, S., 92, 100
Wulf, V., 143

X

Xia, Y., 61

Y

Yang, K., 140
Yap, K.L., 108f, 144
Yetisir, E., 42
Yin, H., 120, 125, 144
Yoo, C.S., 132, 144
Yosinski, J., 101
Young, M., 79

Z

Zaldivar, A., 92, 100
Zaslavsky, N., 94f, 101
Zeiss, C., 169
Zenil, H., 85f, 102
Zhang, J., 117f, 144
Zhang, W., 117f, 144
Zidek, A., 78
Zimmermann, K.F., 119, 139
Zorzato, L., 262, 284
Zuckerberg, M., 127f
Zytkow, J., 100

Subject Index

A

Abduction, 8–9, 235–238, 240, 243, 247–250, 253f, 255
  common cause, 254
  creative, 239f
  eliminative, 239
  selective, 239f
Abductive
  imagination, 241
  methodology, 223, 238
  reasoning, 223, 251
Accommodation, 249, 254
Acupuncture, 24
Agency, 95
Agnosticism, 261–262, 273
Algebra, 166, 176, 183f
Alien reasoning, 6, 81, 87–90, 96–98
Alpha
  machines, 89
  models, 89
AlphaInfinity, 98
AlphaZero, 82f, 88–90f, 98
Anti-realism, 2f, 262–263, 267, 281
Apps (practical applications), 6, 103–106, 108, 111, 113, 115n–116, 118, 121, 125–127, 129, 131–132, 134, 136–138
Archeology, 155
ARPANET, 121–122
Artificial
  general intelligence (AGI), 86, 96
  Intelligence (AI), 1, 3, 5–6, 65–68, 70–71, 73f–74, 89, 115–116, 127f–128, 130, 165, 235, 239
  neural networks (ANNs), 5, 81–91, 93–97
  neurons, 83

Astronomy, 166, 170-171f, 173–174, 205, 263–266, 279
Atmospheric pressure, 177–179, 182
Automated reasoning, 88, 90f
Averaging, 45, 50–52, 57, 59

B

Back propagation, 73, 75
Background knowledge, 21, 25–26, 28-29f, 40–42, 71, 75–77, 85f, 88f, 92, 98, 239, 249–250
Barometer, 167, 177–178, 180, 182
Basic research, 8, 191
Bayesianism, 2, 9, 21, 66, 153–154, 156, 160, 208, 228, 235, 240f, 243–244, 247–248, 250–251, 253–257
Bayesian networks, 72, 94, 160–161
Bayes's Theorem, 243, 247, 251, 255, 258f
Bias(es)
  attrition, 22
  co-intervention, 22, 27, 34–35
  differential subsidiary treatment, 23
  expectation, 22–23
  non-compliance, 22
  observer (or reporter), 23, 27, 34
  outcome assessment, 35
  response, 22
Big Data, 5, 81–82, 85, 104, 116, 118–119
Big Pharma, 39, 198
Biochemistry, 193
Biodefense, 197
Bioinformatics, 119
Biology, 74, 97, 124, 151–153, 155, 193, 197, 213

Black-box problem, 6, 81, 87, 90–92, 95–97
Blinding, 3, 15–20, 22–42

C
Causal modeling, 87, 92, 94
Causality, 83, 153, 268
Causation, 82, 87, 92, 151
Chess, 68, 71–72, 77, 82, 84-86f, 88–89
Climate
    change, 1, 3–4, 45–47, 60f
    models, 52f
    sciences, 155
Clinical trial, 3, 15–28, 38–39, 41
Cloud computing, 6, 103–106, 108–109, 111,
        113, 115–116, 119, 121, 125, 127–132,
        134, 136, 138
Cluster analysis, 7, 147, 153f–155, 162
Cognitive limitations, 82f
Coherence conditions, 249
Collaborative creativity, 122
Communication
    sciences, 104, 107f, 116f, 122f, 124
    technologies, 109, 122f, 124f
Complexity
    dynamic, 4, 6, 103, 107, 112, 120, 123–124,
        126, 128, 130, 132–133
    epistemological, 120–121, 130
    modes of, 120–126, 130
    ontological, 120, 122, 130
    pragmatic, 4, 107
    structural, 4, 6, 103, 107, 112, 120–124,
        128, 130, 132–133
Computational power, 86f
Computer
    science, 71, 87, 97, 129f
    simulation, 49
Computing
    analog, 92
    digital, 92
    quantum, 92
Concept explication, 8, 207, 209, 211, 221,
        224–225, 227–228
Condition
    target, 208–209, 211, 219–221, 227f
    DSD-, 208–214, 216, 219–220, 222, 227
Confidence approach, 45, 52f, 57, 60
Confirmability, 148
Confirmation
    abductive, 249, 251
    inductive, 249
Connectivity, 111, 113, 117, 121, 123, 128–129

Consequence condition, 225–226, 251
Consilience, 243
CONSORT guidelines, 19–20, 24, 39–41
Content delivery networks (CDNs), 127f
Context
    of discovery, 148–149, 237
    of justification, 148–149, 237
Contextualism, 268
Converse
    consequence, 226, 249, 251
    entailment, 249–250
Copernican system, 174f
Correlation, 82, 87, 91–92
Covid-19, 4f–5f, 11, 107f, 112, 119, 125f, 130f
Craig's theorem, 261–262, 271–272, 276
Cryptocurrencies, 127f

D
Data, information and knowledge, 118f–119f
Decision-making, 3–5, 45, 48, 51–55, 57,
        59–60, 120f, 160, 197
Decision theory, 52f, 246
Deduction, 8, 31, 88, 235–237, 241, 270, 272
Deep
    learning, 5–6, 72–73, 75, 77, 81–86, 88,
        90f–93, 95–98, 116
    reinforcement learning, 72, 84, 94
Deep Blue, 68, 71–72, 77, 84
DeepMind, 72, 84, 89
Defeasibility, 249
DENDRAL, 68–69
Denialism, 261–263, 267–269, 271, 273, 276,
        279, 281
Design, 5f–6, 129, 135
    engineering, 93
    research, 8, 207–209, 211, 213–216,
        218–219, 221, 229
    science, 107f, 111f, 114, 117, 123f, 127,
        131, 133, 135, 139
    theory, 135, 213f, 215–216, 218–219
Determinism, 152
Deterministic chaos, 174
Double blinding, 3, 15–20, 22f–26, 28, 40

E
Economics, 87f, 97, 104-105f, 107f, 116–118,
        120f, 124, 128–130, 133f, 152–153, 155
    applied, 130f
    financial, 116–117, 128
Embodied cognition, 95

Empiricism, 6, 85, 92, 147, 165, 167–168
  classical, 167-168f
  constructive, 192–193, 261
  logical, 6, 147–149, 161, 224
  probabilistic, 7, 147, 150–152, 161
End-of-trial tests, 3, 15, 18–20, 23–24, 30–41
Environmental studies, 5
Epistemic contact, 166–183, 186
Ethical values, 8, 131, 134, 136, 191
Experimentation, 7, 147, 151–152, 161, 241
Expert systems, 68–69, 72, 87
Explainable AI (XAI) problem, 6, 81, 91, 93,
    95, 97
Explanation, 4, 8, 20–21, 23, 28, 37, 76, 93, 98,
    152–153, 158, 160–161, 219, 235–258,
    270, 272
  approximate, 242, 257-258f
  causal, 246, 265–266
  deductive, 236, 245–246, 248, 253
  deductive-nomological (DN), 236
  idealized, 242f
  inductive-probabilistic (IP), 236, 251
  likely, 250
  lovely, 243, 250
  mechanistic, 82
  non-deductive, 245
  probabilistic, 244–245
Explanatory
  power, 238, 243–245, 248–249, 251, 254, 256
  success, 251–252, 255–257
  syllogism, 236
Explicative research, 207–208, 219, 223, 229
Extreme weather events, 4, 45–47

F
Fallibilism, 239
Falsification, 65, 70, 75, 77
Factionalism, 269–273
Forecast, 105, 129, 139f
Forensic science, 155, 158
Future studies, 6, 103-104f, 135, 137

G
Genetically Evolving Models in Science
    (GEMS), 82f
Genetics, 193, 196–198
Geometry, 166, 172, 176–177, 183–184,
    186–187, 263–264
Geophysics, 180
GOLEM, 69, 71–75

Gothic painting, 170–171
Greenhouse gas emissions, 47

H
Historicity, 111–112, 124–126, 129–130, 137
History of science, 2n, 136, 199
Human Genome Project, 196
Hurricane
  insurance, 47–52, 56
  models, 4, 45, 49f
  physics of, 49
  risk, 47, 49f
Hyperparameters, 84f, 88f, 91
Hypothetical reasoning, 8, 235, 238

I
ID3, 69, 72
Idealization, 7, 165–166, 182–188, 226, 228
Image processing, 72
Imitation game, 88
Incommensurability, 194
Induction, 5, 8, 65–67, 69–70, 74–75, 77, 153,
    235–237, 248–250
Inductivism, 248
Inference
  to the best explanation (IBE), 8, 235,
      238–250, 253–256
  to the only explanation (IOE), 239, 246,
      253, 258
Information
  and communication technologies (ICT),
      122f, 124f
  science, 104, 110, 128–130
Innovation, 81, 87, 106, 111–112, 116, 121,
    124–125, 132, 137–138
Instrumental
  contact, 168–169
  synthesis, 185
Instrumentalism, 8–9, 201, 261–264, 267–269,
    273–283
  denialist, 268–269, 273, 278–279, 281
  factionalist, 267–271, 273, 281
  historicist, 262
  irenic, 261, 274
  non-cognitivist, 269
  scientific, 267
  sophisticated, 278–280
Interdisciplinarity, 107, 111, 116–118, 132, 135
Intergovernmental Panel on Climate Change
    (IPCC), 46–47

Internet, 1–3, 6, 103–139
    as activity, 113, 134
    Protocol (IP), 113, 122f
    of Science, 119
    of Things, 114f, 117–118, 137
Internet science, 107, 115–116, 130

K
Knowledge engineering, 87

L
Language game, 183–185, 187
Law(s), 71, 86f, 151–152, 155, 167, 173–174,
        176, 179–181, 185–186, 193, 213,
        222–223, 236
    causal, 269
    empirical, 173, 180–181, 185, 208, 222, 228
    mathematical, 181
    physical, 176, 179
    statistical, 236
Legisimilitude, 211–212
Likelihood, 46–48, 235, 243–255
Likelihood ratio (LR), 156–158
Linguistic reduction, 183
Logic, 2, 16, 69–70, 72, 167, 209, 235,
        237–239, 273
    Boolean, 92
    deductive, 66
    of discovery, 5, 96, 237
    formal, 167f
    inductive, 66–67, 70, 241, 253f
    symbolic, 87
Logical positivism, 148
Lottery paradox, 247f

M
Machine
    learning, 2, 65–67, 69–77, 81–95, 97, 108,
        115, 122
    speed, 5, 81
Macroeconomics, 133f
Mathematical
    entities, 270, 278
    models, 119f, 154, 264–265
Mathematics, 1, 3, 6–7, 16, 153, 160, 166, 264,
        270
Measuring device(s)/instruments, 168, 176,
        178, 180, 182, 184, 193, 200
Mechanical engineering, 213
Mechanics, 186–187
    classical, 176

Hamiltonian, 170, 173–174
Lagrangian, 173–174
Newtonian, 173
quantum, 186-187n, 198
Medical evidence, 3–5
Medicine, 3, 16–17, 19, 39, 41, 76, 155, 199
    evidence-based (EBM), 3, 15–17, 19–20, 23
    philosophy of, 1, 3, 16–17
Meteorology, 48
Methodological
    pluralism, 2–3, 123
    pragmatism, 2
Methodology
    of historiographical research programs
        (MHRP), 136
    of scientific discovery, 81
    of scientific research programs (MSRP), 136
Microeconomics, 134f
Micro-physics, 151
Mobile Internet, 6, 103–106, 108, 111, 113,
        115f–116, 118-119f, 121, 123, 125–127,
        129, 131, 134, 136, 138
Multi-agent systems, 113, 133
MYCIN, 68–69

N
Naturalism, 1–2
Natural sciences, 2n, 111, 192, 266
Neo-Pythagoreanism, 278
Network of networks, 2, 6, 103–138
Network science, 104, 107, 115–117, 130
Neural networks, 72–74, 76, 81, 91
Neuropsychology, 151
Nomic research, 207, 218
No miracles argument, 257
Non-cognitivism, 269, 274
Non-eliminativism, 262, 268–271, 273
Nonhuman intelligence, 96–97
Non-pharmacological trials, 24
Novelty
    horizontal (longitudinal), 124, 126, 128, 138
    vertical (transversal), 124, 126, 128, 137–138

O
Objectivity, 168, 179
Ocean cleanup, 207–208, 213n, 217
Ontogenetic learning, 84n

P
Perceptual psychology, 153
Pessimistic induction, 2f, 201

Pharmacy, 213
Physicalism, 148, 272
Physics, 42, 49, 86, 95, 148, 151–153,
        178–180, 185, 198, 264–265,
        275–276
    fundamental, 97, 193
    Newtonian, 172
    quantum, 97, 153
Pictorial form, 165, 172f, 174–177, 181f, 183
Placebo, 17–18, 22–23, 25–28, 33–35
Plausibility, 248, 253–254
Pluralism, 7, 147, 151–153, 161
    about the sciences, 153
    integrative, 152
    methodological, 2, 6, 123
Polanyi's paradox, 86n
Policy-making, 4–5, 130f, 133
Positive Relevance criterion, 250–252
Post-randomization
    biases, 32, 40
    confounders, 21, 23f
    differences, 21
Pragmatism, 2, 191, 195, 206, 237, 268
Prediction, 2f, 4, 6, 33, 51–52, 54, 66, 71, 98,
        103–105, 107-108f, 112, 114, 118–120,
        123f–133, 135–137, 155, 185, 237, 245,
        249, 251, 254, 262, 264, 267–271,
        274–275
Predictive
    power, 85f, 254, 277
    test, 51, 57
Prescription, 6, 103, 105, 107, 112, 114,
        118–119, 127–137, 139
Probabilism, 6
Probabilistic inference
Probability, 1, 3, 5–7, 21, 31, 49–50, 52, 54–57,
        59–60, 65–66, 70, 91, 147, 150–155,
        157–160, 224, 235, 237, 243–248,
        251–254, 256
    conditional, 224, 243
    frequency interpretation of, 153
    posterior, 157, 243, 246–248, 250,
        252–255, 257, 275
    prior, 21, 243, 247–248, 250, 253–256
    theories of, 2
Problem solving, 87, 98, 117, 119, 136, 152,
        194, 200–201, 228
Progress
    in design, 209, 211, 213–214, 216, 221
    empirical, 8, 207–208, 213, 220–223, 228
    explicative, 8, 207, 209, 220–221, 223,
        227–228
    formal, 212
    moral, 199, 207, 209, 229–230
    pragmatic, 7, 191, 194–195

qualitative, 8, 207–209
quantitative, 230
scientific, 1, 3, 7–8, 83, 110, 191–192, 194,
        197–200, 211, 221
technological, 169, 207–208, 214, 229
teleological, 192, 194
Propensities, 244-245f
Prosecutor's fallacy, 158
Psychologism, 167
Pure knowledge, 199

R

Ramsey-sentence RS(T), 276–278, 280–281
Rationality, 5–6, 152, 154, 194
    bounded, 87
    ecological, 87, 95f
    evaluative, 137
Realism, 2, 9, 192f–193, 257–258, 261–263,
        272–280, 282–283
    instrumental, 7, 165–169, 171f, 173,
        179–180, 186
    mathematical, 270
    scientific, 1–3, 7–9, 81f, 98, 165, 171f, 192f,
        243, 257–258, 261, 267, 273
Received view, the, 148–149, 151–152
Renaissance painting, 170–172, 174–175
Revolution, 124-126n, 137
    anti-causal, 83
    causal, 83
    computational, 83
    conceptual, 125n
    methodological, 5–6, 81–82
    scientific, 81–82, 97, 125, 192, 201
Robotics, 92, 95
Rule-based systems, 69, 71

S

Satisficing and optimizing, 240
Scalability, 111, 113, 117, 121, 128
Science(s)
    application of, 4, 6, 103, 105, 107, 110–116,
        128, 130, 132–133, 135, 139
    applied, 4, 6, 103, 105, 107, 110–115,
        117–120, 128–133, 135–137, 139
    basic, 4, 8, 105, 110, 191, 193, 199–200
    constitutive elements of, 120
    data, 104, 117–118, 129–130, 138
    formal, 127f
    of the Internet, 6, 103, 111-112f, 114,
        116–117, 123–124, 127–128, 130,
        132f–133, 135–136, 138
    network, 82, 104, 107, 115–117, 127, 130
    open, 119

Science(s) (*cont.*)
   special, 193, 223
   technology and society, 4f, 104
Sciences of the artificial, 3f, 5, 107f, 111, 115,
     117, 135, 139
Scientific
   creativity, 111, 116f, 120, 131–132,
     137–138
   design, 129, 132, 136
   instruments, 7, 165–169, 177, 181f–183
   reasoning, 8, 235, 239, 243
Scientification, 114
Security, 92, 105, 111–113, 117, 128–129
Semiotics, 237
Significance graph, 195, 199
Skinnerian learning theory, 84f
Social concern about science, 1
Social sciences, 2n, 107f, 111, 117f, 119, 139n,
     153
Socio-informatics, 138f
Statistical methods, 7, 147, 150, 152, 154–156,
     158–159, 161–162
Statistics, 7, 26, 48–49, 52, 72, 83, 147,
     150–151, 153–156, 158–161, 240f
Superhuman performance, 89
Synthesis, 7, 165, 183, 185
   compositional, 7, 165, 185, 187f
   deductive, 7, 165, 185–188
   instrumental, 7, 165, 185
   relational, 7, 165, 185

T
Technological innovation, 81, 111f–112, 116f,
     120, 124, 132, 137–138
Technology, 2, 5, 71, 95, 112, 120f, 122f–123f,
     127f, 131–133, 138, 165, 194, 198, 214
Telescope, 166, 168–169
Testability, 235, 238, 243, 248
Theory
   change, 281–283
   formation, 151–152, 212
   of mind, 90f, 94

Thermodynamics, 178, 182
Transdisciplinarity, 116, 118, 132
Transfer learning, 85f
Truth
   approximation, 8, 207–208, 210–211, 213,
     219, 221–223, 228-229f
   deterministic, 212f
   probabilistic, 212f
Truthlikeness, 211, 224, 235, 242, 256–258
Turing machines, 90f
Typhoon, 4, 46

U
Uncertainty, 3–4, 45, 48, 50–52, 59–60, 187f,
     244f
Unity of science, 148, 151–152
Unobservable entities, 269, 272, 275, 278, 280,
     282–283
Unsupervised learning, 83
Utility, 53, 55
   epistemic, 243f, 245–246
   expected, 53, 55–56, 246
   maximin expected, 53, 55, 59–60, 245
   minimum expected, 55
   theory, 218

V
Verifiability, 148
Verificationism, 89f
Verisimilitude, 211, 242, 256-258f
Vienna Circle, 148
Virtual Reality (VR), 5, 65, 67, 95

W
Web, 6, 103–119, 121–127, 129–132, 134,
     136–138
Web science, 107, 110–111, 115–117f, 121,
     123–124, 127, 130, 133-134f, 139
White-box models, 85f

Printed in the United States
by Baker & Taylor Publisher Services